FASTER THAN LIGHT

Warp Drive and Quantum Vacuum Power

By H. David Froning Jr.,
Edited and Foreword by Tom Valone, PhD

Adventures Unlimited Press

Other Books in the *Lost Science* Series:

- The Anti-Gravity Handbook
- Anti-Gravity & the World Grid
- Anti-Gravity & the Unified Field
- The Free-Energy Device Handbook
- The Time Travel Handbook
- The Anti-Gravity Files
- Tapping the Zero Point Energy
- Quest for Zero Point Energy
- The Energy Grid
- The Bridge to Infinity
- The Harmonic Conquest of Space
- Anti-Gravity Propulsion Dynamics
- The Fantastic Inventions of Nikola Tesla
- The Cosmic Matrix
- The AT Factor
- Perpetual Motion
- Secrets of the Unified Field
- The Tesla Files

FASTER THAN LIGHT

Warp Drive and Quantum Vacuum Power

An aerospace expert explains the science behind breakthrough

Faster Than Light

By H. David Froning, Jr.

ISBN 978-1-948803-16-8

Published by:
Adventures Unlimited Press
One Adventure Place
Kempton, Illinois 60946 USA
auphq@frontiernet.net

AdventuresUnlimitedPress.com

10 9 8 7 6 5 4 3 2 1

FASTER THAN LIGHT

Warp Drive and Quantum Vacuum Power

An aerospace expert explains the science behind breakthrough

ABOUT THE AUTHOR

H. David Froning, Jr. worked at the leading-edge of aeronautical and spaceflight science and technology for over 50 years, including much of the exciting 1945-1985 time-period when the greatest increase in flight speed and distance above Earth occurred. He performed research and development work for the U.S. Air Force, Boeing, McDonnell Douglas, and his own Company, Flight Unlimited. Examples of his pioneering work include design of the configuration concept of the world's first antiballistic missiles and vertical launcher missiles that helped lead to today's vertical launching systems on naval ships. And He was first to explore use of zero-point fluctuation energies of the vacuum for space power and propulsion. His own company, Flight Unlimited, performed work for the U.S. Air Force; National Aeronautics and Space Administration; European Space Administration; ANSER; Allegany Ballistics Laboratory; Thiokol Corporation. He is a Past Fellow of the British Interplanetary Society; Associate Fellow of American Institute of Aeronautics and Astronautics; and Past Member of the International Academy of Astronautics. He served on various panels and was a founding participant in the NASA "Breakthrough Propulsion Physics" Program.

COVER CREDIT

The spacecraft called the “Luminaris” on the front cover and inside this book is reproduced with permission from the artist Igor Puskaric. It is an accurate artistic rendering of a suitable spacecraft for Froning’s *teardrop design* of the engine which is explained in the later chapters of this book. Froning proposes that such a force-producing coil shape would naturally be the requisite foundation of a faster-than-light spaceship to advantageously stress the fabric of space. Therefore, logically it must comprise the length and breadth of this suitably designed ship. The implication is that this spaceship conforms to the shape of the coil and is most compatibly teardrop shaped as well. Froning’s coil powers the ship beyond lightspeed.

We can only imagine how soon such a Luminaris will grace the skies of earth to carry its passengers to places far beyond. – Ed. Note

TABLE OF CONTENTS

ACKNOWLEDGEMENTS

No one can do everything on their own. Most things require help of others. So I must acknowledge some of those others who helped make possible this book.

I must begin with Mom and Dad, Mary and Herman Froning, the very best parents one could ever have. They have now been gone for a while, but I can never forget their un-flagging love and support. Next, my wonderful wife, Irina – whose rare intelligence and beauty would have made her famous in many careers. But luckily for me, she chose a more humble-helping thing. So her hard work made possible all I wrote about in this book. Next, my best and longest friend: Garrett Stone. Fortunately for me, Garrett spent a small part of his final years in unselfish helping of me to edit this book – before being called on to much higher things.

I must thank my great professors at the University of Illinois: Jake Hauser, Harry Hilton, Jon Cromer, Paul Torda, and specially M. Z. Krzywoblocki who kindled my passion for aerodynamics. And special thanks to Professor John Schetzer at the University of Michigan, a wonderful, insightful man who fanned it further.

Spence Robinson was my supportive Nike Zeus Supervisor at Douglas. Ebullient Spence was a joy to work for and travel with. Intelligent-warm-funny, there was no office or place (even in the "deep-south") where he did not make new friends. So, when killed in an accident, Spence's large black church was packed with tear full white-folk who came to pay that great man homage. I want to do it again.

I now realize the best part of aerospace life was not designing and building amazing flight machines. It was the joy of doing it in a fellowship of amazingly brilliant men - with all of us believing we were part of a worthwhile quest. So, thanks to all I didn't mention. It was my great privilege to have worked with you.

Finally, I want to thank Mom again. She enrolled me in a Sunday school which was part of a Church that followed Jesus' command to "heal the sick" (not just preach the gospel). For this got me turning to God first when faced with physical challenges before immediately resorting to medical means. This also got me turning to God when other problems arose. Thus, every good thing mentioned in this book was from this turning. So, at the risk of annoying devout atheists who might read this book, I must acknowledge good-God most of all.

H. David Froning, Jr.

FOREWORD

For years I have been amazed at the detailed physics, untethered vision, and scientific optimism that is always conveyed with Dave Froning's presentations at our Conferences on Future Energy. Such excitement filled me with each of his discoveries, such as the close correlation between the equations for the speed of sound and the speed of light. This singular discovery of these similar and analogous variables may be the greatest find that Froning made during his life. We now have a reason to hope that traveling beyond the speeds of sound and light are actually quite similar. This also makes me recall the concept of systems theory which is applicable here and has great validity in academic pursuits. Dave's basic premise for this book seems to be his second slide from the Eighth Conference on Future Energy (COFE8) where he states:

> **'Specially-conditioned EM fields can "warp" gravitational metric and "polarize" zero-point quantum vacuum.'**

This is where we find the most value and intrigue for the future of space travel. His teardrop shaped electromagnetic toroids (shown in later chapters) are critically pivotal for specially conditioning the electromagnetic (EM) fields. For those not familiar with the quantum vacuum or zero-point energy, we would like to refer you to my easy-to-read book, *Zero Point Energy: The Fuel of the Future*. There one will find a succinct explanation of the "polarization of the vacuum" which is based on a modification of the virtual particles emanating from it. This outcome of realizing the import of Dave's one-liner above is that the hope for Alcubierre's warp drive may be as simple as the Froning toroid with custom EM excitation. An encouraging evidential set are the extensive simulations run for Dave on a computer by colleague Robert L. Roach, one of which is shown below. The color coding shows the vacuum compression regions running from red to blue, with the highest compressed area being right in front of the leading edge of the craft (shown as a black outline edge on) that optimistically will be saucer-shaped with the toroid inside. As the speed of the craft exceeds light speed c, the equipotential lines start sweeping toward the back of the craft with

more complex patterns at higher multiples of c (yes... several times the speed of light).

This book is unique and a real breakthrough for the physics of space travel but it would not be as convincing nor believable if Dave had just started directly with Chapter 5 or 6. Without any autobiographical background of his lifetime career in aerospace engineering with major companies, who often had secretive contracts with the military, this book would have less credibility. The first half dozen chapters are therefore quite a revelation providing a candid disclosure of the clandestine rocket development that for decades, proceeded mostly without media coverage. Froning met and worked with so many impressive people that I have taken an editor's license and underlined as many as I could find, for emphasis. One of the most prominent is Dave's meetings with the famous Col.

Chuck Yeager, the first man to fly faster than the speed of sound. I was fortunate to hear a presentation from Col. Yeager back when I was a grad student in the State University of NY at Buffalo and was impressed with how he overcame his fear of the unknown supersonic realm that everyone thought was impenetrable and unreachable. For those not curious about the historical military context from which the present civilian faster-than-light science emerged, they can skip the first four or five chapters and get right into the later ones where the good stuff is!

It has been a pleasure to personally edit this collection for friend and colleague, Dave Froning. I have underlined most of the famous aerodynamicists that Dave worked with to make it easier for the reader to note them for reference, even during speed reading. It is my belief that this book will help future generations decipher the secret to faster-than-light travel. You, the reader, on the other hand, here in the present day, will no doubt find this information a valuable companion to your library of future science that is emerging today. Onward and upward to the stars (ad astra)!

Thomas F. Valone, PhD, PE

Integrity Research Institute

PREFACE

It is difficult to single out one epoch in history when humanity made its most progress in advancing flight. But I believe it was between 1945 and 1985 when the greatest surge of progress in flying faster and farther occurred. This progress didn't happen from grand, noble purpose – but mainly, from fear and pride. This epoch began in 1945 when the World's first nuclear explosion occurred above New Mexico desert and a perilous 40-year "cold-war" between Russia and America began. And 1945 was the year the design of the first aircraft to break the sound barrier began. This epoch and the cold war began to end in 1985 as Russia's Mikhail Gorbachev initiated a period of openness "glasnost". Coincidentally, the Pioneer 10 space probe has crossed Neptune's orbit at this time and now is flying past the end of our solar system into interstellar space.

This period, which I call the "Halcyon Years "of air and space flight, happened because of the perceived need by Russia and America to gain air and space supremacy over each other. During this period the " sound" and " heat" barriers, which were preventing faster-longer flight, were overcome – as speed increased almost 50 times and flight distance almost 100,000-fold. Digital electronics and jet engines revolutionized airline travel, and rocket engines and heat protection systems enabled humans to land-on and return from the moon Both Russia and the U.S developed space stations; spaceships that flew somewhat like airplanes and space probes to visit every major planet and the Sun.

I was fortunate my aerospace life encompassed all but the first 5 years of this remarkable period. And though I wasn't one of its famous participants, I had interesting encounters with some who were. On the other hand, colleagues and I played our own small parts in helping overcome some flight barriers facing America during this exciting time. These flight barriers were overcome in classified military programs that weren't publicized like NASA's well-known space programs were. These military programs are no longer classified, but little is still known about them. So, some things in this book may add a small amount of aeronautical and spaceflight history about some of the U.S. military vehicle advances that were made during the halcyon years of spaceflight. spaceflight.

The halcyon years also perfected technologies that revolutionized airline travel by making it much swifter, safer, more comfortable and less expensive. These technologies included jet engines whose performance wasn't limited by the maximum allowable speed of propeller blades. And this created the stupendous air-travel infrastructure of aircraft, airlines, airports, hotels, travel organizations that have given millions of new careers and experiences to the people of Earth.

Rocket and satellite technology also advanced enough during the halcyon years to begin a world-wide commercial space industry of communication satellites that orbit Earth and provide the internet and telecom networks that have changed life for almost every human upon it. Finally, the halcyon years were also the golden years of planetary exploration – when NASA took advantage of a rare (once every 175 years) alignment of Jupiter-Saturn-Uranus-Neptune to visit them all – exploring more outer planets than all space probe explorations since.

The next period (1985-2015) lasted to the end of my aerospace life. Less bold than the previous halcyon one, it emphasized different aerospace advances than speed and range. It also revealed the painful truth that formidable barriers still obstructed our next big steps in space. This book mentions some of these now well-known obstacles, like excessive vehicle fuel consumption and cost and insufficient safety for human spaceflight. Regretfully, there is no vehicle today that comes close to overcoming these obstacles. But examples of the kinds of future vehicles that might do this in the future are shown. These examples are not likely to be just like those that finally revolutionize spaceflight. But they illustrate the kinds of advances in future vehicle designs that may be needed.

I was also introduced to new ideas like "space-time dilation near the speed of light" and bizarre ideas like the "zero-point energy" in space itself and faster than light travel by "warping" such space. And pursuing these strange ideas began with an extraordinary experience in an unfriendly communist land.

In summary, the book looks back on aerospace life, during the halcyon age of air and spaceflight, when the swiftest surge of aerospace advancement that ever happened (so far) on Earth took place. It then looks back on less-than-halcyon aerospace years that followed - sometimes spent on seemingly fruitless research on ways to overcome the formidable barriers to spaceflight that still remain. But although this work, done with brilliant-talented-dedicated people, wasn't as successful as we might have liked, hopefully, it and future work by more brilliant-talented-dedicated people will set the stage for a new 'halcyon age of air and space-flight advancement – one that will even take us to the stars.

Finally, I discuss possibility of Earth societies becoming space-faring ones that expand beyond our solar system. Past studies show, even slow sub-light speeds would allow self-replicating starships would allow their self-replicating people to eventually colonize our entire galaxy. But other studies show seemingly impossible socio-political obstacles facing civilization that would become space-faring ones. So the book ends with a view on the possible futures for spaceflight.

H. David Froning, Jr.

1 OVERCOMING THE "SOUND BARRIER"

Although many people knew that dense-sturdy rifle bullets flew through air faster than sound-waves do, very few aeronautical experts in the early 1940's believed inhabited aircraft could ever fly faster than the speed-of-sound. For such craft would have to safely accelerate through a "sound barrier", a wall of high-pressure air which forms, as shown below, about aircraft whose speed equals the speed of sound in the air. So this barrier was generally believed unsurmountable in 1946 when I began aeronautical engineering at University of Illinois. For brave British test pilots had already lost their lives in fast craft that had become uncontrollable - disintegrating even before reaching the speed-of-sound.

FIGURE 1 CONDENSATION OF AIR IN THE REGION ABOUT AN F-18 AIRCRAFT – SOUND-SPEED IS REACHED

I joined the quest to overcome the "sound-barrier" long after it had first begun in the early 1940's. I was in graduate school at the University of Michigan and became fascinated with transonic aerodynamics – study of the radical change in airflow about a vehicle and the swift change in forces and moments acting on it as it approaches, exceeds and surpasses the speed-of-sound. After graduation I served a short stint in the U.S. Air Force and then resumed my aeronautical engineering career at Boeing, the Company I had first joined after graduating from the University of Illinois. I had returned to Boeing because it had just developed the very first large transonic wind tunnel that could obtain excellent flow quality at transonic speeds. I reasoned that working in this tunnel would let me be among the first to learn of the actual forces and moments acting on aircraft and missiles as they reached and exceeded sound-speed. And the brand- new information from these tests would give me a good head start in the design of different future high-speed craft. And this indeed happened as month-after-month and test-after-test solved more of the many mysteries of transonic flight.

FIGURE 2 TODAY ABOUT 60 YEARS AFTER THE AUTHOR WAS THERE, THE BOEING TRANSONIC WIND TUNNEL IS STILL GENERATING VITAL DATA FOR FUTURE FLIGHT.

In 1955 I received an extremely tempting job offer from Douglas Aircraft in Santa Monica California. They were very interested in my particular combination of academic, military and industry experience, so, it seemed time to make a change. Then, while driving south to California, I remembered I had to locate a new Air Force base to serve my two weeks of reserve active duty each year. The very best possible air base would be the Air Force Flight Test Center at Edwards California, where all new Air Force aircraft were tested. But a chance for a lowly reserve officer like me to serve at such a place seemed remote. But, Edwards wasn't much out of my way to Santa Monica. So I decided to give Edwards a try.

I told my purpose to a lady at the Flight Directorate Building and was amazed when she immediately said, "You should talk to Colonel Yeager" and led me into his office. I had very few aviation heroes, but Col. Yeager was one of them. I still remembered my excitement while at the University of Illinois in 1947 when I heard the electrifying news that Charles (Chuck) Yeager had become the *first human to fly faster than the speed of sound.* No details of his flight had been released, yet. But I was aware of his epic achievement as an aeronautical engineer and pilot, Colonel Yeager seemed interested in what I had learned about transonic stability and control at the University of Michigan and the

Figure 3 The Bell, X-1 Rocket Plane. This plane was flown by Charles Yeager to break the "sound barrier" for the very first time. This photo was taken at about the same time as Yeager's famous flight. Today, this area is part of today's Edwards Flight Test Center – then, usually only called "Muroc".

Boeing Wind Tunnel. Apparently, new Air Force aircraft were encountering new flight control problems at transonic speeds. So I imagine that he was probably keen to glean all he could on this topic – even from a lowly Air Force reserve officer like me. After a very enjoyable talk, I was overjoyed when my hero shook my hand on parting and told me to report to him at his office as soon I came to Edwards to begin my first two weeks of Air Force reserve duty.

My first job at Douglas was refining aerodynamic characteristics of missiles launched from aircraft, to improve their flight behavior. So, being an aviator with aircraft experience at Boeing and the Air Force was helpful. Similarly, my previous Boeing experience and current Douglas experience would prove useful in my Air Force Reserve work at Edwards, whose major activity in much of the 1950's was developing the first generation of Air Force supersonic aircraft.

A military aircraft reached higher speeds, they began suffering buffeting and reduction of stability and control as they approached sound-speed in steep high-speed dives. Then huge increase in air pressure and drag occurred as craft reached almost the speed-of-sound (about 1200 kilometers per hour) was reached. And my first transonic testing of supersonic vehicles in the Boeing transonic wind tunnel was revealing adverse flight behavior continuing until well after sound-speed was surpassed. So, gaining much more knowledge of transonic aerodynamics and flight dynamics was very vital for overcoming the sound-barrier in 1955. Therefore, it was a major challenge at Edwards Air Force Base when I reported to Col. Yeager for my first period of Reserve Duty.

Since I last met him, Col. Chuck Yeager had been promoted and sent to Europe. So, I reported to Colonel Jack Ridley, who was almost as legendary as Yeager. Ridley had been Yeager's main consultant during his famous flights and reputedly had an uncanny ability to make right technical decisions during those pioneering days of high-speed flight. Ridley is also credited with organizing Edwards into the Flight Test Center of Excellence it is today. And it was he who had to decide what to do with the lowly reserve officer who showed up at his office one day.

At our first meeting Ridley saw that my work at Douglas Aircraft could be useful to his Directorate. For, my Douglas aerodynamic and flight simulation work was revealing adverse yaw-roll coupling for the yaw and roll behavior of some classes of vehicles at transonic-supersonic speeds – to cause catastrophic divergence of their flight. This kind of problem was a major concern n in his Directorate's development of supersonic aircraft. So, he decided each year of my reserve duty would support a different aircraft project. Roughly 1/3 of my time would be study of aircraft project reports. Another 1/3 would be support of aircraft flight tests; and the final 1/3 would be to share my applicable aerodynamic and simulation experience with appropriate project members. I think this Ridley plan worked out well during my entire time at Edwards. In 1956 Colonel Ridley was promoted and transferred to another command, and in 1957 I was shocked and saddened to hear that he died in a C-47 crash in Japan.

At Edwards I learned of American and British military test pilots who were lost in early attempts to overcome the sound barrier. American pilots were Richard Bong in a P-80; Milo Burcham in a YP-80; Glen Edwards in a YB-49; and George Welsh in an F-100. And over my 5-year period of my reserve duty at Edwards, two more test pilots lost their lives there. They were Mel Apt in an X-2 research plane and Ivan Carl in an F-104 test aircraft. I had never worked with either one but knew them well enough to have a feeling of "loss". And I remember their deaths causing a feeling of mild self-condemnation in me, for the relatively safe aeronautical contribution my science and engineering life was making towards supersonic flight – as compared to the ultimate sacrifices these men had made.

After I left Edwards in 1960, two more test pilots: Joseph Walker (F-104 in 1966) and Michael Adams (X-15 in 1967) also perished. So I estimate that *at least 7 Air Force test pilots were lost* in the dangerous, challenging quest for safe super-sonic flight. Similarly in Apollo, three test pilot astronauts perished in a ground accident and 4 test pilot astronauts (and also 8 mission specialists) were lost in two Shuttle crashes. So it would seem similar numbers of military test pilots were lost in Air Force development of supersonic flight and in NASA's development of supersonic flight during the first epochs of human spaceflight.

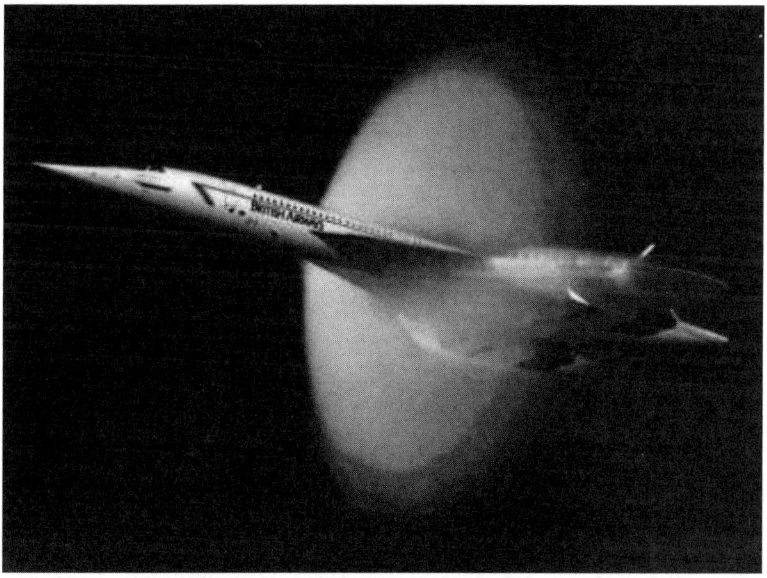

Figure 4 British Airways Concord supersonic airliner accelerating through the condensation shock associated with the speed-of-sound, on its way to higher velocity (twice sound-speed}. The Concord fleet of British Airlines flew passengers through sound-speed almost 100,000 times– accelerating and then decelerating through sound-speed (Mach 1.0) during each flight.

By about 1975, ample engine thrust and much more theoretical-experimental knowledge had given aircraft developers enough propulsive power and aeronautical understanding to design and fly aircraft that could swiftly and safely "slip-through" the sound barrier. So, accelerating or decelerating through the sound-barrier became routine for military craft, and also for the crew and passengers of British and French "Concord" supersonic airliners carried millions of passengers beyond sound-speed for 27 years between 1976 and 2003.[1]

[1] **Ed. Note:** During this time, the US also developed its own navy blue-colored supersonic transport (SST) airliner at a cost of $1 billion (to contrast with the Concord which was white). Due to environmental groups lobbying the government, it never received the further support to become an active part of any airline. During the 1980's this editor attempted to find any investor or benefactor who could save this unique historical vehicle from destruction for pennies on the dollar while it was housed for over ten years in a building the was eventually bought by a church. The church held Sunday services right next to the SST until the means for its disposal manifested, which doomed this SST and postponed the US development of any viable SST for another forty to fifty years by any reasonable estimate.

2 BREAKING THE "HEAT BARRIER"

Now, even before the sound barrier was fully overcome, the "heat barrier" rose up to make things difficult for vehicles that flew faster than about twice the speed of sound. For at higher speed than Mach 2.0, aerodynamically-heated air soaking into aluminum skins was hot enough to seriously weaken them. And at speeds 5 times faster, air became hot enough to melt strong steels.

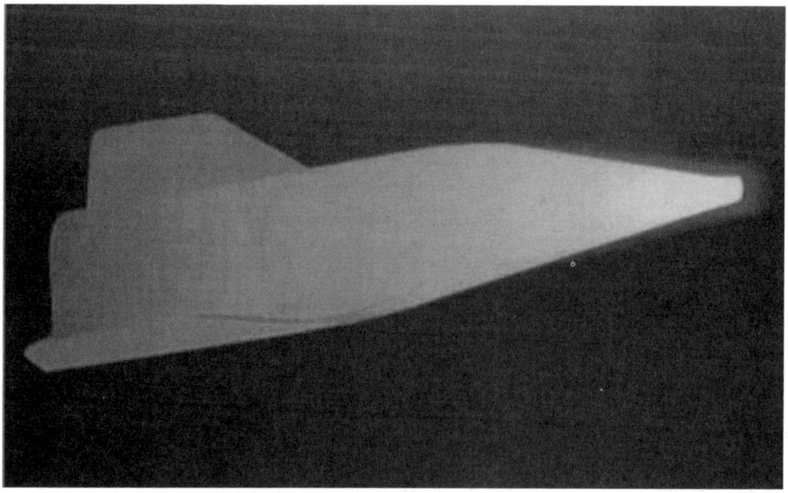

FIGURE 5 NOSE AND SKIN OF A VERY HIGH-SPEED VEHICLE BECOMING HOTTER THAN MELTING TEMPERATURE OF SOME STEELS. SUCH VEHICLES WERE NEEDED FOR HIGH SPEED WEAPON DELIVERY, DEFENSE AGAINST INTERCONTINENTAL MISSILES AND RETURN OF PEOPLE SAFELY TO EARTH AFTER SPACE

In the 1960's researchers developed high-temperature titanium structures that could withstand temperatures to about

500 degrees centigrade (500°C) and allow development of the XB-70 bomber and SR-71 spy plane which exceeded Mach 3.0. Metallurgists then developed INCONEL – a nickel, chromium and iron alloy that withstood 700 C temperatures. This resulted in its use on the Mach 6.0, X-15 research plane. We found that steels with columbium were found able to withstand 1400 degree temperatures to allow speeds 10-20 times those of sound.

Low-temperature metal structures, protected by higher-temperature plastic, glass, or ceramic materials were found best for re-entry vehicles. Here, metal structure was protected by thick" heat-sink" or "ablative" materials, which by melting or vaporizing, could safely absorb or carry away harmful heat.

Heat-sink materials were used on early ICBM nose cones. But much lighter ablative materials were soon found superior. Thus, ablative materials were used in later military systems and

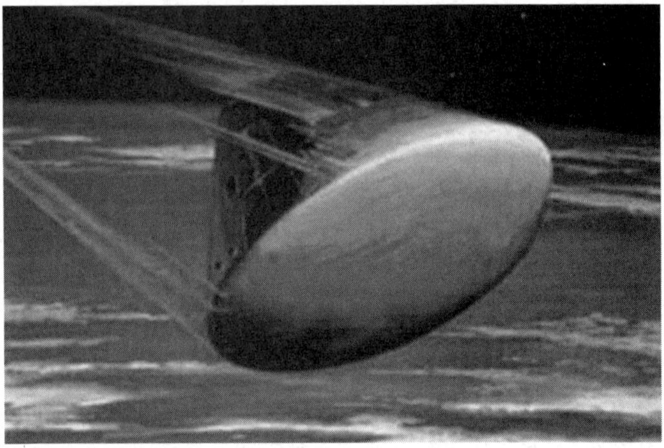

FIGURE 6 APOLLO REENTRY VEHICLE HEAT SHIELD PROTECTING THREE APOLLO ASTRONAUTS FROM SEARING HEAT DURING EACH SAFE RETURN TO EARTH.

with all NASA re-entry systems of the Mercury, Gemini, Apollo programs. Such materials were also used on hyper-sonic un-piloted missile systems which traveled faster than about Mach 6 in air. Typical ablative materials were phenolic fiberglass, nylon and

Teflon. NASA cited no serious heat-shield development problems on their programs, possibly, because its broad, high-drag heat shields suffered relatively mild heating intensities.

The Douglas Nike Zeus Anti-Missile Missile was a second generation hypersonic system whose "sharp" nose had to withstand higher pressures and heating rates than the large, blunt noses of military and NASA re-entry vehicles. Zeus also had to travel faster through higher-pressure and lower-altitude air. So, all its surface areas tended to experience higher pressures and heating rates. Zeus was also guided and controlled in both air and space, requiring control surfaces driven by stainless steel shafts, which had to be shielded from very hot, high-pressure air.

Overcoming Nike Zeus's hot heat barrier was vital for me. I had been an initial architect of this 3-stage missile – whose four movable control surfaces steered all 3 missile stages. They were located on the missile's most forward stage to allow removal of much wing weight and drag from the missile's second stage. This increased its speed, altitude and area it could protect. And things went well at first. Early hypersonic wind tunnel tests revealed the missile was flyable, even though vortex wakes shed by the four movable control fins were found to cause rapid changes in missile pitching and yawing moments during maneuvering flight. This behavior challenged missile autopilot designers, but was acceptable.

On the first flight, separation of the Booster stage by igniting the second stage rocket was not achieved – only a very large explosion. A small task force headed by Norman Augustine (who later became Assistant Secretary of the Army and, finally, the CEO of Lockheed-Martin Corporation) found that the drag of the two forward missile stages was much more than that of the separating booster. So, the middle and booster stage remained close together long enough for explosive rocket pressures to build up in the region between them. This problem was solved by holes in inter-stage structure – to let some gases from the burning second stage rocket escape as remaining rocket gases pushed the booster away.

On the next test flight successful staging was achieved. The missile accelerated further and higher, but then exploded again. I was devastated when we found that each of my four control fin's arrowhead-shaped leading edges (which were shaped to reduce the very high hinge moments Zeus's hydraulic actuators had to overcome) were also channeling unexpectedly large amounts of hot, high-pressure air into the V-shaped regions existing between each control fin and the missile's conical nose. And this hot air was destroying each stainless steel fin control shaft. Unfortunately, neither myself, nor those experts who reviewed my fin geometry knew enough about supersonic air breathing propulsion to realize the shape of the fin's leading edge, though not of ordinary air-scoop shape, was, in fact, acting like an airflow-gathering supersonic inlet or diffuser.

My poor plan-form shape was soon changed into a conventional one, and tight tolerances between body and fins left only very narrow channels for hot air to approach fin shafts. But all shafts were again destroyed on the next flight. After examination, hot air was still found forming behind complex shock-waves formed by the fins, and this heat was still propagating to the control shafts inside the shallow boundary-layer airflow along the missile's nose. So we had the enormous problem of sculpturing subtle barriers in the vicinity of the four control fins that could prevent formation and flow of hot boundary layer air.

About that time, a second problem-solving person: Dr. Henry Ponsford, a new structures leader, stepped forward to meet the immediate need. He proposed spraying water into the shaft region to maintain shaft temperature below its melting point. This added unwanted bulk, weight and complexity into each test missile. But water-cooling worked, allowing the vitally-needed flight tests to proceed while we desperately tried to sculpture a light, low-drag barrier ("a glove") to effectively shield the shafts from hot heat. A third key person: Robert Womack, a young, bright, intelligent, hard-working Douglas engineer worked day after day, week after week designing and testing dozens of different "gloves". None worked. Then, one night, instead of going to dinner with others, Bob stayed; kept designing one more barrier and tested it. And, lo, it worked!

FIGURE 7 SUBTLE BARRIERS 'GLOVES" THAT FINALLY EFFECTIVELY SHIELDED NIKE ZEUS CONTROL FIN SHAFTS FROM EXTREMELY HOT, HIGH-PRESSURE HEAT DURING ITS VERY HIGH SPEED HYPERSONIC FLIGHT.

It's probably unfair to single out these individuals since many people played vital parts in solving many more formidable heating problems during Nike Zeus development. But these come to mind in the overcoming of Zeus's heat barrier.

Finally, the third generation hypersonic systems embodied re-usable thermal protection system (TPS) materials – like ceramic tiles which were designed to maintain their structural and thermal integrity for the entire life of each Space Shuttle – lives which had to survive many scorching hot trips from space to Earth. Shuttle tiles were numerous enough that some could occasionally fail and be replaced. But none were allowed to melt or to ablate (to lose their mass) when exposed to heat. Success of the Space Shuttle reusable TPS was proven by over 130 safe trips through the atmosphere during its returns to Earth.

FIGURE 8 REGIONS OF VERY HIGH HEAT THAT ENVELOPE THE LOWER (WINDWARD) SURFACES OF THE REUSABLE SPACE SHUTTLE BODY, NOSE AND WING

The Space Shuttle's Thermal Protection System (TPS) worked very well. Its high-temperature "tiles" and multi-layer insulated structure safely survived the heat barrier during over 130 safe re-entries through Earth's atmosphere from space. But in February 2003 a large piece of insulation foam fell from the Space Shuttle Columbia External Tank during its way to orbit, hitting and damaging (piercing) a reinforced carbon-carbon panels that were used on the Orbiter's left wing.

A long investigation concluded that this damage created a flow path for the influx of re-entry heat that caused Space Shuttle Columbia's disintegration high above Texas on the 23rd of February. Before this, most TPS experts had assumed Shuttle TPS could successfully overcome the heat barrier for likely operational situations. But unfortunately, all 8 Columbia Astronauts had to give their lives to provide final information on needed changes in TPS and TPS procedures to insure Space Shuttle protection during return through the heat-barrier to Earth. So, the Space Shuttle TPS can surely now declare victory over the heat barrier. For its now protected vehicles during 133 safe round-trips to space – many more trips than all those trips protected by expendable, melting and ablating TPS.

So, the heat-barrier, like the sound-barrier, has been overcome. But, the high cost of significant TPS weight and hazard remains. Thus, the challenge remains to develop TPS materials and vehicle TPS designs that can reduce this large cost.

3 THE WORLD'S FIRST ANTI-MISSILE MISSILES

Missile developments during the Halcyon Years included Intercontinental Ballistic Missiles (ICBMs) and Anti-Ballistic Missiles (ABMs) to defend a country against another country's (ICBMs. I was at Douglas for a time when assigned to its ABM program: the **Nike-Zeus Anti-Ballistic Missile**.

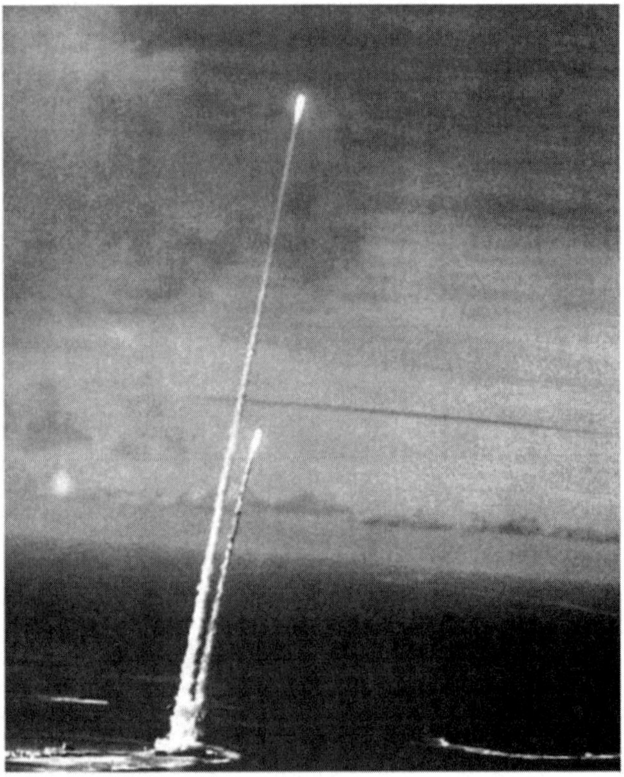

FIGURE 9 SALVO OF ZEUS/SPARTAN ANTI-MISSILES LAUNCHED DURING CBM INTERCEPT TESTS

Everything about the Nike-Zeus ABM System was splendid. It was named after "Nike," the Greek Goddess of War and "Zeus," the supreme ruler of all the gods. And its mission was the most important one I could imagine in 1960 – defense of the United States against devastating nuclear destruction carried by ICBMs from the Soviet Union, which carried huge nuclear warheads at speeds of about 25,000 miles/hr. over thousands of mile distances on high arching paths above Earth towards the United States. And since every part of ABM systems (target detecting and discriminating and tracking phases and missile guiding and flying phases) was then declared impossible by powerful and prestigious scientists, this added more challenge-motivation-resolve to all on the Nike Zeus ABM team.

The Nike-Zeus missile was one of the three elements of the Nike-Zeus Ballistic Missile Defense System being developed by Bell Laboratories and the Western Electric Company for the U.S. Army. One element was the Phased Array Radar (PAR) that would detect incoming enemy ICBMs as they crossed high over the North Pole. Another was the Missile Site Radar (MSR) – a structure that looked much like a Mayan pyramid in both size and shape. Using PAR data, the MSR established precise tracking of all incoming threat objects and guided swiftly, moving Nike-Zeus missiles on appropriate trajectories to intercept the ICBMs sufficiently high and far away so any detonating nuclear warheads would not subject defended areas to catastrophic nuclear blast and radiation damage. As developer of the "transistor," Bell Laboratories had pioneered the integrated circuits that give birth to today's solid state digital world, so they clearly led this technology. So Nike-Zeus PAR, MSR, and their signal and data processing systems were at the leading edge of radar and electronic state-of- art.

Engagement time and space provided by the Zeus PAR and MSR required Zeus missiles to race over hundreds of miles of distance and climb about as high in the sky in less than a few minutes of time, requiring The swift, agile Nike-Zeus missile had been designed years before I joined the Nike-Zeus Aerodynamics

Group and was to have its first flight test soon. Zeus's main mission was intercepting high-speed ICBMs but it also had to intercept enemy aircraft at vert long range and low altitude, where missile speed was slowest and large wings were deemed necessary to provide maneuvering force. Zeus missiles reach speeds more than twice those of the fastest surface-to-air missiles that were then flying.

But extremely swift missile speed through Earth's atmosphere caused the metal skin of the Zeus missile to become hot enough to melt. Thus, layers of high-temperature, Teflon material was used to prevent very hot heat from reaching and weakening missile structure. Shown below is the metal skin of

Figure 10 Original winged Nike-Zeus missile

the original winged Nike-Zeus missile totally covered by heavy Teflon material.

During development, aerodynamic heating was found higher than expected, so more Teflon thickness and weight was needed to protect missile structure. And this added weight was reducing missile speed and, hence, the protection that it could provide. So, when I joined the Nike-Zeus Aerodynamics group my first job was to explore reducing missile weight by reducing the area of the Zeus wings.

Less wing area reduced missile lift, so some lessening of maneuvering against evading aircraft was expected. But less anti-aircraft ability was acceptable if higher speed with smaller-lighter wings, increased Zeus's anti-missile capability. Two junior engineers helped me estimate Zeus missile weight, lift, and drag changes by slight reductions in Zeus wing sizes. Then, on a computer, we flew the modified missiles on their long air defense mission – and were surprised to find no loss in Nike-Zeus anti-aircraft capability at all. I then recalled that a vehicle's lift is roughly proportional to the square of its speed. So, possibly, lift loss from wing removal was being compensated by lift increase from the

Figure 11 Later version of Nike-Zeus missile

higher speed that less wing weight and drag was allowing. I decided to see what would happen if all wing weight, lift and drag was removed. Both anti-missile and anti-aircraft capability significantly increased. *My hunch was right*.

Then the idea came that all wing and control fin weight and drag on the second stage could be eliminated by locating a single control system (which would provide aerodynamic control in air and thrust-vector control in space) on the most forward stage. After some work, my re-designed missile was found almost 10%

lighter; much faster in speed; and with more capability than had originally been promised to the Army. Shown on the previous page, is the final version of the new missile (called "Spartan"). It differs from my original re-design by having small fixed fins on the two rear stages, and a different platform for the control fins on the forward stage, which provide aerodynamic and thrust-vector control.

But now all missile groups and senior designers were called in to search for any technical flaws in the Zeus re-design that my tiny group of relatively junior engineers had done. I became very anxious. Would my attempt to design something new be public failure and end my future at Douglas? Thankfully, no major flaw was found. The next step was convincing the U.S. Army to make a radical change in the Nike-Zeus missile. The missile's guidance hardware and two powerful solid rocket motors remained the same, but everything else was changed. My idea of a single aero-reaction control system to control all three vehicle stages would enable huge savings in missile cost and complexity, but probably longer development. The astute Vice President of the Douglas Missile and Space Division (Robert L. Johnson) presented our Nike-Zeus missile modification to Bell Laboratories and Army officials on a Friday. From past experience with very minor missile changes, no decision was expected for weeks. So, imagine our surprise and amazement when full go-ahead on the changed missile was received the very next Monday morning. The features and benefits of our revised missile were apparently so overwhelming that usual chains of reviews and approvals were done away with.

But challenges soon arose. The most serious were due to my 4 forward-located control fins. Swift fin rotations quickly generated aerodynamic control forces and moments in air. And, when a small central rocket engine ignited, swift fin rotations quickly generated propulsive control forces and moments from by rocket nozzles, embedded in each fin. Such aero-propulsion control hadn't been attempted before and was made more difficult by each high internal and external pressure and temperature every control component had to withstand.

Previously mentioned were the many test failures and laborious research to perfect a way to protect stainless-steel control surface shafts from the erosive hot heat of high-speed airflow. Much effort was also required to perfect high temperature materials to protect the swift swiveling propulsion control nozzles from erosive internal heat from 90 degree turnings of rocket exhaust flow. And another challenge was development of an extremely responsive autopilot to quickly-precisely steer the very high-speed missile – and this despite the violently changing aerodynamic forces and moments caused by the strong, vortices shed over missile rear body and fins by the 4 deflecting control fins.

Thus, I was frustrated when asked to give up my Nike Zeus aerodynamics group responsibilities and head a new anti-missile missile project in a different organization. For almost all of the problems caused by my re-designed Nike Zeus missile were far from being solved. But though I left the Zeus program, I remained in close contact with its people. And I became much more comforted and less concerned as I noted "right persons" were always seeming to come forth at the right time to solve each Zeus problem in a timely way. Thus, every problem I had once despaired over was finally overcome, and the next page shows successful launch of a finally-perfected, properly-performing missile.

In February, 1967 a salvo of missiles like the one below launched against an incoming Atlas ICBM launched from the United States. One missile failed in mid-flight, but the other missile worked perfectly (along with all the other system elements) to become the world's first missile to successfully intercept an ICBM.

Figure 12 Test launch of Nike-Zeus missile

Visionary people at Bell Telephone Laboratories had been thinking about a small companion to the Zeus Missile. It would accelerate faster to about the same speed over a short distance to provide "last-ditch" protection of vital targets by achieving top-speed speed in only several seconds. I was selected to head-up a small team to investigate the feasibility of a "Sprint missile". At first, it didn't seem possible. Rapid achieving of very high speed at very low altitude caused vehicle stresses-pressures-heating-rates far beyond any ever conceived for high speed flight. And we found stupendous uncontrollable aerodynamic moments developing over cylindrical Sprint missiles during high acceleration at very low altitude and high-speed. Finally, I had a long dialog with Jack Hines, a very expert Douglas aerodynamicist, about missile shapes that would minimize excursion of vehicle center-of-pressure during accelerated flight. Jack finally said a cone shape was best. But Jack, who wasn't shy about suggesting things, did not urge it – possibly because cylinder shapes had been always been satisfactory for missiles since the dawn of rocketry. A thought then came to me that a cone shape might also minimize missile center-of-gravity excursion during rocket burn. This convinced me to attempt an "all-cone" Sprint missile.

After some grumbling, the team went to work on it. Things worked out better than expected. Tapered rocket motor cases and rocket grains caused by cone shape were found superior in their ability to survive high acceleration. And very innovative component packaging kept center of pressure and center of gravity excursions so small that missile moments never exceeded flight control abilities. We also found that the needed sharp nose could withstand stupendous heating rates during short duration missile flights. And a cone shape spread high flight pressures and heat flux over the missile in the best possible way. Thus, we were able to

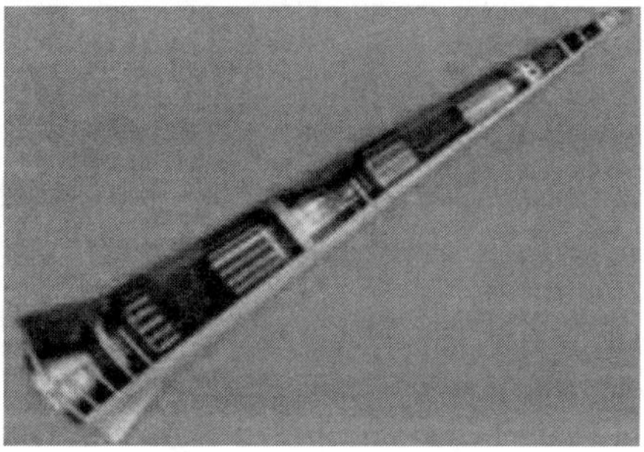

FIGURE 13 ORIGINAL DOUGLAS SPRING MISSILE DESIGN RECOMMENDED TO BELL LABS AND US ARMY

design a two-stage 'Sprint Missile' companion to Nike-Zeus (below) that met all Sprint missile system requirements.

Unexpectedly, the U.S. Army opened the Sprint missile procurement to other bidders, and our all-cone missile design [2] was in the data package all bidders received. The Douglas Sprint proposal team was led by new people of high rank and sometimes of different technical opinion than mine. So the new submitted Douglas design had a cylindrical lower stage, which made it somewhat more ordinary than the original Douglas all-cone design. Martin Marietta won Sprint. A main reason was superior structural integrity of the rocket grains their rocket contractor provided. Months later, still depressed, I was called into our Chief

Engineer's office. He had a drawing of Martin's award-winning two stage all-conical Sprint missile. It was identical in cone angle to my team's design and only about one inch different in total missile length. The differences were: a somewhat longer nose and movable control fins on the Martin upper stage vs. movable control flaps on our original upper stage; no fins at all on the Martin lower

FIGURE 14 SPRINT MISSILE DEVELOPED BY MARTIN MARIETTA COMPANY. IT WAS USED WITH THE NIKE-ZEUS/SPARTAN MISSILE IN THE US ARMY SAFEGUARD ANTI-BALLISTIC MISSILE DEFENSE SYSTEM

stage; vs very small fins on ours. As I left, our Chief Engineer smiled, saying, "So, stop being sad Dave! Martin is building your all-cone Sprint design".

In 1975, The Safeguard Anti-Ballistic Missile (ABM) System began to be installed near Grand Forks, North Dakota. The system included: a north-facing Perimeter Acquisition Radar; Missile Site Radar and Zeus/Spartan and Sprint missiles. And all Anti-Missile Missiles were in underground silos as shown below.

The Safeguard Ballistic Missile Defense System was the very first to defend the United States and U.S. ICBMs against enemy attack. It has now been replaced by newer anti-ballistic missile systems. But its existence helped maintain peace between the United States and Russia during the Cold War by helping to deter enemy nuclear attacks during its existence. It also resulted in the

FIGURE 15 US ARMY MISSILE TEST FACILITY

first Strategic Arms Limitation Treaty, which caused dramatic reductions in nuclear warheads and ballistic missiles of the United States and Soviet Union. Though I never voiced it, I always hoped that at least once in my aerospace career I might be in at the birth of a new vehicle. So I am certainly grateful for the opportunity to have had a part in developing the World's first two defense-oriented anti-missiles. And I know my Zeus/Spartan/Sprint teammates feel just the same.

Note: technical Information and photographs of the Nike-Zeus/Safeguard and Sprint missiles are from www.wikipedia. www.google.com.images and full-size Zeus and Spartan and Sprint missiles are on display at the US Space and Rocket Center in Huntsville Alabama in the U.S.A.

4 LAUNCH AND MISSILE SYSTEMS TO PROTECT NAVAL CRAFT

FIGURE 16 AIR AND SEA WAR ENVISIONED BY SOME IN 1963-WAR THAT COULD THREATEN ABILITY OF LARGE AND SMALL SURFACE NAVY COMBATANTS TO SURVIVE FUTURE AIRCRAFT AND MISSILE ATTACKS.

The Halcyon age of civil and military light was in full-bloom, with receptivity to bold, compelling ideas for improving it. And this included vehicle launching systems. Zeus vertical launcher development for the Army greatly improved surface missile launching rate, and I felt that surface-missiles of the Navy Sea Systems Command could also benefit from such vertical launching.

FIGURE 17 VERTICAL LAUNCHING OF A NIKE ZEUS MISSILE FROM UNDERGROUND LAUNCHING CELL

I was given modest resources to explore the possibility of naval ships having defense against both aircraft and ballistic missiles and very high launching rate against both. At this time (1963) no naval ship had been sunk by a missile, so there was some apathy towards protecting ships from missiles and there was much resistance to our idea of swiftly launching many missiles from inside ships.

Just as many ready-to-launch US ballistic missiles or anti-missiles can be quickly vertically-launched from underground "silos" if under attack by many enemy threats, we found many ready-to-launch navy missiles could be quickly launched against many threats from inside below-deck vertical canisters that also served as missile storage and shipping containers. By contrast, existing Navy launchers swiftly sent missile bodies, wings and fins from below-deck magazines to above-deck trainable launchers where they were quickly assembled into ready-to-launch missiles. These launchers were mechanical marvels in swiftly transporting and assembling missiles, but their transfer and assembly still used-up precious seconds between each launch. But, throughout all naval history fears were felt over below-deck ignition of explosive ordnance. So, the Surface Navy had long fearfully forbidden igniting of explosive ordnance (like missile rocket motors) below the main decks of all large naval combat ships.

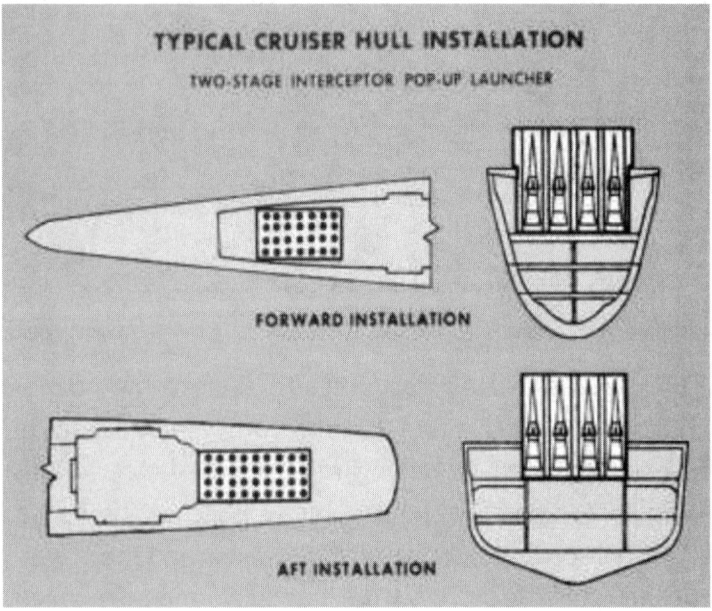

FIGURE 18 TYPICAL CRUISER HULL INSTALLATION

This installation above was proposed in 1963. It shows below-deck missiles protecting any navy ship with an Aegis SPY-1 radar against ICBMs. Installations of missiles more like existing navy missiles were also proposed to cope with aircraft and cruise missiles.

So there was strong initial Navy resistance to below-deck launching of missiles. Overcoming this resistance was difficult. But many Navy people began accepting the need to defend ships against both aircraft and missiles and launch missiles more rapidly from the ships. One was a Navy Captain on an international panel concerned with protecting small naval craft from air attack. He invited me to brief his panel on our ideas for this. This led to a Douglas team that evolved the system shown on this page, to protect small ships against air attacks.

FIGURE 19 US NAVY SMALL NAVAL SHIP

Existing small naval ships had insufficient depth for installation of fixed vertical canister launchers. So, missiles in cannisters were installed within a trainable launcher. But the idea of the missile and its shipping-storage-launch canister being a "ready-round-of ammunition" was retained. Our system happened to be of special interest to the West German Navy, since we had used their 250 ton "Jaguar" **fast patrol boat (FPB)** as a typical small ship platform for our defense system. So, in about 1985 I was invited to a Jaguar FPB Base in Flensburg Germany on the Baltic to learn a little more about Jaguar FPB operations.

As I entered the German base with very high hopes of learning about FPBs , I recalled earlier, as a 15 year-old school boy during World War II, of being very interested in the German battleships Bismarck and Tirpitz because of their very extraordinary sea battles with the British navy. But, these hopes soon sank when I met sullen, scowling, imperious "Frigaten Kaptain Mueller", the FPB Base Commander. It was apparent He was a senior German naval officer who had probably fought in bloody battles against English naval ships in World-War II and, possibly, still disliked any English-speaking person he met. He was probably obligated to tell me something. But I felt that he wished me gone as soon as possible as He said nothing, letting me struggle to formulate my first question.

While struggling, I happened to glance at a big picture of a warship on the wall behind him, and recognized it as the "Tirpitz", one of the two German battleships I admired as a boy. Before realizing it, I blurted out all my school-boy knowledge and admiration for this ship. He seemed dumb-founded at what I knew about it. He had been its Executive Officer until it was finally sunk by waves of British heavy bombers in a Norwegian Fiord. And his scowl changed to almost a smile during a very animated discussion of English and German battleships. Then he stood up, politely bowed, and invited me to join him on an FPB operation!

He assembled a crew and we were off on a wild ride over the Baltic for most of the afternoon. But during all of racing around, four "blips" remained on the FPB's radar screen. There were East German and Russian MIGs following us from high above. Pointing at the four blips, Kapitan Mueller quietly said: "This is my problem. I hope your system can help me". I left him motivated to do just that.

After leaving with resolve to give Kapitan Mueller a system that could fight off the formidable enemy foes above him and the Baltic Sea, I realized its development would require its key elements undergo handling and launching trials on a ship somewhat like an FPB. So imagine my surprise on returning home, to find such a ship was owned by Douglas and its founder Donald Douglas Sr. It was a Navy craft now called "Pacific Surveyor". It was smaller and not as sleek as a German FPB. But it had room for an FPB launcher and sailed the seas of the Pacific Ocean. Almost the instant that interest in the Pacific Surveyor was shown, Donald Douglas Sr. (its co-owner) wanted to hear about my project.

He, of course, was that aviation pioneer whose most successful aircraft was probably the DC-3, which many still consider the greatest airliner advance in history. I was invited to lunch on his yacht "Ladyfair" in Long Beach Harbor and told him of the new missile and launcher we were developing to protect small naval ships against aircraft and missiles. Being both a flyer and a sailor, he was enthused over developing both flight vehicle and naval launching technology with his "Pacific Surveyor". No longer responsible for daily Douglas direction, Mr. Douglas, not only agreed to our use of his boat, he also gave himself the job of drawing up deck modifications for installing our launcher on Pacific Surveyor.

FIGURE 20 INSTALLING MISSILE SHIPPING-STORAGE LAUNCH CANISTERS IN THE LAUNCHER ON THE DOUGLAS SHIP

The next page shows the launcher already installed on Pacific Surveyor's modified deck and missiles in circular canisters being lowered into it. Not shown is the smaller volume square canister that was also to be successfully launched from.

Our missile-canisters were viewed as "ready-rounds of ammunition" – same in concept to what we proposed to the navy for missile transport-storage-launch on larger U.S. Navy ships. So, missile transport-storage-launch from canisters was vital in sea trials. Mr. Douglas visited some trials as the below photo shows.

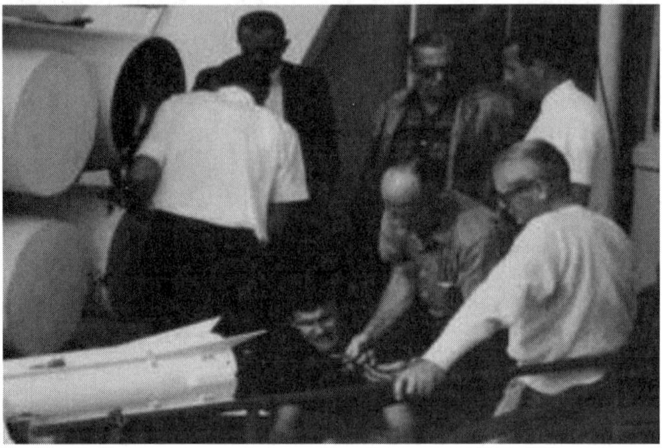

FIGURE 21 MR. DOUGLAS AND MR. CONANT PREPARE LIVE MISSILE ON SHIP

We completed handling trials and constructed two flight missiles that would be propelled by solid rocket motors to supersonic speed in several seconds. Navy cameras on San Clemente Island in the Pacific would track our missiles after their launch from our Douglas ship. Mr. Douglas provided both his ships for our firing trials: his Pacific Surveyor to launch our missiles and his Ladyfair for support. Two days before launch, Pacific Surveyor departed from the Navy Seal Beach facility with two live missiles in the launcher. Ladyfair departed from Long Beach harbor with: Mr. Douglas and Mr. Conant; (a very good friend and former Vice President) and a corporate person and me, who made up the crew.

The first day we were tossed about vigorously while plowing through a mild Pacific storm. We finally reached Catalina Island that night dropped anchor off a remote part of the Island called "Cat Harbor". I ate very little and finally fell into fitful sleep despite the wind and rolling-yawing ship. But loud shouting brought me suddenly awake. Heavy seas and strong currents had caused the anchor to start dragging during the night and we were now slowly moving toward the rocky shore. The other crewman and I were somehow able to quickly raise anchor as Mr. Douglas and Mr. Conant rushed to start the engine and man the tiller. But, just as the engine started up, the hull loudly scraped over unseen rock or reef and our rudder sheared off. And we could soon see surf breaking over jutting rocks and dark rocky shore that was rapidly moving closer. Things were happening so fast I had no time for panic and fear, though later I realized the dire peril our rudderless boat had been in as we were swept towards rocks. Strangely, time seemed to slow enough to listen and pray. In an instant I recalled the Bible account (Matt 14: 24-32) of Jesus walking on the Sea of Galilee on a stormy night to frightened disciples in a small boat; and Jesus saving Peter from sinking and wind ceasing when Jesus came aboard the ship.

The wind did not cease when Jesus came aboard our ship – but He guided Mr. Douglas and Mr. Conant to do what was needed to save our lives. We had no rudder, but we still had an engine with twin throttles and propellers which meant we had a crude means of accelerating and turning ourselves with propulsion steering. Mr. Douglas and Conant, being expert sailors besides being expert aviation men, quickly learned how to play with engine power to each propeller. Their skillful engine throttling slowly turned and accelerated our boat away from the rocky shore. Then, after a very tense time we were finally able to slowly work our way back to a safe distance from the shore. We then let down the anchor again, making sure this time it was firmly in place. What a relief!

The next morning the storm had weakened and an ocean-going tug to tow us back to Long Beach Harbor was called for. But rain and poor visibility were now predicted for Pacific Surveyor on the launcher day. This would prevent good Navy photographing of missile launches and flight. This was important for compelling proof of our system's naval potential. So I immediately "scrubbed" our missile firings until the next fair weather day. Mr. Douglas said nothing, but seemed to like the decision. The voyage back to Long

Beach under tug power was pleasant. I learned some about Mr. Douglas's early life – like when he was 14 in Washington DC, and took the trolley all the way to Fort Meade in 1908, to see Orville and Wilbur Wright both fly their Wright Pusher.

When we parted in Long Beach late in the day I took a chance with Mr. Douglas and joked that: "I hope you are not going to deduct too much from my wages for my project resulting in so much damage to your Ladyfair". He laughed, saying "Ladyfair is insured, so I won't deduct any of your wages this time Dave". "Furthermore, next time we will go out for pleasure (not work) as soon as she's fixed." This promise wasn't fulfilled. But I was relieved. We were still friends.

Weather was perfect on the re-scheduled firing day. I had been flown to the Navy control and observation area on San Clemente Island and the countdown to Missile 1 launch was in its last 10 minutes. Now, my enthusiasm and zeal had driven any suggestion of failure from thought during the program's entire duration. But right then, a negative thought occurred. I suddenly remembered being once told that, historically, the first launch of every Douglas missile had always ended in a failure and fire; furthermore, my program's first launch would involve, not only an untested missile and motor, but un-tested launcher as well. And a motor or launcher failure could endanger our small vessel and its crew. My future again seemed in peril as it seemed to be a week ago. But the suggestion was immediately replaced by the assurance that all would be well.

Missile 1 flew from its launch cell like an arrow, making a thunderous sonic boom as it exceeded sound speed in less than two seconds and sped from sight. Missile 2, launched from a narrower cell, also flew like an arrow, generating another thunderclap and disappearing from sight. Missiles, rockets and launcher performed flawlessly and every person on my team had done a perfect job!

The launch trials were a complete success with modest

FIGURE 22 MISSILE LAUNCH PHOTOS FROM HIGH SPEED CAMERAS ON THE SHIP AND ON SAN CLEMENTE ISLAND

canister temperatures and pressures and very acceptable noise and vibration measured during missile acceleration out of the larger volume circular canister and the lesser volume of the preferred square one. In all cases, measured values agreed well with Predicted values – so that we could scale up our data to predict results for larger missiles on larger U.S. Navy ships. Shown are missile launch photos from high- speed cameras on the ship and from Navy cameras on San Clemente Island.

Mr. Douglas seemed ecstatic, having me give a luncheon talk on the tests to his top executives. Navy resistance to vertical launching of missiles from inside ships noticeably diminished as I showed the very mild canister pressures and temperatures experienced during successful launching of our missiles from a naval craft. Soon we received news that the U.S. Navy was beginning a preliminary vertical canister launcher design, similar to our vertical vented Nike Zeus Launcher. We were also told their design included our missile-canister concept (wherein the missile canister serves as shipping-storage-checkout and launching system). The German Minister for Research and Development was also pleased that my German Jaguar FPB ride had resulted in

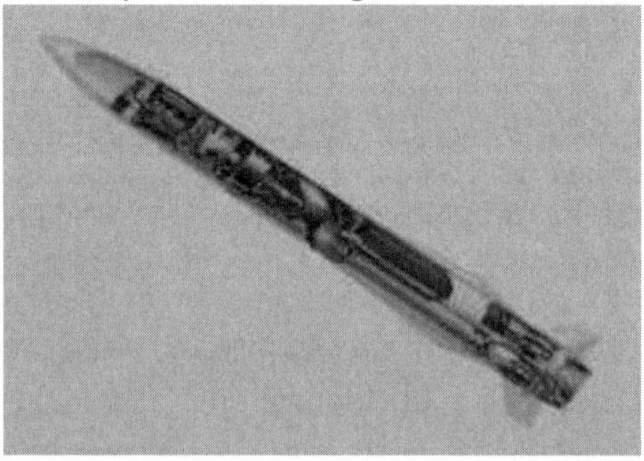

FIGURE 23 SEEKER, GUIDANCE, WARHEAD, PROPULSION AND FLIGHT CONTROL SUBSYSTEMS OF MISSILE ELEMENT OF THE DEFENSE SYSTEM DESIGNED BY DOUGLAS AIRCRAFT TO PROTECT WEST GERMAN NAVY SHIPS

our missile handling and firing trials that Donald Douglas, himself, had participated in. Within weeks the West German Ministry of Defense awarded Douglas a large contract to perform a detailed study of a small ship defense system for its Navy.

Our small ship protection system mission required no advancement in vehicle speed or acceleration, or propulsion, structures, control system state-of-the art. But it required significant advances in electromagnetic (EM) wave propagation ability of ship radars (to properly illuminate attackers with EM radiation). Also needed were advances in emitting target-illuminating EM wave energy from the ship. This energy is reflected by the target and "homed in on" by an advanced radar receiver and guidance system in our missile. Ship radar contractor (HSA) and missile radar contractor (Motorola) said their work on our Study gave them opportunity to advance radar system state-of-the-art. Their research also increased my knowledge of EM energy generation-modulation-detection – knowledge u s e d later in study of spaceships, propelled by EM fields.

Our system definition study for the German Navy gave me an excuse for more contacts with Mr. Douglas during those

FIGURE 24 NEARLY SIMULTANEOUS MISSILE LAUNCH FROM FORWARD AND AFT EX-41 VERTICAL LAUNCHING SYSTEMS

occasions when he visited his office in Santa Monica and requested that I brief him on our German study progress. The German Navy was very enthusiastic over our defined system, and lobbied very intensely for funds to develop it. Unfortunately, the German Ministry of Defense could not obtain these funds. Thus the German Navy and Douglas had no opportunity to develop our system. By then, U.S. Navy assessments confirmed our claims that current above-deck trainable launching systems could not launch missiles rapidly enough against heavy air attacks. And industry was invited to develop an EX-41 Vertical Launching System for U.S. Navy ships.

By then Douglas and McDonnell Aircraft were merged into a single McDonnell Douglas Corporation. Donald Douglas Sr. had retired and the merged corporation chose not to compete for the EX-41 vertical launcher. Shown below is a missile launching from it.

In March, 1983, just before the Halcyon years of air and

FIGURE 25 SM3 MISSILE LAUNCHED FROM A CANISTER IN AN EX-41 LAUNCH CELL INSIDE AN AEGIS BMD SHIP. THIS SYSTEM IS SOMEWHAT LIKE THAT PROPOSED TO THE US NAVY BY DOUGLAS AIRCRAFT (3) ABOUT 50 YEARS AGO. IT ALSO USED A SPY1 RADAR; CANISTERIZED MISSILES AND FORE AND AFT VERTICAL LAUNCH CELLS.

spaceflight were over, U.S. President Ronald Reagan announced his Strategic Defense Initiative (SDI) to develop new technology to intercept enemy nuclear missiles. The Navy now had significant potential against such threats with their Aegis/SPY-1 radar and their new EX-41 vertical launching system. So, one selected SDI "Star Wars" system used these elements and a new anti-missile. Navy Aegis BMD system development continue to this day, using the AN/SPY-1 radar and also the EX-41 vertical launching system (VLS) and 2-stage SM3 missiles with "kill-vehicles" that can destroy enemy warheads by high impact energy. Wikipedia describes the Aegis BMD tests and says up to 80 defense systems might be installed on ships.

In a way, I have nothing at all to show for all those years spent on naval defense work. It didn't result in any Douglas system ever protecting any German or U.S. Navy ship; nor result in a merged McDonnell Douglas Company ever making a vertical launcher for any Navy. But, still, it was satisfying to do pioneering work that may have helped provide needed momentum for developing the launching systems that all major navies now use. It was also satisfying having the chance to plant the first suggestions that naval ships could provide BMD protection for themselves and others. In any case, it also gave people exciting, technology-advancing work. And it even helped my personal development somewhat. Foreign travel for 3 years forced me to learn to relate better with other people and other cultures – and this somehow developed more empathy in me for others – not just those I already knew and liked. So I am very grateful for my naval adventure. It surely enriched and deepened my human experience.

5 AIR BREATHING AND NUCLEAR FUSION POWER AND PROPULSION

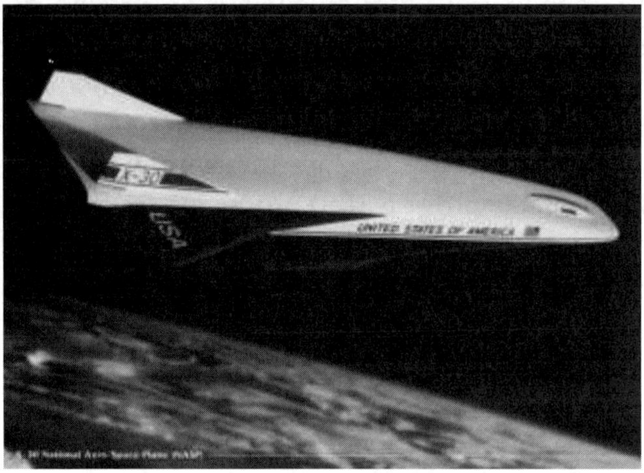

FIGURE 26 THE SELECTED NASP VEHICLE REFLECTED THE McDONNELL DOUGLAS NASP DESIGN

My arrowhead-shaped Zeus control fins, which funneled scorching hot air toward each control fin shaft, had revealed my woeful ignorance of supersonic combustion ramjet-scramjet) propulsion. So I resolved to remedy this as soon as possible. This resolved was aided by an opportunity to compare missiles propelled by solid-rocket propulsion with those propelled by ramjet-scramjet propulsion. This comparison enabled me to work with the nearby Marquardt Company – one of the few U.S. ramjet engine makers. And this led to meeting and working with ramjet-scramjet people at McDonnell Aircraft after it took over Douglas Aircraft. This led to working on a Defense Advanced Research Projects Agency (DARPA) and Air Force program called "**National AeroSpace Plane" (NASP)**. It was to dramatically reduce travel cost between Earth and space by means of a reusable airplane-like vehicle propelled by ramjet-scramjet engines and rocket engines. Various aerospace companies were competing for the development of NASP, including the McDonnell Aircraft

Division of McDonnell Douglas. And I was invited to join the McDonnell NASP team from my prior working with vehicle and propulsion people on this team.

NASP was an exciting and challenging program. But excessive weight and volume of its cryogenic liquid hydrogen fuel was preventing the NASP vehicle from achieving the smaller size and lighter weight needed to make it attractive.

So somewhat on my own, I began looking for ways to reduce NASP propellant. Dr. Frank Mead[2], a visionary scientist who then headed much of the Air Force advanced propulsion research was pursuing possibilities for much more intense heating of hydrogen propellant than by the usual way of mixing and burning it in oxygen or in air. One possibility was intense heating of hydrogen by energy released from "clean" nuclear fusion of hydrogen and boron 11 nuclei (p-B11). Another was energy release from mutual annihilation of very small amounts of matter and anti-matter (protons and anti-protons). I also received anti-matter help from David Morgan and Steve Howe of Lawrence Livermore and Los Alamos National Laboratories; and fusion propulsion help from Dr. George Miley at University of Illinois Fusion Lab and Robert Bussard of the EMC2 Company.

Bob Williams, the NASP program manager, became interested in my NASP nuclear fusion and matter/anti-matter annihilation work. At one meeting in Washington DC, he invited another visitor, Apollo astronaut John Young, who first flew the Space Shuttle, to hear my briefing. They both encouraged me to keep going on this "more far-out NASP work". I also enjoyed similar discussions with Pete Conrad, another Apollo Astronaut and then a

---

[2] **Ed. Note:** Dr. Frank Mead became the Director of Research and Development for the Air Force in the 1990's and early 2000's but was prevented from providing any notable innovation to aerodynamic propulsion by the Air Force brass, as related to this Editor in person. Apparently, he came too close to classified technology more than once and in the end, Frank said to me, "I am just a storefront" when we talked at JPC 2001.

Vice President in our Commercial Aircraft Division. Sadly, Pete was killed years later in an accident.

Now, by about this time the 'Halcyon Years" were almost over. Bob Williams was replaced by less bold NASP management and later NASP was finally ended when its technical-performance goals were deemed unachievable. But my work on NASP-like vehicles powered by advanced propulsion continued at McDonnell Douglas until I retired. And my company, Flight Unlimited, continued this work.

Both nuclear fusion and matter-antimatter annihilation reactions were found able to heat hydrogen fuel to such high engine exhaust temperatures and velocities that needed NASP liquid hydrogen fuel mass could be reduced by factors of 4 to 5. And vehicle cost analysis showed that lighter NASP-like vehicles propelled by matter/anti-matter annihilation could achieve vehicle Life Cycle Costs as little as only about one twentieth of those of the Space Shuttle Space Transportation System costs [4]. But anti-matter work was finally discontinued because of uncertainty in needed shielding mass to absorb intense gamma rays emitted by total annihilation of matter, and also uncertainty in the high-cost to manufacture anti-matter. Our work finally focused on clean fusion of hydrogen {specifically its proton nucleus) and boron 11 ions (p-B11 fusion) – with energy from their fusion heating airflow in either a nuclear rocket or ramjet-scramjet engine. Unfortunately, p-B11 fusion[3] required much more input energy and higher fusion temperatures than other fusion reactions. But other fusion reactions emitted high-energy neutrons that caused much radioactivity. By contrast, p-B11 fusion only emits fairly few low-energy neutrons that caused very

[3] **Ed. Note:** This form of fusion promises to be four times the output and much safer than any other source. See COFE3 presenter, Eric Lerner's website: https://focusfusion.org/category/eric-lerner/ or https://lppfusion.com/ for more info.

little radioactivity.[4] So, all of our final NASP-like vehicles embodied high-speed air breathing ramjet-scramjet engines and rocket engines that heated hydrogen by nuclear-fusion. Shown in the next figure is such a vehicle – a collaboration between my company and aerospace students at Saint Louis University [5]. The vehicle used an "Inertial Electrostatic Confinement" (IEC) system[5] conceived in 1967 by Dr. Robert L. Hirsch, and then considerably advanced by R. W. Bussard over many years [6].

We found igniting fusion power systems before takeoff required large amounts of ground electric power and this significantly increased ground operations costs. But air breathing spaceplanes didn't need fusion rocket propulsion until high-speed and altitude are reached.

[4] The reaction is p + 11B → $3He_4$ (8.7 MeV) and "aneutronic" with no dangerous, radioactive neutrons emitted. Also see https://focusfusion.org/ .

[5] **Ed. Note:** History has proven that Dr. Philo T. Farnsworth also built and demonstrated an IEC for Bell Labs in the 1960s. Most references now credit both with the nomenclature "Farnsworth-Hirsch Fusor" as noted in Dr. George Miley's text, *Inertial Electrostatic Confinement (IEC) Fusion: Fundamentals and Applications*, Springer, 2013 and many other online sources. IEC has been identified in recent times as an ideal fusion power unit because of its ability to burn aneutronic fuels like p-B11 as a result of its non-Maxwellian plasma dominated by beam-like ions. This type of fusion also takes place in a simple mechanical structure small in size, which also contributes to its viability as a source of power.

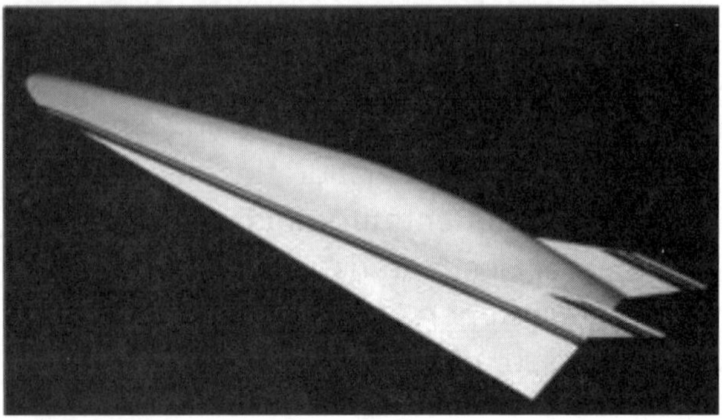

FIGURE 27 A NASP-LIKE VEHICLES EMBODIED HIGH-SPEED AIR BREATHING RAMJET-SCRAMJET ENGINES AND ROCKET ENGINES THAT HEATED HYDROGEN BY NUCLEAR-FUSION

So, an on-board power system to provide enough electric power to ignite a fusion reactor at high altitude and speed was found to be very desirable. In this respect, the Russian Leninetz Company showed possibility of MHD (magneto-hydro-dynamic) interactions gathering significant electrical energy from hypersonic airflow passing through jet engines by an MHD interaction. And the extracted electrical energy could be re-inserted into nozzle airflow to increase jet thrust. Leninetz's claim of increasing jet engine thrust was not confirmed. However, groups at Princeton and elsewhere confirmed the possibility of efficient MHD (magneto-hydro-dynamic) extraction of electrical energy from engine air flow above Mach 7 speed.[6] So, an MHD power system was designed to extract electricity from scramjet airflow and then conditioned it into electric power for igniting a fusion reactor.

[6] Mach 7 = seven times the speed of sound, or 5371 mph – **Ed. Note**

Below is an MHD scramjet concept by Professor Paul Czysz of Saint Louis University. Seen, are both electrodes and magnets within scramjet engine walls.

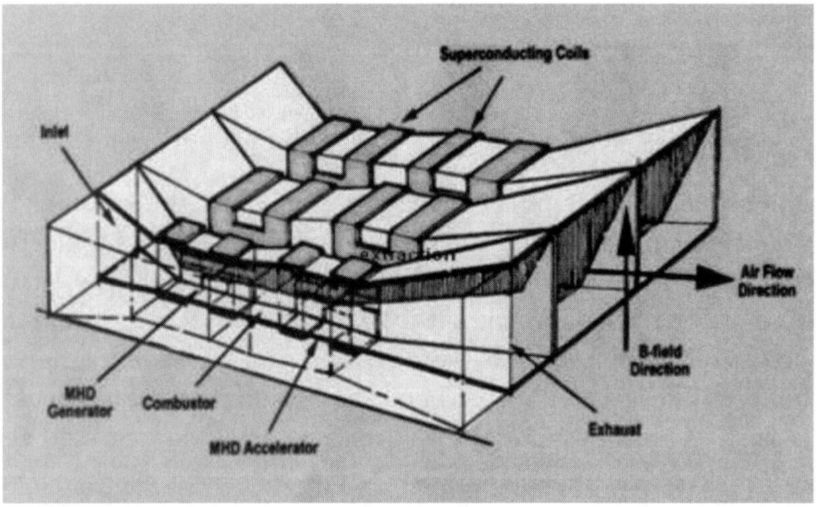

FIGURE 28 AN MHD-SCRAMJET ENGINE DESIGN BY PROF. PAUL CZYSZ

For our system, power extraction from scramjet engine airflow started at about **Mach 10**, when extracted electric power was about 10 MW and enthalpy loss was only about 10%. Ordinary hydrogen burning in the scramjet's combustors continued. After the fusion system was ignited, its energy was deposited into the air-breathing engine's exhaust by electron beams. And ohmic heating of the engine exhaust flow generated engine thrust with negligible consuming of mass.

NASP-like vehicles that were powered and propelled by MHD air breathing and fusion rocket engines, were studied by my company and the University of Illinois [7] for the Air Force Research Laboratory (AFRL) in 2002 to 2004. Shown, below is the Dense Plasma Focus (DPF) fusion power system used in the study.

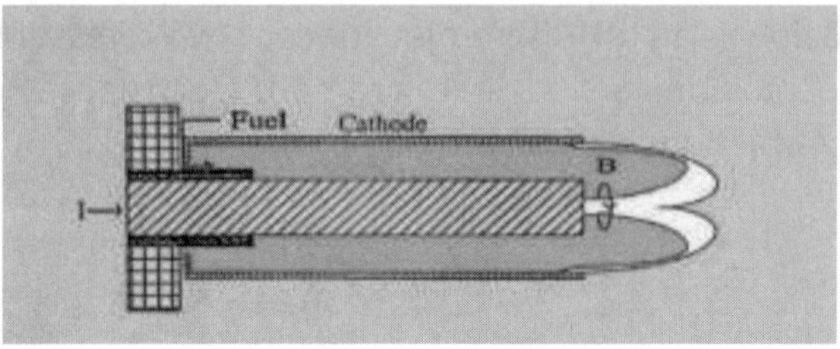

FIGURE 29 MAGNETIC CONFINEMENT FUSION OF HYDROGEN AND BORON 11 IONS IN DENSE PLASMA FOCUS DEVICE (P-B11)

The capacitor system for a Dense Plasma Focus rocket generated intense current discharges to transform gaseous nuclear fuel into a p-B11 plasma within concentric electrodes. The plasma accelerates itself along the cathode; turns 180 degrees; and is then powerfully compressed by very strong induced magnetic fields of hundreds of Teslas[7] until p-B11 nuclear fusion occurs. Shown below is the air-breathing vehicle this fusion system was integrated with.

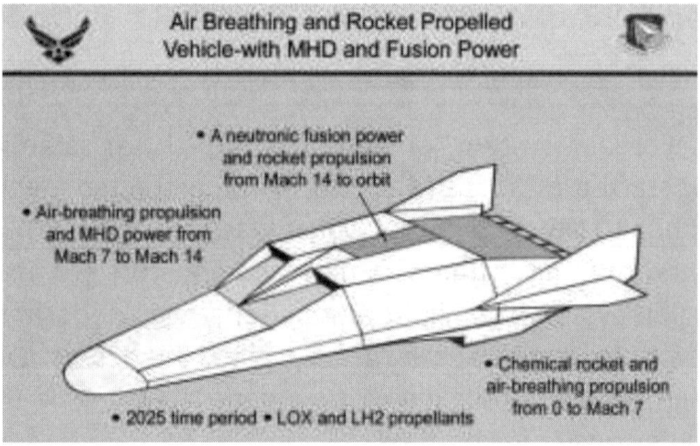

FIGURE 30 AIR-BREATHING VEHICLE INCORPORATING A FUSION SYSTEM

This vehicle embodies advanced power systems (MHD and fusion); ramjet and scramjet engines; rocket engines (chemical and

[7] Ed. Note: One Tesla = 10,000 gauss. A kilogauss magnet is usually a neodymium-iron-boron "rare earth" magnet since ceramic "refrigerator" magnets are only on the order of a few hundred gauss. However, a magnetic field of "hundreds of Teslas" can only be made with very high current, usually requiring superconducting coils, thus being at the top of the strongest manmade magnetic fields on earth.

fusion); flow path shared by a combined rocket and jet engine and flow path shared by scramjet and fusion engines. It consumed only 1/5 the fuel an ordinary scram-jet/rocket vehicle would consume, and its estimated life-cycle cost was one-twentieth that of the Space Shuttle. Large cost reductions resulted from large reductions in vehicle fuel, and use of strong-light structures formed from rapidly-solidified titanium and silicon carbide – to reduce fuselage, wing, tail weight about 30%. Weight, height, length and span of 2025 vehicles were less than current B-1, B-2, C-17 aircraft. So, B-1, B-2 or C-17 Air Bases could be used by these vehicles if they could provide liquid hydrogen storage and fueling for a fleet of vehicles.

Show below is a summary of vehicle weights, engine types and speed ranges.

Vehicle Weight At	Engine Type	Speed Range
Takeoff...... 174 t		
Propellant .... 74 t	Air-Augmented Rocket for thrust and TVC	Mach 0 -26
Air Flight 47 t		
Space Flight 27 t		
Empty .... 100 t	Supersonic Combustion Ramjet	Mach 7-14
Payload .... 18 t		
Structure .... 28 t		
Propulsion .... 42 t		
Rockets 21 t	Nuclear Fusion Rocket	Mach 7-26
Scramjets 21 t		
Systems .... 12 t		
Dry Mass .... 82 t		

FIGURE 31 CHART COMPARING VEHICLE WEIGHTS (TONS), TYPE, AND SPEED RANGE

The 2025 vehicle would drastically reduce Earth-to-orbit flight costs. And the redundancy from its multiple engines and flow paths would increase flight safety somewhat. And lower speed-temperature-pressure ranges for its engines should increase its reliability somewhat. But a major uncertainty was the high confinement energy and temperatures needed for p-B11 fusion.

Unfortunately, we had no time or resources to explore fundamental fusion physics problems associated with the very high confinement energy needed for p-B11 fusion.

In this respect, many fusion physicists assume there is little new to learn about electric or magnetic fields beyond the electromagnetic (EM) fields that were formulated by James Clerk Maxwell over a century ago. But, using modern group theory and topology, Dr. Terence W. Barrett [8] has formed more complex EM fields by special conditioning of ordinary Maxwell EM fields.[8] Moreover, he has shown ordinary EM fields of lower U1 Lie symmetry can be transformed into more complex EM fields of higher SU2 Lie symmetry in several ways. Barrett's SU2 EM fields are described by more complex tensor mathematics. And EM interactions involving these more complex SU2 fields contain added "A" tensor fields which combine with electric "E" tensor fields and magnetic "B" tensor fields in various ways.

Barrett also shows ways of transforming ordinary U(1) EM field energy into SU(2) EM field energy. One is polarization modulation of phase-modulated EM wave energy propagating in waveguides or optical cavities. Such polarization modulation is described in [9] and is summarized by Fig. 32 on the next page.

[8] **Ed. Note:** Though this "group theory" may seem complicated, a *measurable* vector potential A is in the SU(2) group, which is better than U(1) where A is *not measurable*. See Barrett's Table 1 on the next page and his article with the author on their EM theory of FTL flight in the Appendix.

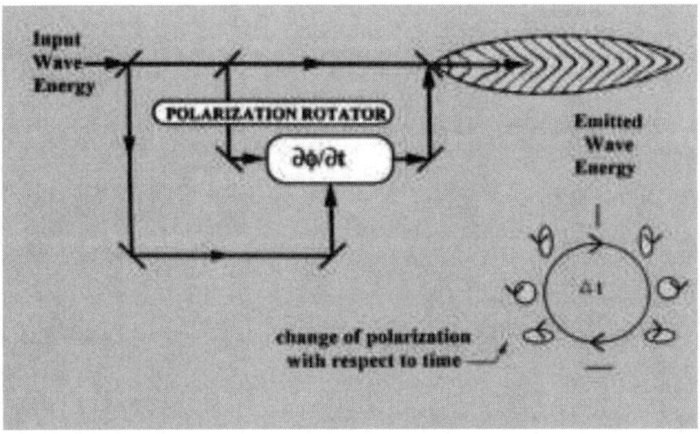

FIGURE 32 EXAMPLE OF SPECIALLY CONDITIONED SU(2) EM RADIATION

Use of beams of SU(2) radiation to confine plasmas in fusion reactors like the "Inertial Electrostatic Confinement" (IEC) reactor (below) was studied in [10]. Fusions occur where fusion ion beams collide. So we were encouraged that [10] indicated that irradiation of the IEC fusion region with added beams of ion-confining SU(2) or

Table 1

	U(1) Symmetry Form (Traditional Maxwell Equations)	SU(2) Symmetry Form
Gauss' Law	$\nabla \bullet E = J_0$	$\nabla \bullet E = J_0 - iq(A \bullet E - E \bullet A)$
Ampère's Law	$\frac{\partial E}{\partial t} - \nabla \times B - J = 0$	$\frac{\partial E}{\partial t} - \nabla \times B - J + iq[A_0, E]$ $- iq(A \times B - B \times A) = 0$
	$\nabla \bullet B = 0$	$\nabla \bullet B + iq(A \bullet B - B \bullet A) = 0$
Faraday's Law	$\nabla \times E + \frac{\partial B}{\partial t} = 0$	$\nabla \times E + \frac{\partial B}{\partial t} + iq[A_0, B] =$ $iq(A \times E - E \times A) = 0$

SU(3) radiation could greatly increase the numbers of fusion events and therefore the strength of the energy output.

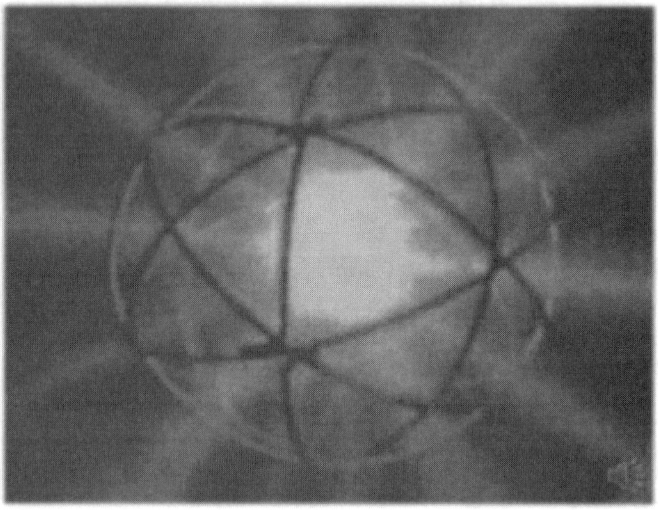

FIGURE 33 THOUSANDS OF FUSIONS/SEC. OCCURRING IN THE CENTER OF AN IEC REACTOR AT THE UNIVERSITY OF ILLINOIS, FORMERLY DIRECTED BY PROF. GEORGE MILEY, AUTHOR OF *LIFE AT THE CENTER OF THE ENERGY CRISIS: A TECHNOLOGIST'S SEARCH FOR A BLACK SWAN* (WORLD SCIENTIFIC, 2013), A BOOK ADVOCATED BY THIS EDITOR WHICH SUMMARIZES YEARS OF PROF. MILEY'S FISSION, FUSION, AND LASER RESEARCH.

6 FIELD POWER AND PROPULSION FOR FUTURE FLIGHT

Since its beginning in 1903, air travel progressed with propeller propulsion. But it wasn't revolutionized until the advent of jet propulsion, invented in 1937 by Frank Whittle and Hans von Ohain, for jet engines greatly reduced air travel time and cost. But this revolution also required invention of the transistor by John Bardeen, William Shockley, Walter Brittain in 1948, for this began solid-state electronics and digital computers for needed guidance-navigation-control.

And spaceflight by rocket propulsion progressed steadily since its invention by Robert H. Goddard in 1928. Examples of this progress are the many communications and navigation satellites that orbit Earth today. But spaceflight still has not been revolutionized like air flight was. For radical reducing of its cost and risk has not yet been accomplished. So, despite conscientious efforts by government and industry, human spaceflight is a taxpayer burden, not a commercial opportunity.

Our Air Force study described in the last section significantly reduced vehicle flight cost by significantly reducing its propellant consumption by use of fusion propulsion. We endeavored to increased its safety and reliability by having most systems able to "back-up" ones that failed and by reducing speed and pressure and temperature ranges over which each single system had to work. So, integration of propulsion and power in the 2025 vehicle provided significant flight cost reduction and modest flight safety increase – compared to rocket systems operating over very wide speed-pressure-temperature ranges.

Part 2 of this Air Force vehicle study used the 2025 Vehicle defined in Part 1 as a starting-point to explore a very advanced "2050 Vehicle" that: (a) flew more futuristic military space missions than the 2025 vehicle; and (b) embodied new systems and technologies to further the cost and hazard of Air Force aerospace mission that orbited Earth. At that time NASA was interested in "returning to the moon to stay". So an added air force mission of the 2050 Vehicle was rapid delivery of emergency or rescue services from Earth to people on the Moon. This mission required vehicle takeoff and payload weights and volumes similar to those of the 2025 Vehicle. But more than twice its 11 km/s impulsive velocity was needed. This required much of this higher velocity to be by the actions and reactions of fields; not only by combustion and expulsion of matter.

The 2050 vehicle on the next page had some systems similar to 2025 vehicle systems (air-augmented rockets; scramjet engines; MHD and fusion power). But science and technology assessments of even more advanced systems were also made by physicist Dr. Eric Davis, of the Institute for Advanced Study at Austin. Eric examined issues of certain advanced systems that were believed possible for Air Force missions by 2050.[9] This work fed into selection of certain new systems for the 2050 vehicle. These were: antennas for emitting special EM field energy; zero-point energy extraction system for inertia-reducing or thrust producing field propulsion; and carbon 60 nanotube-reinforced structures of very high strength-to-weight ratio.

---

[9] **Ed. Note:** Davis' latest study, "Faster-Than-Light Space Warps, Status and Next Steps" won the American Institute of Aeronautics and Astronautics' (AIAA) 2013 Best Paper Award for Nuclear and Future Flight Propulsion. See https://www.space.com/21721-warp-drives-wormholes-ftl.html for a summary of that paper by Space.com. His amazing articles are also available on https://www.researchgate.net/profile/Eric_Davis6 . His coworker, Dr. Hal Puthoff, has also contributed a Defense Intelligence article on Advanced Space Propulsion in the Appendix.

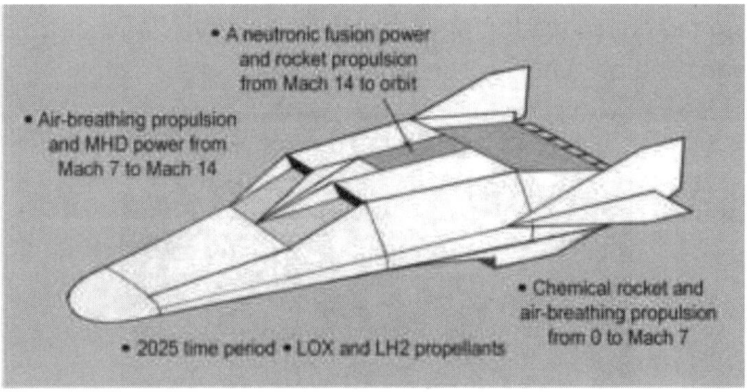

FIGURE 34 2050 VEHICLE: FOR FUTURE AIR FORCE AEROSPACE MISSIONS IN A LATER PERIOD AND 2025. SOME VERY ADVANCED MILITARY PAYLOADS DEEMED POSSIBLE IN THIS LATER PERIOD WERE ALSO STUDIED.

Developing 60 percent more impulsive velocity than the 2025 vehicle's rocket and jet engines with similar payload and takeoff mass required both jet propulsion and a favorable interaction between vehicle-emitted EM fields and the ambient fields in the medium being flown through. One such medium is the "electromagnetic zero-point quantum vacuum" which pervades the stupendous vastness of cosmic space. Here, Haisch, Rueda and Puthoff [10] suggested that a body's acceleration through electromagnetic zero-point quantum vacuum gives rise, at least in part, to its "inertia" – force that instantly opposes a body's acceleration or deceleration to higher or lower speed.[10] Extending this idea, we (Froning and Roach [11]) explored the possibility of vacuum-disturbing electro-magnetic (EM) discharges from an accelerating vehicle nose reducing the resistance of the quantum vacuum to it. A preliminary approximation of this by computational fluid dynamics (CFD) techniques is on the next page. Shown in Fig. 35, is the *favorable radiation pressure gradient* formed about a ship that is moving much slower than light and emitting a vacuum-disturbing EM beam. The reduction is from the discharge favorably

---

[10] **Ed. Note:** This famous paper was cited by Arthur C. Clarke in his last book, *3001 – The Final Odyssey*. It is also the first paper available on Dr. Puthoff's webpage, https://www.researchgate.net/profile/Harold_Puthoff .

modifying vacuum (reducing vacuum permittivity and permeability in forward regions) about the ship.

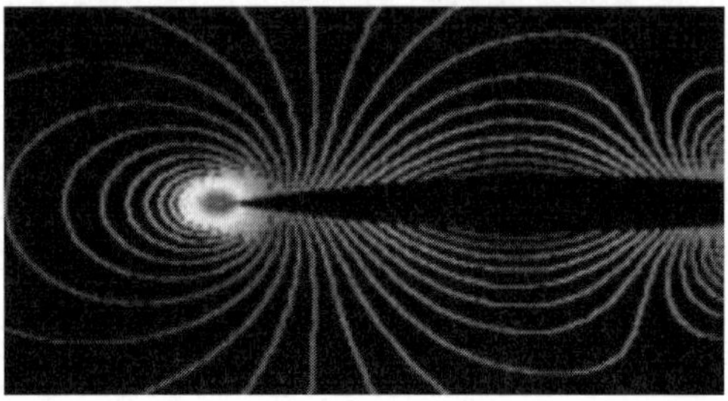

FIGURE 35 EM DISCHARGE DIMINISHING RESISTING PRESSURES EXERTED ON SHIP THAT IS ACCELERATING AT SLOWER-THAN-LIGHT (V=0.3C) SPEED.

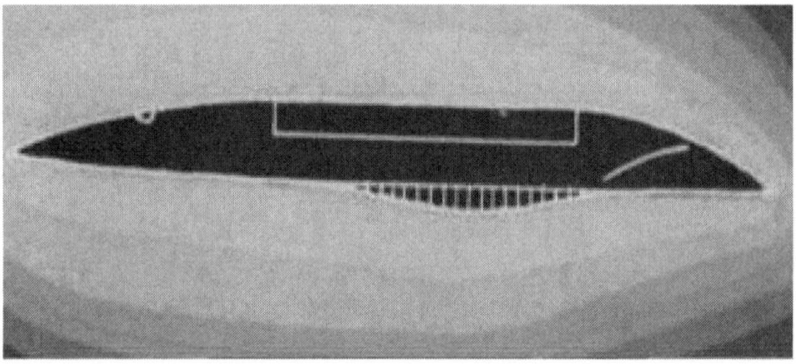

FIGURE 36 FIELD PROPELLED SHIPS SHOULD SUFFER LESS AIR PRESSURE AND AIR HEATING IN PLANETARY ATMOSPHERES. BUT THIS FIGURE INDICATES A POSSIBILITY OF SHIP GENERATED EM DISCHARGES EXCITING SURROUNDING AIR ATOMS AND MOLECULES—TO CAUSE AN AURORA BOREALIS-LIKE APPEARANCE

Inertia-reduction or gravity-reducing field propulsion is usually associated with Star Trek-like "warp-drives" used for faster-than-light starships. But it may be needed long before star flight to reduce heat, pressure, and friction-like wear and tear on vehicle components like bearings; vehicle skins, combustion chambers, and turbine blades. Such adverse environments required enormous amounts of maintenance on the Space Shuttle which prevented the Shuttle's frequent and economical use. By contrast: airliners suffer much less environmental wear and tear; are maintained at much more modest cost; can accomplish many more flights and fly much longer. For airlines suffer much less heat, pressure and frictional wear on airframes, engines and components. So, field propulsion was selected for the 2050 vehicle, not only for needed flight performance for rapid transits between Earth and Moon; but to reduce flight costs and increase safety by reducing the heat, pressures and stresses its airframe, propulsion, and power systems must repeatedly withstand. Embodying field power and propulsion in the 2050 vehicle, while maintaining its payload and takeoff mass the same as the 2025 vehicle's, required significant reduction of vehicle propellant and significant reduction in its structural weight. In this respect, the figure below by the U.S. Army Research Laboratory indicates that nanotube structures (which have 100 to 300 times greater tensile strength-to-weight ratio than the strongest high-carbon steels) possess very high buckling strength as well.

This encouraged us to strive for very thin, light and strong aerospace materials that could be reinforced by an ultra-strong and ultra-light matrix of carbon nanotubes. For calculations indicated

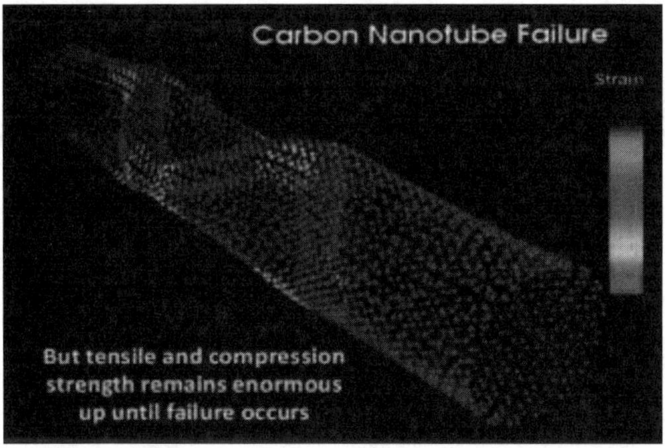

FIGURE 37 CARBON NANOTUBE FAILURE MODE OF BUCKLING

such structures could double the strength and halve the structural weight of a 2050 vehicle airframe.

After the Air Force study ended, we used 2050 Vehicle technologies in a vehicle for sub-orbital space-tourism or for two-hour hypersonic airline flights between major cities of Earth. Shown below are: vehicle weights; propulsion systems and their operating ranges; and passengers that could be profitably flown to space.

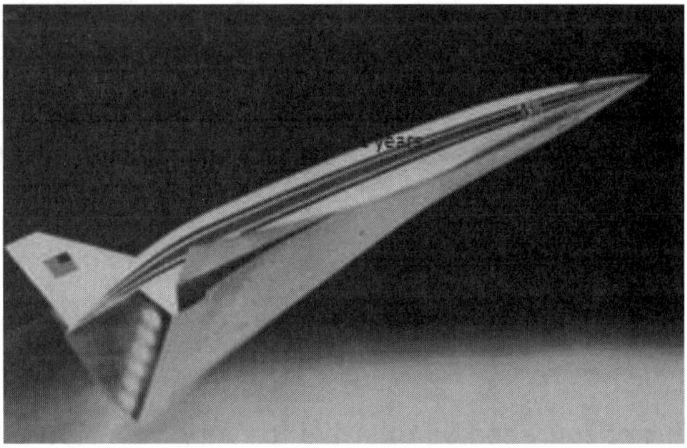

FIGURE 38 SPACE-TOURISM OR HYPERSONIC-AIRLINER DERIVATIVE OF THE FIELD + JET-PROPELLED 2050 VEHICLE -- TO GIVE THOUSANDS OF CAREERS AND SPACEFLIGHT EXPERIENCE TO THOUSANDS OF PEOPLE ON EARTH.

Finally, cost analysis in [13] indicated a space tourism variant of the 2050 vehicle (a fleet of five jet and field-powered space tourism vehicles) could profitably fly 440,000 people to and from space by 2,000 flights in 11 years. The following information summarizes the costs of development and launch for various solid rocket motors and various known rockets.

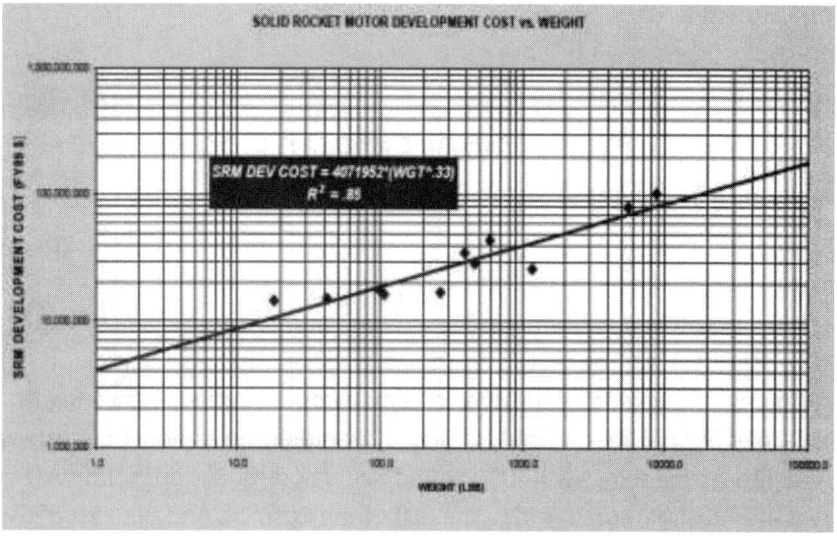

FIGURE 39 COST ANALYSIS OF SOLID ROCKET MOTOR (SRM) DEVELOPMENT[11]

	GLOW	RATE/YR	COST TOTAL	COST/LAUNCH	FIXED COST	VARIABLE COST
TITAN 34D	1900000	2	$73.472	$36.736	$63.206	$5.133
		4	$83.738	$20.935		
		6	$94.004	$15.667		
		8	$104.270	$13.034		
		10	$114.536	$11.454		
		12	$124.803	$10.403		
SCOUT	47200	2	$5.701	$2.850	$4.889	$0.406
		4	$6.513	$1.628		
		6	$7.326	$1.221		
		8	$8.138	$1.017		
		10	$8.951	$0.895		
		12	$9.763	$0.814		
DELTA II	428000	2	$28.702	$14.351	$24.662	$2.020
		4	$32.742	$8.186		
		6	$36.782	$6.130		
		8	$40.822	$5.103		
		10	$44.862	$4.486		
		12	$48.902	$4.075		
ATLAS II	353000	2	$21.977	$10.988	$18.877	$1.550
		4	$25.076	$6.269		
		6	$28.176	$4.696		
		8	$31.275	$3.909		
		10	$34.375	$3.437		
		12	$37.474	$3.123		
ATLAS CENTAUR	360600	2	$25.621	$12.810	$22.011	$1.805
		4	$29.230	$7.307		
		6	$32.839	$5.473		
		8	$36.449	$4.556		
		10	$40.058	$4.006		
		12	$43.667	$3.639		

FIGURE 40 COMPARISON OF NASA LAUNCH OPS COST (FY88$ SHOWN)

[11] **Ed. Note:** Presented at the 2008 SCEA-ISPA Joint Annual Conference and Training Workshop - www.iceaaonline.com – approved for public release

7 RELATIVISTIC SPACEFLIGHT FOR TRAVEL TO DISTANT STARS

I watched with interest as the U.S. and Russia began competing with each other to be first to do things in space and to place men on the Moon. But I wasn't as enthusiastic as many others about space flight. I assumed our entire solar system would eventually be explored by rockets and interesting things would be found on its bodies. But our solar system was infinitesimal compared to the rest of the Cosmos. And this rest of the cosmos was separated from Earth by stupendous gulfs of time and space. For light, moving at the ultimate speed limit of our universe (186,000 miles/sec.) takes 30,000 years to reach the center of our Milky Way galaxy and millions of years to reach a neighboring one.

Then I stumbled upon a translation of a German paper by the famous German rocket scientist, Eugen Sänger [14]. I was stunned by it. Sanger showed a ship moving at nearly light speed would, in accord with Einstein's Special Relativity (SR), contract space ahead of the ship and slow tempo of time inside it (as shown below). Thus, distant stars could

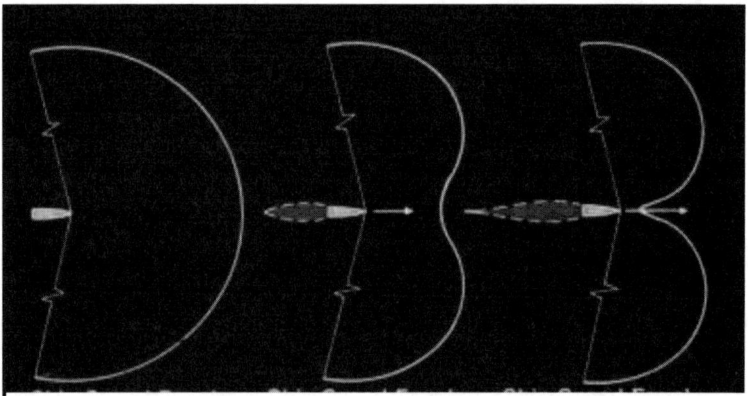

FIGURE 41 MORE AND MORE SPACE CONTRACTION AS A SHIP GETS NEARER AND NEARER TO LIGHT SPEED (RED EXHAUST REPRESENTS THRUST) AND TIME DILATION

be reached in lifetimes of ship crews if they stayed close to the speed of light.

Using Special Relativity, Sänger showed that relativistic space-time dilation would cause human aging in ships to slow more and more as they reached speeds that approached nearer and nearer to the speed-of-light as that occurred on longer and longer interstellar trips. Using one earth gravity acceleration during the initial half of a trip, followed by equal deceleration during the final half, Sanger proved, as shown below, that the **nearest star would be reached in 6 earth years**, as only 3.5 years or less passed on the ship. He estimated our Milky Way galaxy's center would be reached in 19 ship-years as 30,000 years passed on Earth and the M-31 galaxy was reachable in only 29 ship-years as 2.2 million years passed on Earth. And he even showed the universe's circumference (then estimated to be 3.0 billion light-years) could be traveled in 42 ship years, Sänger thus showed we may not be forever confined to our currently very small part of the cosmos.

Interstellar Destination	Elapsed Tme On The Earth	Elapsed Time On The Ship
Nearest Star	6.0 Years	3.0 Years
Center of Our Galaxy	30,000 Years	19 Years
Andromeda Nebula M-31	2.2 Million Years	29 Years

FIGURE 42 TRIP TIMES ESTIMATED FOR 1.0 G ACCELERATION AND 10 G DECELERATION IN A MOVING SHIP FRAME.

Sänger had suddenly made spaceflight more interesting to me. But he assumed the use of an ultimate rocket – a "photon rocket" which converted matter into energy with perfect efficiency by annihilation of matter with anti-matter.[12] Also, the weight of his ultimate rocket

[12] **Ed. Note:** Dr. Brian von Herzon from UCLA worked on antimatter propulsion years ago and told this Editor that besides antiprotons or anti-hydrogen, which can be stored in a

included no structural mass but still consumed prohibitive fuel amounts even for trips to nearest stars. I soon concluded (perhaps too hastily) that too much rocket fuel would have to be carried within interstellar rocket ships.

A conceivable source of fuel was the tenuous clouds of hydrogen existing throughout galaxies at densities of 100-10,000 atoms per cubic centimeter. And being familiar with missiles with a rocket lower stages and ramjet upper stages I envisioned a lower Sanger-like anti-matter rocket stage for initial "boost" to nearly light-speed, and then a ramjet-like upper stage to ingest a very long swath of tenuous interstellar material which would exist along the ramjet's long interstellar route. Then, I envisioned a **fusion reactor** in the ship "burning" (by nuclear fusion) hydrogen atoms or molecules the ship ingested. So, the best flight trajectories, I thought, would traverse a galaxy's densest clouds of interstellar hydrogen if burning fuel is necessary.

FIGURE 43 HORSE-HEAD NEBULA WHICH CONTAINS MANY TENUOUS REGIONS OF INTERSTELLAR GASES

Magnetic funneling of very large volumes of interstellar material into ramjet-like ships was needed at slower speeds, while smaller volumes were needed at very high-speed for adequate ingestion and

magnetic bottle, lots of water is needed to act as the propellant. His estimate was that one gram of antimatter could propel the space shuttle into orbit but its production is expensive.

combustion – as nearly light-speed. Shown next is an "interstellar ramjet" moving at nearly the speed-of-light.

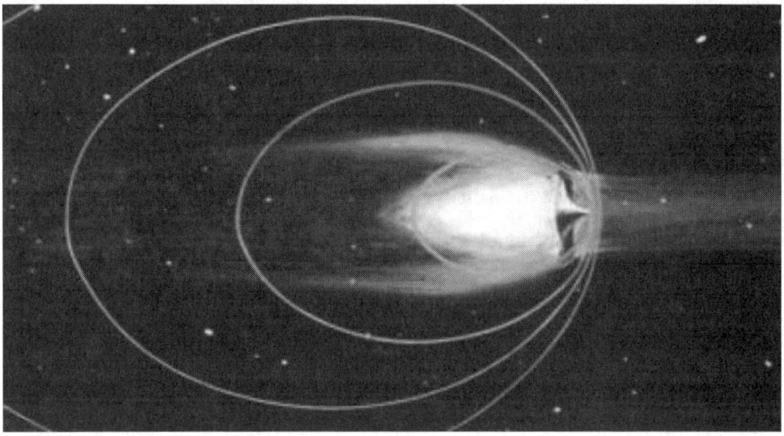

FIGURE 44 MATERIAL INGESTED AND FUSION-COMBUSTED BY AN INTERSTELLAR RAMJET AT NEARLY LIGHT-SPEED.

At that time (1967) my only experience was in aeronautics (not astronautics) so I didn't know if my interstellar flight ideas would interest any astronautics people or how feasible they were. To find if there was any interest, I summarized my early ideas in a paper [15]. It reviewed Sanger's work that revealed the possibility of traversing much of the universe in relativistic ships during lifetimes of their crews; and I showed that ships with both rocket and ramjet-like propulsion might be needed for this.

I submitted [15] to the Papers Committee of the International Astronautical Federation (IAF). Its Committee Chairman, Harry Ruppe (a former Werner von Braun V2 rocket scientist) apparently liked the paper and invited me to present it at the 1967 IAF Congress in Belgrade Yugoslavia. At first there was some consternation in the Douglas Astronautics Division, as a non-astronautical person would be talking about an advanced astronautical topic at a space forum. It ended when my paper was favorably reviewed by some experts at Douglas Astronautics. *So my first interstellar flight foray was allowed to begin!*

The Congress was being held in Yugoslavia (now Serbia) in 1967 during the cold war and almost all attendees were from the Soviet Union or Communist bloc countries. Stars of the Congress were Soviet Union participants. Russia was then the world leader in spaceflight, always achieving new space firsts before America (for example, first satellite-man-woman-dog in space; first walk in space; and first un-manned probes to the moon, Venus and Mars). But America was beginning to catch up. The trip began badly. My flight arrived so late at Belgrade airport, my hotel and flight to Dubrovnik (where I was going to stay before the congress) was cancelled. Unlike in America and Europe, people at the counters weren't inclined to help me at all, as I met seeming indifference and coldness at every turn. I finally had no choice but to stay in the terminal all night, for airport desks were now closing and employees leaving the terminal for the night. So, it looked like my only company would be gun-carrying soldiers at the terminal doors. Luckily, I was able to find a small corner area with uncomfortable chairs where I wouldn't be noticed by soldiers. I got little rest that night but things got better next day. I caught the Dubrovnik flight and the man next to me on the flight had a sister there who rented rooms. And at his sister's home I had a good stay and good talks with the lady's son who loved American Basketball.

But things seemed to get grim again as the Congress began. Its sessions were held in various buildings, many of which housed drab-looking Communist Party offices. The photograph on the next page shows one the different buildings where different congress sessions were held. The largest "Opening Session" of the Congress was in a large auditorium in a building with many Yugoslavian government and Communist Party offices. As Congress delegates had to shuffle past many of these offices on our way to the big auditorium, I remember feeling an increasingly, heavy sense of depression with every step that I took.

FIGURE 45 YUGOSLAVIAN GOVERNMENT AND COMMUNIST PARTY OFFICES OCCUPIED THE SAME BUILDING WHERE THE 1967 INTERNATIONAL ASTRONAUTICAL CONGRESS WAS HELD

I received earphones in the auditorium, settled down, and set my earphones to English. After many welcoming, ceremonial speeches which I heard clearly in English, the featured speaker, Dr. Leonid Sedov, from the Soviet Delegation began the most important address of the day: the latest and the most complete summary of all Soviet space science accomplishments during their first 10 years of space exploration. But, for some reason, his particular talk was not being translated into English. So there I sat, frustrated and angry and surrounded by a sea of seemingly unfriendly faces, and unable to comprehend any of the spoken word.

Since Dr. Sedov's talk was not illustrated, I had nothing to look at or to listen to. But with nothing technical to listen to, I now had plenty of time to think about other things. I was somehow able to slowly still the anger welling up within me over not hearing what I had eagerly looked forward to hearing. A strange thought came to think on all the good things of my past that I was grateful for, and all the kindness and help I had received from wonderful and unselfish people. A surge of gratitude and affection seemed to suddenly go out from me to all those wonderful ones of my past who had blest me. I saw no vision but slowly began feeling presence and power reigning within me, and distinct awareness of harmony, goodness and love encircling that auditorium and me – just like encircling spiral arms that surround a giant galaxy's center. Figure 46 is my favorite galaxy reflecting this concept.

Another surge of gratitude surge flooded my consciousness and I tingled as if high voltage was flowing through me. Amazingly, right there in that Communist Party Auditorium, I suddenly found myself loving everyone that surrounded me.

As time suddenly seemed to resume, I no longer cared that I wasn't hearing anything about spaceflight. But I became aware of someone several rows ahead stirring vigorously. A man suddenly turned, looked back at me, and passed a paper to me through many people. It was an English translation of "**Ten Years of Space Exploration in the Soviet Union**," by L. I. Sedov – the very talk then being given.[13]

From then on my experience was completely transformed in that Communist land. Changed consciousness transformed depression and frustration in what had seemed a gloomy, unfriendly land into joy, humor, idea sharing and starts of friendships with receptive people from Soviet Union and Communist-bloc countries. And though there were few English or American people to talk with, I don't recall serious language barriers, for missing words or phrases seemed to always get "filled in" during every conversation. This was long ago, but I still remember that thrilling, soul-satisfying experience when, for a change, I let all my own wishes and desires go and rejoiced over all the good I had been given.

---

[13] **Ed. Note:** The International Astronautical Federation still holds an annual conference called the International Astronautical Congress today. Up to date information is available at http://www.iafastro.org/ .

FIGURE 46 HUBBLE SPACE TELESCOPE VIEW OF THE ANDROMEDA GALAXY 2.2 MILLION LIGHT YEARS FROM EARTH

8 QUANTUM ZERO-POINT FLUCTUATION ENERGY FOR PROPULSION

Ordinary company work kept me too busy to do more than a short paper [16] on interstellar ramjets. Then, a rude shock. I found that a respected fusion physicist, Robert Bussard, had conceived of a fusion-propelled interstellar ramjet [17] long before me! And, to make things worse, I found his analysis, in many aspects, was much more complete than mine. But even his more complete analysis had critics – mainly those who were very strongly skeptical of the feasibility of magnetic funneling of enormous volumes of tenuous interstellar material into fusion reactors. As more and more experts refuted feasibility of magnetic funneling of huge gas volumes into interstellar ramjets, I finally abandoned tenuous interstellar matter as a starship fuel source.

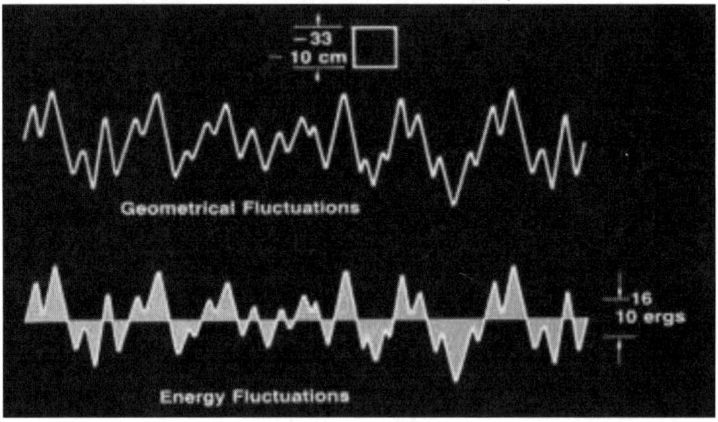

FIGURE 47 SHOWN ABOVE ARE GEOMETRICAL (10^{-33} CM) AND ENERGY FLUCTUATIONS (10^{16} ERGS) OCCURRING DURING THE PASSAGE OF TIME IN THE SMALLEST REGION QUANTUM PHYSICS ALLOWS.

About then, a company physicist referred me to the work of Professor John Wheeler of Princeton. Wheeler showed the incredible possibility of what seems inert and empty space actually teeming with energy over sub-microscopic scales of time and distance, "fluctuating in accord with quantum physics and resonating at the scale of the Planck length (10^{-33} cm) between configurations of varied curvature and topology" [18].

Such fluctuations in the geometry of space cause "quantum zero-point energy fluctuations" as is symbolized in Fig. 47.

The average "expectation value" of energy (10^{16} erg) in this tiny region is stupendous.[14] It Is 20 orders of magnitude greater than the energy in a proton, whose volume is 80 orders of magnitude greater than that of this tiny region's 10^{-33} cm "Planck length". Shown below in a different way, is the distribution of energy fluctuations in a spatial region of the quantum vacuum at a given time. Energy fluctuations, like electromagnetic energy pulsations, are visualized as different lightning flashes of different wavelength or frequency. So, lower energies are associated with longer wavelength or lower frequency. Thus, lower energy quantum fluctuations are symbolized by lightning flashes of cooler color (like: blue, yellow and green), and higher energy quantum fluctuations are symbolized by lightning flashes of warmer color (like: orange, pink and white).

Now space is generally viewed as an energetic medium, a "quantum vacuum" that includes the four massless quantum fields associated with the zero-point ground-state of the four forces of nature (gravity, electromagnetism, strong force and weak force). Quantum vacuums also include virtual particles – such as electrons, positrons and mesons, and some believe another quantum field is associated with the universe's accelerated expansion by unseen "dark energy".

I used Bussard's geometrical and mathematical representation of an interstellar ramjet's relativistic interaction with interstellar matter to make a similar one of a quantum interstellar ramjet's relativistic interaction with the long swath of quantum fluctuations the ship's frontal area would sweep-out during a long journey through space. This is what is shown below.

---

[14] **Ed. Note:** 1 erg is equal to 10^{-7} joules or 100 nanojoules (nJ) and 1 J/sec = 1 Watt

Show below at a given instant is a short portion of the very long swath of quantum electromagnetic energy fluctuations to be swept-out and ingested by a ramjet-like ship along its long interstellar route. Electromagnetic energy fluctuations of different intensity occur within the swept-out swath of zero-point EM fluctuations at a given instant. As in the previous figure, they are symbolized by many different lightning flashes of different frequencies (different colors).

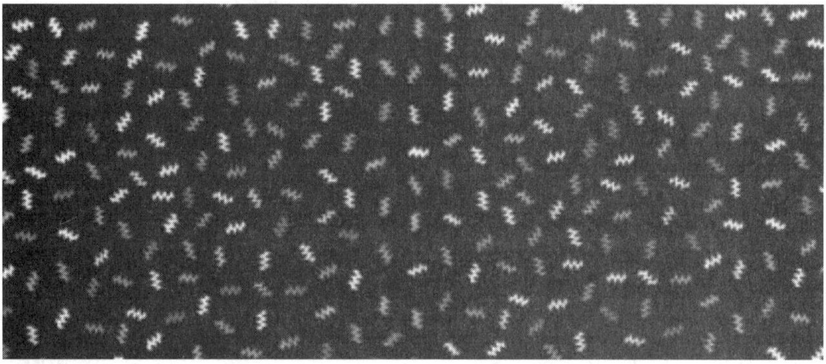

FIGURE 48 QUANTUM FLUCTUATION ENERGIES WITHIN A SHORT PART OF THE VERY LONG SWATH OF SPACE THAT IS SWEPT OUT BY RAMJET-LIKE STARSHIP CAN THESE ENERGIES BE HARVESTED OVER FEMTOSECOND SCALES OF TIME AND NANOMETER SCALES OF DISTANCE FOR POWER AND THRUST?

I was able to relate "expectation values" of extractable zero-point energy density in the swept-out swath of quantum vacuum with ship power and thrust that can be generated. This includes the efficiency (α) that the ship's energy ingestion system extracts and harvests zero-point energy (ZPE) from the vacuum; and this was done as a function of the **efficiency (β)** with which extracted energy (from processes like electron-positron pair annihilation) could be converted by the ship propulsion system into the useful kinetic energy of a thrusting beam of light.

I didn't identify a specific way to extract quantum zero-point energy from space, or a specific way to transform the extracted energy into laser-like light. But my representation enabled computing ship performance as: a function of time and distance over which energy could be interacted with and extracted with efficiency (α); and the efficiency (β) by which the extracted energy is converted into the useful kinetic energy of an emitted laser-like beam. An artist conception of a "quantum ramjet" accelerating to nearly the speed-of-light is on the lower part of this page.

My idea of a **quantum interstellar ramjet** is shown below. Presented in London, It was published in the ***Journal of the British Interplanetary Society*** in 1980 [19] and was the first to consider the possibility of *using the quantum vacuum as a source of energy and space power and propulsion.*

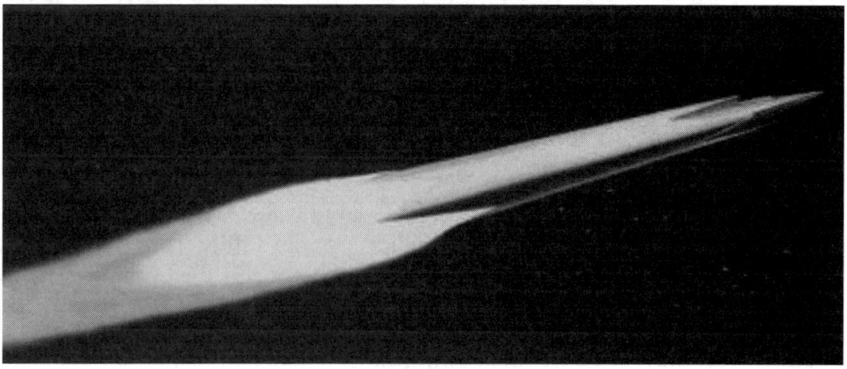

FIGURE 49 QUANTUM INTERSTELLAR RAMJET: FOR REACHING NEARLY THE SPEED C FLIGHT BY HARVESTING THE QUANTUM VACUUM ENERGIES OF "EMPTY SPACE" ITSELF

My paper was not an optimistic vision of using fluctuation energies of space for energy needs. It showed that quantum interstellar ramjets faced the extremely difficult problem of interacting with energy fluctuations lasting less than one femtosecond of time (10^{-15} seconds) and existing over distances as short as a nanometer in length. Unfortunately, in 1980, instrument response times were millions of times slower and thousands of times longer than such short temporal and spatial scales. But now, advances that have created narrowly-focused femtosecond petawatt (1 million GW) lasers should soon allow the quantum vacuum to be probed and interacted with over times and distances approaching these values.[15]

---

[15] **Ed. Note:** The detection, measurement, and verification of quantum vacuum fluctuations that Froning refers to here have been made. The book, *Practical*

Years later at a special gathering, I found myself seated at dinner next to Robert Bussard (who had conceived of an interstellar ramjet before I had). I told him how much I admired his Interstellar Ramjet paper and how I borrowed some of its excellent methodology for a quantum interstellar ramjet analysis that I had made. He smiled broadly and said he had actually read my quantum Interstellar ramjet paper and liked it well enough to give it to his wife for use on one of her advanced university student projects. So he considered us "even" in the borrowing of each other's ramjets. We soon become good friends and his larger company and my much smaller one enjoyed years of collaboration on fascinating fusion power and propulsion energy research for NASA and ESA.

Although no quantum interstellar ramjet is likely to fly in my lifetime, it is my only claim to any kind of public fame. Science fiction writer Arthur C. Clark used my quantum ramjet as a huge, heroic, life-saving starship that carried one million people safely from dying Earth to a new world in his 1986 science fiction novel, "***The Songs of Distant Earth.***"[16] Now, more than 30 years later, science fiction vehicles are still saving Earth and other worlds in innumerable ways. But I'm still pleased that my Earth-saving quantum ramjet was among the first.

More importantly, Air Force scientist, Frank Mead became interested in my zero-point vacuum energy ideas. He directed his brilliant scientific consultant, Robert Forward, to work with me on how it might be applied to spaceflight power or propulsion. At first, Forward told me, "Right now I see no way to get a handle on that energy, it may be as impossible to use as heat in the ocean."

But Forward became very interested in quantum energy and he gave me much support at conferences, especially when I was attacked by skeptics about its existence. And he soon learned of the "Casimir Effect"

Conversion of Zero Point Energy from the Quantum Vacuum for the Performance of Useful Work by Thomas Valone cites many of the best papers answering this prediction with certainty.

[16] **Ed. Note:** Arthur C. Clarke also used a similar work "Inertia as a ZPE Effect" by Puthoff, Rueda, and Haisch (Phys. Rev. ,1995) that he called it the "Space Drive" for his book *3001: The Final Odyssey*. Hal Puthoff was pretty happy about it.

wherein closely-spaced metal plates exclude the longer wavelength quantum fluctuations between them to cause inward pushing force from outside fluctuations to compress the plates.

Forward showed that this could drive the plates together without effort and the EM energy of all excluded quantum fluctuations could be drawn-off like electric current from a battery. This resulted in a famous paper by Forward on a vacuum fluctuation battery [20]. Unfortunately, his battery didn't generate net electrical power because needed input energy to force open the closed plates to repeat the process would be more than the released energy. But his work showed the vacuum could be interacted with *to cause force and release energy*. So it greatly increased scientific interest in the nature and behavior of the quantum vacuum.[17]

U.S. Patent #7,379,286 (Haisch and Moddel 2008) describes possible extraction of quantum zero-point energy from gas flowing through many narrow channels. The patent suggests that extracted energy removed from the channels may be zero-point energy removal by a Casimir Effect. It also suggested that removed zero-point energy is replenished by more zero-point energies from space each time depleted gas flow emerges from the channels after each circuit by the gas.

It would be splendid if this is, indeed happening and if the zero-point energy extraction could be enormously increased by much smaller and narrower nano-manufactured channels. Unfortunately, I have heard nothing about the status of this Patent. So apparently there is much still to be done before, what would be the most incredible energy source of

---

[17] **Ed. Note:** See *Zero Point Energy: the Fuel of the Future* by Thomas Valone for a discussion of Forward's battery, his AFRL report on ZPE the Casimir effect, and the energy density of the quantum vacuum. Frank Mead also received a US Patent for his transducer of vacuum fluctuations, which is analyzed by this editor in both books just cited in these footnotes.

all – energy from the vastness of empty space itself – can be harvested for Earth's future energy and space-flight needs.[18]

[18] **Ed. Note:** Dr. Moddel and a grad student presented their findings regarding their patent at COFE7 and expressed the opinion that a prototype of the nanoporous cube would require significant funding to accomplish. Their initial findings looked promising. Prof. Moddel is the Director of a Quantum Energy Research Lab http://ecee.colorado.edu/~moddel/QEL/ZPE.html . Also worth mentioning is the Arthur Manelas Device presented by Dr. Brian Ohern at COFE10 is a stand-alone, overunity strontium ferrite electrical generator that also was attributed to vacuum energy because of a nanocrystalline structure (his slides and DVD available at www.futurenergy.org).

9 FASTER-THAN-LIGHT FLIGHT IN HIGHER DIMENSIONAL SPACES

It's fortunate that time slows on ships traveling at nearly light-speed so they can reach distant stars before much aging of their crews can occur. But time doesn't slow on Earth. So many Earth generations would have lived and died by then. So it would be even better if ships could travel faster-than-light relative to earth, to reach and return from stars in short times on the ships and Earth. For the crew could then share trip results with those who had remained on Earth.

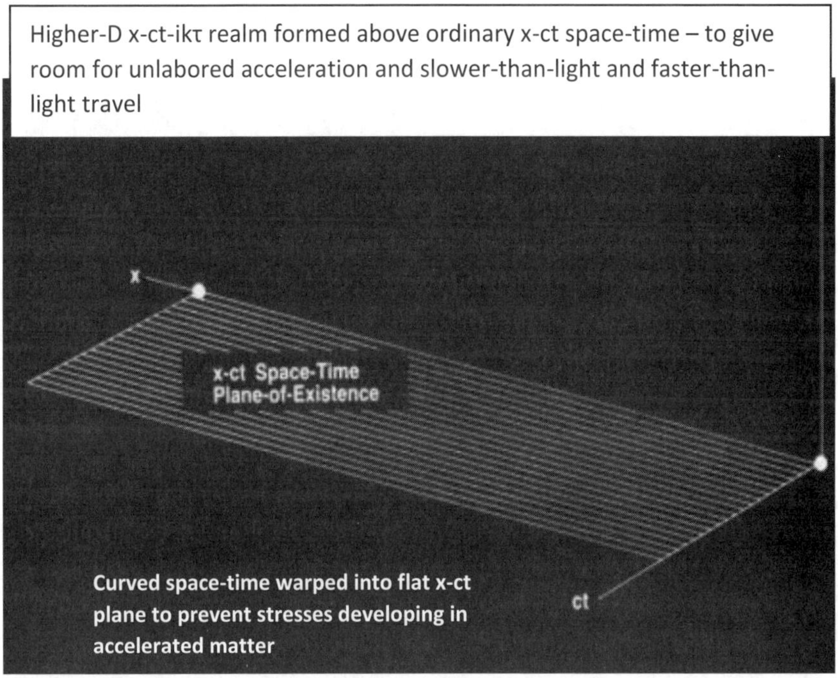

FIGURE 50 LOWER D X-CT PLANE OF SLOWER-THAN-LIGHT TARDYONS- EMBEDDED IN A HIGHER-DX-CT IKT REALM THAT BOTH SLOWER-THAN LIGHT TARDYONS AND FASTER-THAN LIGHT TACHYONS TRAVEL IN.

Lower-D x-ct plane of slower-than-light **"tardyons"** – embedded in a higher-D x-ct-ikτ- realm that both slower-than-light tardyons and faster-than-light **"tachyons"** travel in.

In the early 1980s speculative articles in physics journals introduced the idea of "tachyons" which could go faster than light (FTL). And some viewed tachyons as faster-than-light solutions of Einstein's Special Relativity (SR). For, SR's Lorentz contraction factor $[(1-(V/c)^2]^{1/2}$ has real values only for speed V slower than light (STL). But, if multiplied by $i = [-1]^{1/2}$, this factor becomes $[(V/c)^2-1]^{1/2}$ which has real values only for FTL speed. After some geometric analysis, I was able to visualize both STL and FTL solutions of SR as occurring in a deeper realm of existence

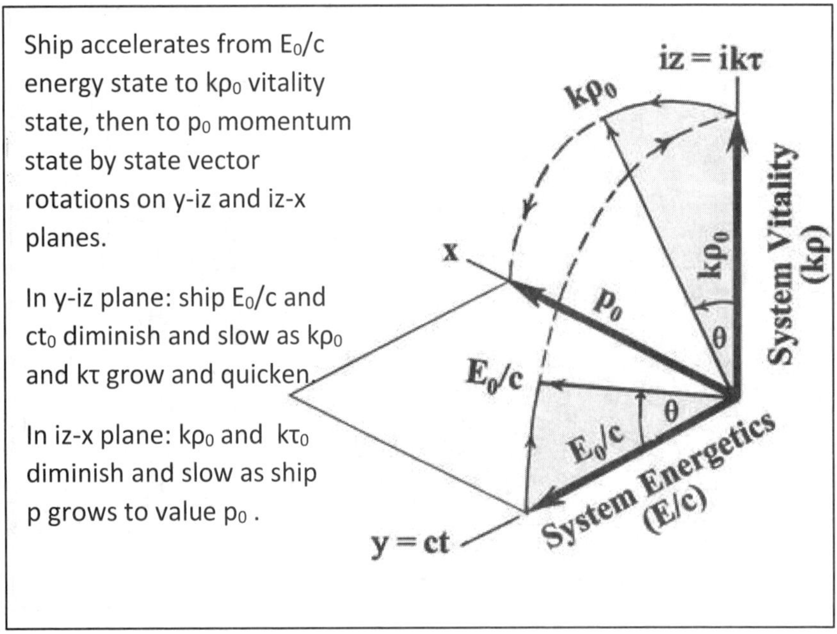

FIGURE 51 UNLABORED ACCELERATION BY SHIP STATE ROTATION IN Y-IZ AND X-IZ PLANES

than the plane of existence of ordinary x-ct space-time. This was with a vertical coordinate ikτ, which is orthogonal to those which describe space travel (x) and time travel (ct) of slower-than-light tardyons on an x - c t space-time plane of existence. And this x - c t plane is seen to be embedded in the volume of higher-dimensional x-ct-ikτ realm rising (like a sky) above it.

Sides of the higher-D x-ct-ikz cube realm below are: x-ct space-time planes; "time-less" space-tau x-ikτ planes and "space-less" time-tau ct-ikτ planes.

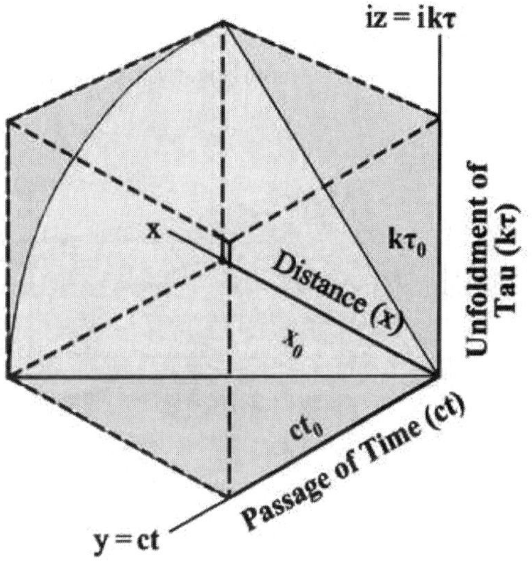

This deeper realm has room for innumerable x-y space-time planes of existence and for innumerable x-iz space-tau and y-iz time-tau planes of being also. Faster speed than c isn't allowed in the realm's conical region. But, in its reaming regions, speeds dx/dt that are billions of times c are allowed.

FIGURE 52 X-CT SPACE-TIME PLANES; "TIME-LESS" SPACE-TAU X-IKT PLANES AND "SPACE-LESS" TIME-TAU CT-IKT PLANES.

Ship energy E/c is related to its time-travel ct; vitality kρ is related to its tau travel kτ; and momentum p is related its space travel x. And shown below is a way for swift, unlabored ship acceleration to very high speed in this deeper x-ct-ik realm.

Just as my slower and faster-than-light flight domain was simplistic, so was my representation of STL and FTL vibratory states within it. Shown below is a very simple de-Broglie wave-packet representation of a ship's vibrational states at STL and FTL speed, wherein a slower-than-light tardyon ripple in x-ct space-time metric becomes a faster-than-light tachyon ripple in nearly orthogonal x-iz space-tau metric. Unlabored acceleration requires constant ship momentum and hence no change in the wave packet wave-length λ. So, λ is shown to be

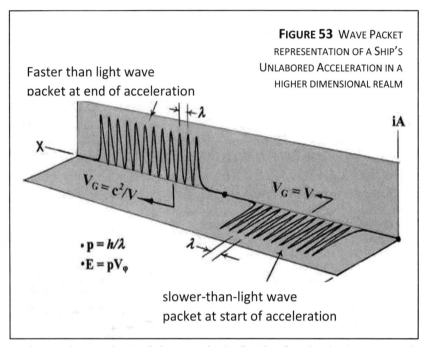

FIGURE 53 WAVE PACKET REPRESENTATION OF A SHIP'S UNLABORED ACCELERATION IN A HIGHER DIMENSIONAL REALM

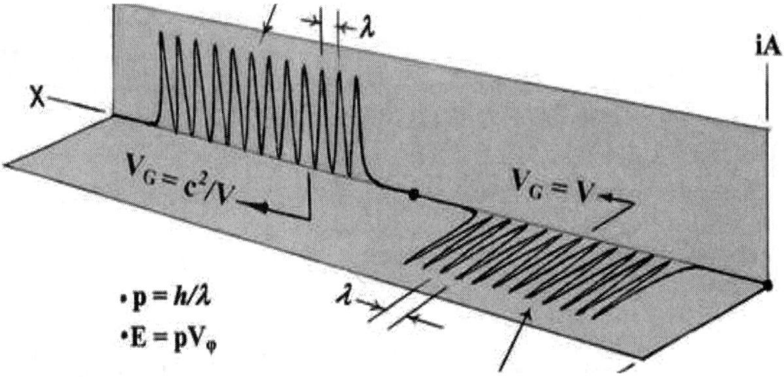

unchanged. Initial FTL "**phase velocity**" Vϕ of individual wave-packet undulations is c^2/VG: where VG is packet's **"group velocity".** Here, packet VG at maximum speed is c^2/ Vϕ where: Vϕ is the same as the packet's initial VG at the start of its acceleration.[19]

Unlabored acceleration requires constant ship momentum and hence no change in the wave packet wavelength λ. So, λ is shown to be unchanged. Initial FTL "**phase velocity**" Vϕ of individual wave-packet undulations is c^2/VG : where VG is packet's "group velocity". Here, packet VG at maximum speed is c^2/ Vϕ where: Vϕ is the same as the packet's initial VG at the start of its acceleration.

The packet vibration plane rotates in the higher x-ct-ikτ realm with change in packet speed VG/c relative to Earth. Angle ϕ of this plane

[19] See "Faster Than Light" Report #708 from www.integrity-research.org for explanation of phase and group velocity, plus a number of physics journal reprints on FTL travel.

relative to the horizontal is $\tan^{-1}(VG/c)$. So undulations are in a nearly-horizontal plane at very slow VG. But it rotates towards vertical with increasing speed. So, vibration state is nearly-horizontal at slowest speed and nearly-vertical at fastest speed.

My view of tardyons and tachyons as existing in orthogonal space-times in a higher-D realm than space-time was published [21] over objections of two reviewers who believed my tachyons were too "metaphysical". But they avoided the "causality" problem all other tachyons suffered. For a tachyon in space-time could be seen by o n e observer as moving forward-in- time, but to another at different speed and distance, it would be moving backward in time.

My wife Irina visited an office with a UFO book [22] on a table, and while idly thumbing through it, she saw the word "tachyon", a strange word she once heard me say. Intrigued, she finally found a bookstore with that book and brought it home to me as a gift. At first I was not grateful since I faithfully avoid UFO books. For, though they often mentioned objects with extraordinary flight behavior, they invariably lacked evidence one could deeply delve into. But this book seemed superior on most others, with compelling photos and descriptions of significant amounts of testing and analysis of the photos and metal samples and sound recordings. So, I decided to look at this UFO book.

It was a Swiss farmer's story [20] about contacts with alleged human-like beings over a several years in the 1975 period. Mentioned, were fairly frequent visits by allegedly space-faring beings whose "beamships" traversed 500 light years of distance (more than 1,000 trillion kilometers) between their home planet (within the Pleiades star cluster and Earth. They made this enormously long trip in only 7 hours, requiring average flight speeds more than 100 million, billion times the

---

[20] **Ed. Note:** Billy Meier is his name and Col. Wendel Stevens was the coauthor of at least three books on this famous farmer and was a frequent presenter at UFO conferences. Billy had only one arm but still took the *most high-resolution photos* of any UFO that are unmatched even to this day. See "UFO... Contact from the Pleiades, Volume 1" by Wendelle Stevens and Lee Elders [22] and also, "UFO...Contact from the Pleiades: A Supplementary Investigation Report" by Wendelle C. Stevens.

speed-of-light. I skeptically assumed the farmer had probably been told this tale by persons versed in science or science fiction – not by authentic astronauts. But the extremely short travel times clamed for the ships implied higher acceleration and speeds than one might expect – with the ships spending most all their time flying over an insignificant part of their flight distance at slower speed – which left them the very difficult task of stupendous acceleration to stupendous speed in only seconds of time. But instead of this surprisingly unexpected flight profile increasing disbelief in the farmer's story, it made me even more intrigued in it.

The alien craft accelerated swiftly to light-speed in 3.5 hours by a "light-emitting drive". A "tachyon-drive" then accelerated the ship over a 250-mile astronomic distance toward the target at much higher rate in only several seconds to speed over a trillion times c. The beamship then slowed at the same enormous rates to cover the remaining 250 light-years of distance to its destination.

Preventing prohibitive stresses developing in the ship and crew during very high acceleration would require appropriate warping of space-time metric or quantum vacuum by the beamship's drives – to prevent ship mass and time dilation as light-speed is approached and surpassed. The farmer was told time passed at the same rate on both ship and Earth. This is consistent with ship drives preventing relativistic slowing of ship time, shrinking of ship length; and magnifying of ship mass, compared to when at rest on Earth.

The visitors said their tachyon drive took only seconds of Earth time to achieve a "hyperspace-state" that, in effect, caused space and time to cease as a ship accelerated to estimated speeds more than a billion times light-speed c. Strangely, the "causing of space and time to cease" was somewhat like what occurred during the ship unlabored acceleration described in my Figs. 55, 56. Shown is space ceasing during initial upward rotation of a ship's energy state vector in an y-iz pane of existence; and then time ceasing (being slowed to zero tempo) as ship energy vector completes a 90-degree rotation into a time-less iz-x plane of existence. And one can visualize only seconds of Earth-time elapsing during swift downward rotation of the ship's momentum vector to a nearly horizontal attitude on the iy-x plane and to a near-zero angle from the x-coordinate of the x-ct plane of existence of ordinary space-time. And this final vector orientation is associated with the ship's maximum speed – a velocity more than a trillion times light -speed.

I was still skeptical that beam-ships and their drives existed. But if they did a drive's emitted EM fields would have to favorably couple with gravity or quantum vacuum fields. But these fields have greater complexity than ordinary EM fields (whose modest mathematic complexity is defined by a "Lie symmetry" of only U1). Thus, ordinary

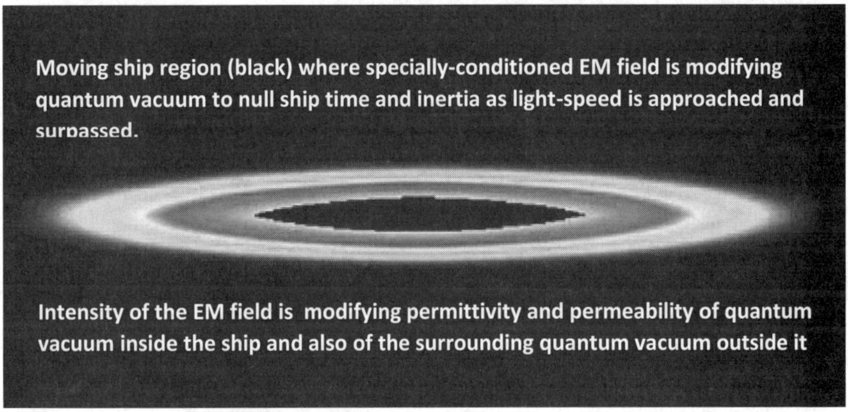

FIGURE 54 A SHIP VISUALIZED AS SWIFTLY ACCELERATING FOR SEVERAL SECONDS BY THE PROPULSIVE ACTION OF ITS 'TACHYON DRIVE' TRAVELING ABOUT A 250 LIGHT-YEAR DISTANCE AND REACHING A MAXIMUM SPEED OF MORE THAN A BILLION TIMES THE SPEED OF LIGHT.

electromagnetism couples weakly with gravity. So, as shown below, beamships would emit special EM fields of higher complexity and symmetry to couple favorably with those that give rise to gravity and inertia.

Achieving of nearly unlabored ship acceleration requires that the ship's surrounding medium present very little resistance to the ship. But this, of course, requires the ship to only slightly disturb its medium. This was quaintly described by the visitors in [22] as: **"the ship is protected by a protective girdle (force field) which lets guide away all interference without pushing it away".** In addition, they defined "hyperspace" as a higher dimensional realm, or different state of being that replaces a spatial sense of distance or temporal sense of time.

The hyperspace condition created by a tachyon drive (described on page 53 of [22]) seemed somewhat similar to the height of my ikτ coordinate that rose above a flat x-ct space-time of existence to form the higher-dimensional x-ct-ikτ realm described here. So, the mathematics and geometry associated with this higher realm could be used to display

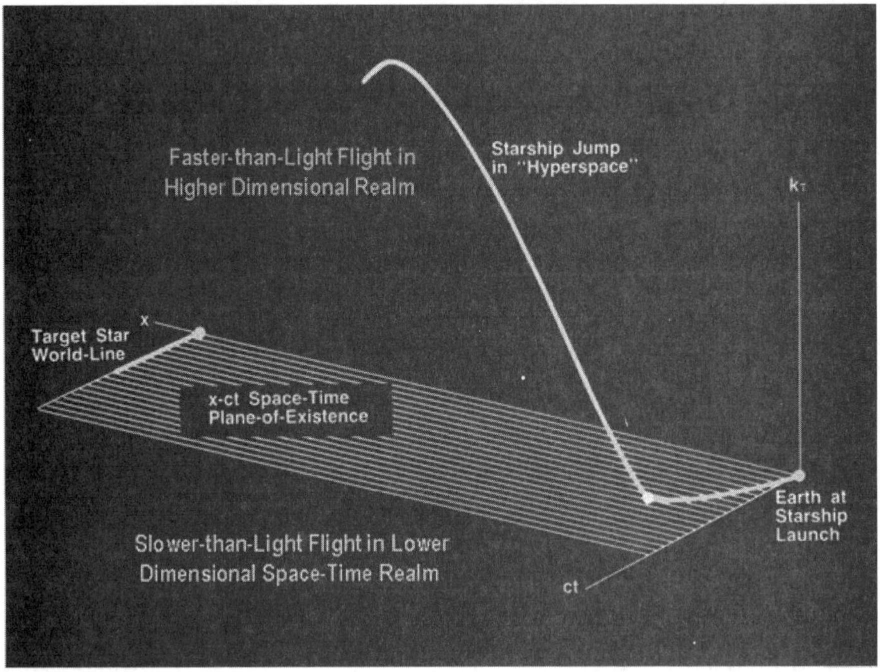

FIGURE 55 TRAJECTORY (WORLD-LINE) OF SLOWER AND FASTER-THAN-LIGHT FLIGHT IN A HIGHER-D REALM THAN SPACETIME

of slower-than-light (STL) and faster-than-light (FTL) segments of alien ship trajectories in my x-ct-ikτ realm which, which was only a 3-D approximation of a higher-dimensional realm.

Shown in Fig. 55 is the path (world-line) of a starship moving at STL and FTL speed in a higher-D realm than the lower-D x-ct space-time of a merely material world. Shown is the first half of the ship trip as it accelerates at a given rate to light-speed and then accelerates at extremely higher rate to very high FTL speed.

Near-horizontal space travel (x) and time travel (ct) occurs close to the x-ct plane until c is reached, then near-vertical climbing flight begins. And as sky gives aircraft room to climb and dive above Earth, so the ikτ coordinate gives ships room to climb and dive above space-time. Shown, is the shorter, slower path traced-out by an accelerating ship in space-time and then beginning of a much longer, swifter arching path

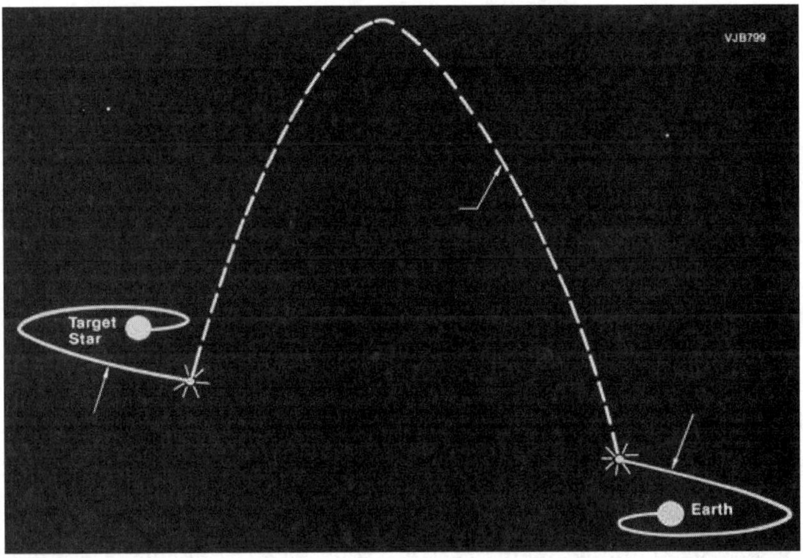

FIGURE 56 STARSHIP ACCELERATES AWAY FROM EARTH, DISAPPEARS AND REAPPEARS IN ONLY SECONDS. BUT DURING THESE SECONDS OF DISAPPEARANCE, THE SHIP, IN EFFECT, LEAPS HIGH ABOVE SPACE-TIME AND OVER STUPENDOUS DISTANCES TO REACH SPEEDS THAT ARE BILLIONS OF TIMES LIGHT-SPEED

traced out by ship in the higher-D x-ct-ikτ realm. Finally, acceleration to stupendous peak speed and altitude (ikτ) and beginning of slowing to lower speed and altitude until the journey finally ends.

Ship flight can also be viewed from the perspective of an Earth observer who is watching a starship fly away – accelerating in the direction of its target (a planet in another soar system) and then vanishing from sight as its initial acceleration ends. The ship then re-appears in only seconds – at the speed it disappeared at. But the ship is now suddenly 500 light-years away – very near to its destination.

My UFO-stimulated work [23] was published by the British Interplanetary Society (BIS). As UFOs are taboo topics in science journals, I was limited to describing FTL requirements in higher dimensional realms. But even with no UFOs in it, my work received enough interest for the BIS to request that I write another FTL article [24].

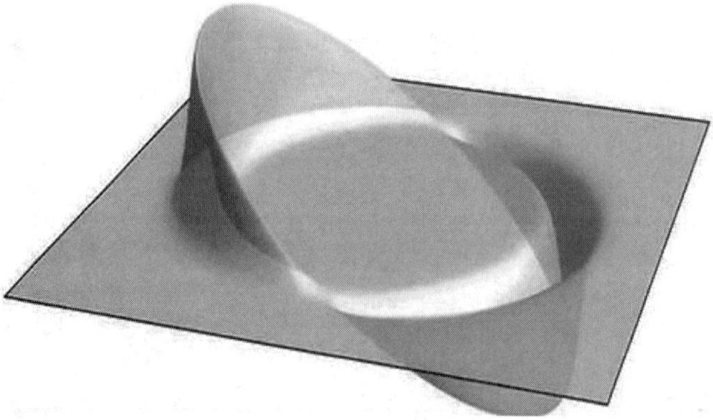

FIGURE 57 A GEOMETRICAL REPRESENTATION OF HIS SPACE-WARPING SOLUTION FROM ALQUBIERRE'S PAPER

In 1972 Miguel Alqubierre [25] defined a space-time warping solution of General Relativity (GR) that allowed unlabored starship acceleration to faster-than-light (FTL) speed. And I believe it is the best work on unlabored FTL flight that has been done so far. A geometrical representation of his space-warping solution is shown here. Seen is ship space-warping field-propulsion accelerating it to the left by "flattening" of ordinarily, curved space in and around the ship, to prevent stresses or forces developing in it during extremely accelerated flight. Also shown is vertical space warping near the ship to cause its impulsion.

Alcubierre's representation above of vertical warping of flat 2-D metric, and my representation of vertical rising above flat 2-D metric are somewhat related. For his shows vertical warping of flat metric in a higher-D realm than that of the warped metric; while my representation shows vertical rising by a metric warping ship in a higher-D realm than that of the warped metric.

Alqubierre's work stimulated much additional work by others for many years. Unfortunately, most of this additional work was discouraging. For it found that extremely weak coupling between gravity and ordinary EM fields required generation of stupendous amounts of EM field energy to warp space-time metric into Alqubierre's topology above, which enables faster-than-light travel.

Some of us believe warping of space-time metric or polarizing of quantum- vacuum with acceptable field energy will require special conditioning of ordinary EM fields into higher-order EM fields of much greater complexity. Examples of higher order EM fields are SU2 and SU3 EM radiation fields that have been studied by Barrett [8,9,10] and have the higher Lie symmetry of the SU2 and SU3 matter fields associated with the "weak" and the "strong" force in atomic nuclei.

My last FTL effort was a preliminary simulation of exceeding light-speed by EM energy emitted from an accelerating ship much like that shown in Figure 56. Shown below is favorable distorting of a ship's surrounding quantum vacuum by favorable modifying of its electrical emissivity and magnetic permeability by EM beams. Not included were EM and vacuum field interactions inside the vehicle itself. All work was by Robert Roach of Georgia Tech University – who used computational fluid dynamics (CFD) and certain assumed similarities in fluidic and electromagnetic energy transfer and disturbance propagation. His work revealed encouraging possibilities, but also the staggering difficulties of properly modeling all of the many critical and currently unknown processes and interactions that would need to be understood for proper modeling and simulation of FTL flight.

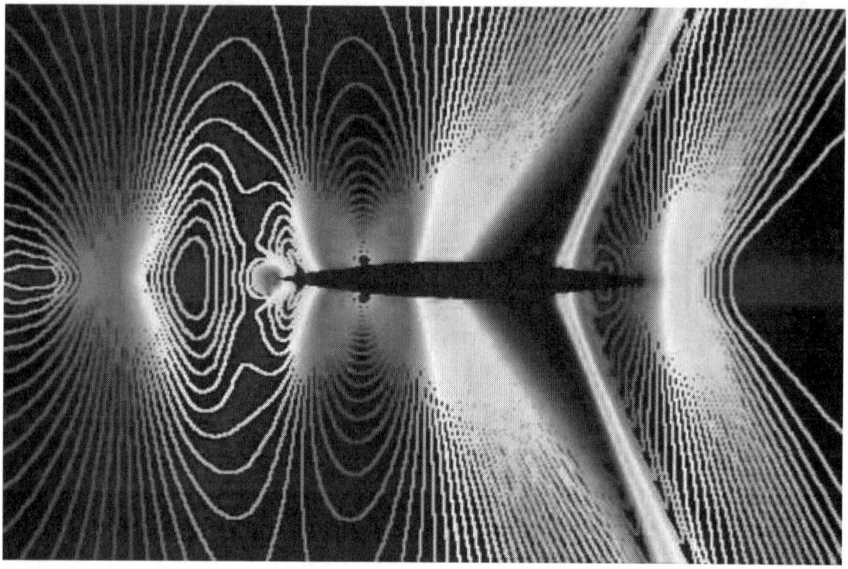

FIGURE 58 SIMULATION OF NEAR LIGHT SPEED TRAVEL (RIGHT TO LEFT) SHOWING SPACE-TIME DISTORTION FROM ZERO-POINT ENERGY PRESSURE GRADIENT **(see back cover for color version)**

Its staggering difficulties, and innumerable uncertainties and unknowns causes FTL flight to be generally rejected by orthodox science, with little in accredited science literature (like *Nature* and *Physical Review* Journals) to encourage or support bizarre FTL ideas like: higher-dimensional realms: warping of matter or space-time metric; extracting of energy from empty space. Unfortunately for FTL advocates, FTL miracles are mentioned mainly in science fiction and UFO literature. But strangely, unusual things somewhat like FTL miracles have been mentioned in religious literature. For the Bible's Old and New Testament tell of the higher representatives of a supreme Mind or Spirit doing extraordinary things thousands of years ago that were somewhat like FTL miracles. Some of these similar things were: materializing and dematerializing matter; parting a sea; walking on water; stilling a storm: and immediate transport to a distant shore.

This chapter would not be complete without mentioning my discovery of the **similarity of the equations for the speed of sound and the speed of light**. Here is one of my conference presentation slides that

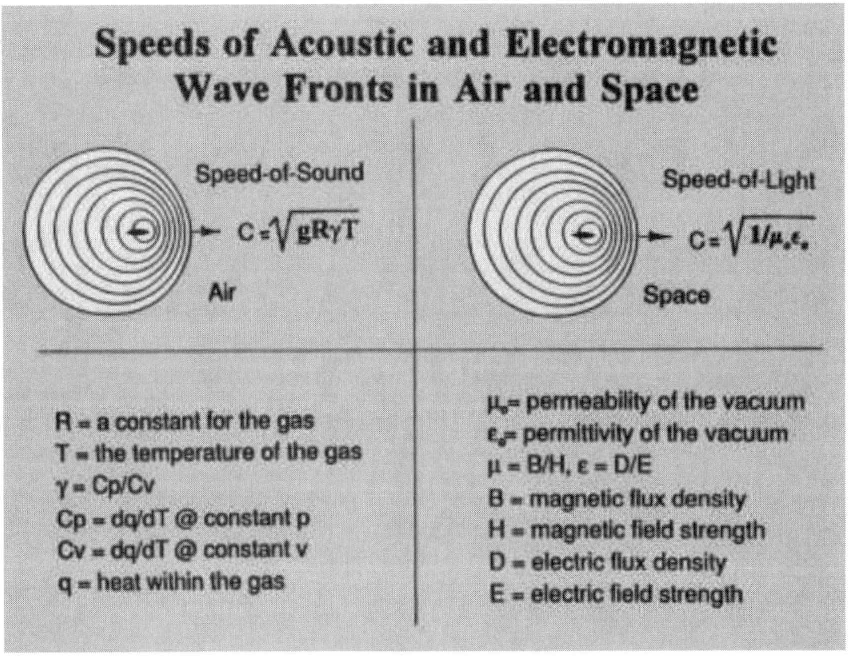

FIGURE 59 COMPARISON OF THE SPEED-OF-SOUND AND THE SPEED-OF-LIGHT EQUATIONS

summarizes these details. Normally, both media (air for sound transmission and the physical vacuum for light transmission) form compressions in front of the craft in the direction of travel, as shown in the diagram "Speeds of Acoustic and EM Wave Fronts in Air and Space".

Another piece of evidence for arguing the feasibility of FTL flight is the comparison between the Prandl Glauert equation of air drag (where here c is treated as the "speed limit of the medium," whether sound or light) and the well-known equation of mass increase (inertial mass) from Special Relativity, which also can be regarded as drag. However, there is the theoretical variation, with the right conditions, for the quantum vacuum to *assist in the forward motion of a craft by creating negative pressure*. Let's see how that can be formulated since the implications for FTL travel are momentous. Since we know how heroically Chuck Yeager proved that faster-than-sound travel was possible, Figures

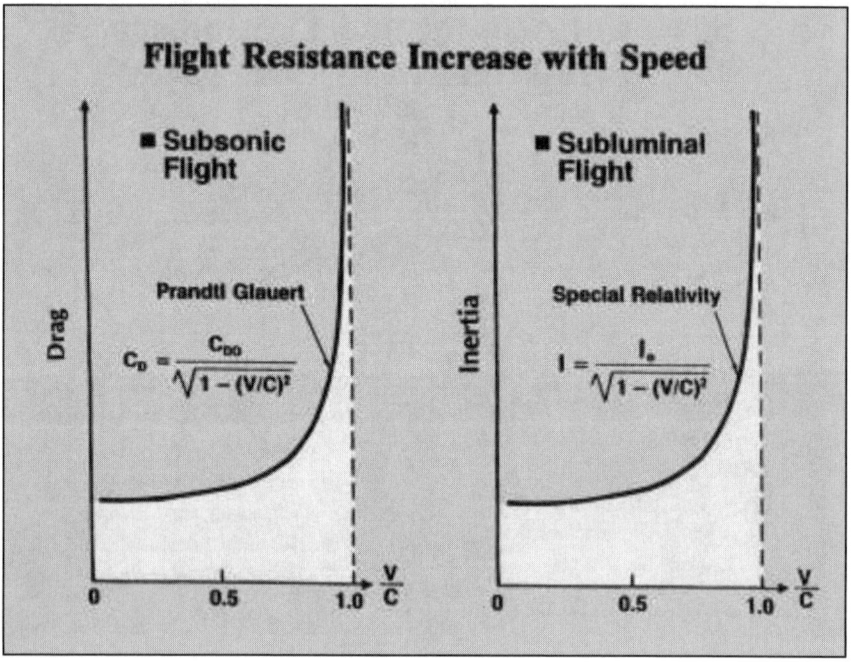

FIGURE 60 SIMILARITY OF THE PRANDTL GLAUERT AND SPECIAL RELATIVITY EQUATIONS, WHERE INERTIA INCREASES IN THE SAME MANNER AS MASS DOES NEAR THE SPEED OF LIGHT

59 and 60 give the best credence available today to the underlying mechanics of FTL travel since the variables are so similar for obviously different media.

Froning [19] was the first known peer-reviewed article to suggest the possibility and problems of extracting zero-point energy from the quantum vacuum for power and propulsion. Figure 61, taken from this article [19] indicates spaceflight as the chosen application. Zero-point energy expectation values given in Wheeler [18] were used, and processes that could: (a) materialize electrons and positrons out of the vacuum for mutual annihilation; and (b) gather the resulting photons into a thrusting beam of light were assumed. The investigation identified the scales of time and distance that zero-point energy must be interacted with and extracted from the quantum vacuum, to swiftly accelerate starships to almost light-speed. It was found that interaction with and extraction of zero-point energy from the vacuum must occur over 10^{-6} to 10^{-7}cm scales of distance – distance 4 to 5 orders of magnitude larger than the diameters of atoms.

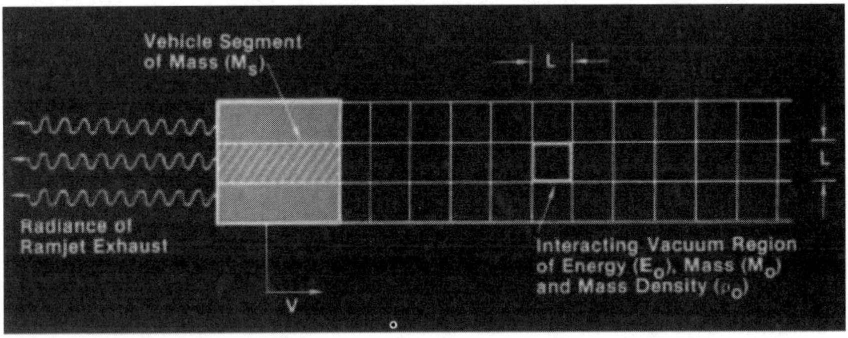

FIGURE 61 SCHEMATIC REPRESENTATION OF A "QUANTUM INTERSTELLAR RAMJET" EXTRACTING ZERO-POINT ENERGY FROM SPACE

Another important part of this later research on FTL is the interesting outcome of the nature of the vacuum interaction with an accelerating spacecraft as compared with the air interaction with acceleration of a moving craft. A repulsive field similar to that generated by electronic interaction between air vehicle skin and passing air can be provided the lower vehicle of Figure 62 – which is accelerating with respect to the quantum vacuum - by embodying space-warping, particle-repelling field generators in its outer skin. But the quantum vacuum has a "negative" pressure which can be viewed as acting in the opposite direction that inward-pushing positive pressure fields act. So, with repulsive field pressures already acting "outward" from the vehicle skin, there is nothing to prevent "outward pulling" (rather than inward-pushing) pressures to act over the entire ship. And this would cause higher-than-ambient, outward-pulling, acceleration-assisting force to act upon the front of the vehicle; and lower-than-ambient, outward pulling pressures to act upon its rear – that also results in an acceleration-assisting force.

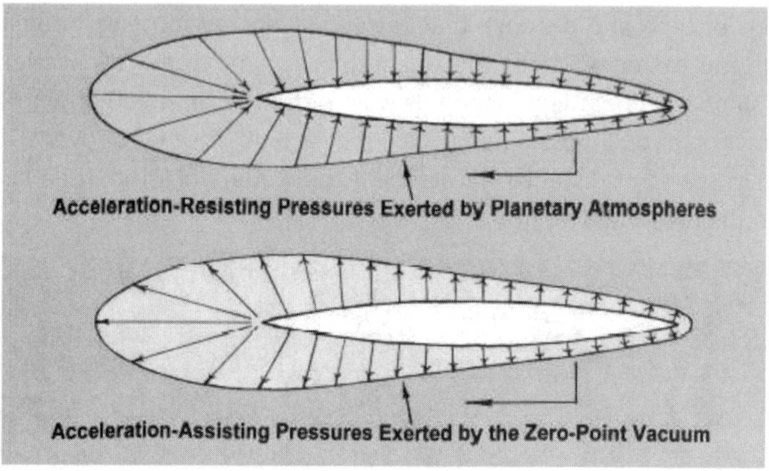

FIGURE 62 "RESISTING" AND "ASSISTING" VEHICLE PRESSURES EXERTED BY PLANETARY ATMOSPHERES AND THE QUANTUM VACUUM

Such acceleration assisting force would maximize in the transluminal speed range where pressure gradients become the most intense. Such force would be diminished by entropy-increasing energy dissipation in shock waves – such as emitted Cherenkov radiation. However, a negative pressure vacuum should still assist repulsive field propulsion systems. It must be admitted that fluid dynamic analogues of vehicle flight through warpable space and perturbable vacuum by simulations using Computational Fluid Dynamics (CFD) are only first-order approximations inadequate for any kind of precise computation for actual vehicle or propulsion system design. But it is believed that they are useful in introducing persons such as engineers to features and problems of field-propelled flight by visualization rather than the complex tensor calculus of General Relativity.

Figures 58 and 63 are examples of such a simulation and subsequent visualization with color-coded pressure regions. Note how the mathematical simulation shown in Figure 63 exhibits a more extreme flattening of the shock waves of the vacuum at twice the speed of light. This flattening trend was also continued and became even more extreme as more computer simulations were run and went up in multiples of the speed c [28].

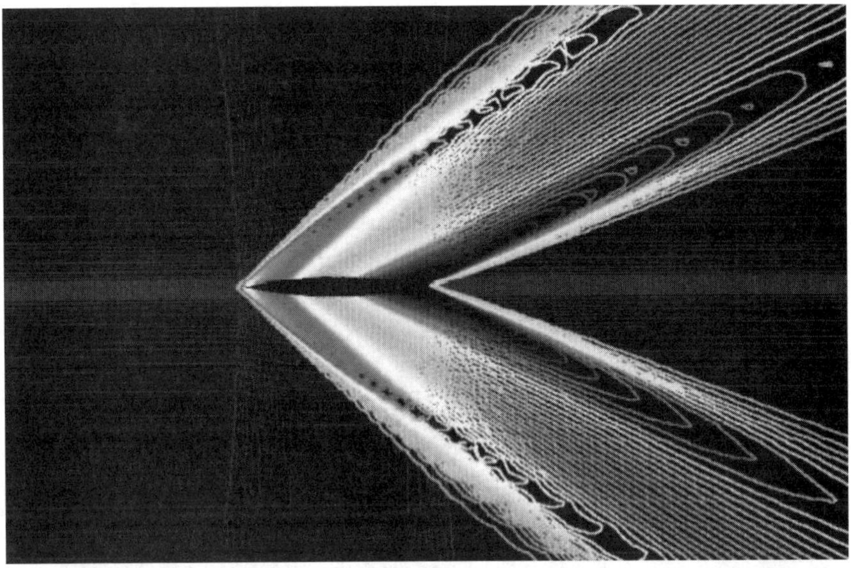

FIGURE 63 ZERO-POINT RADIATION PRESSURE GRADIENT CAUSED BY ACCELERATING SHIP AT TWICE THE SPEED-OF-LIGHT **(shock wave due to no EM Discharge fore or aft)**

The most bizarre possibility for meeting Earth's power and propulsion needs may be the harvesting the stupendous energetics that possibly exist within the quantum mechanical ground states of the vacuum of "empty" space. These energetics are symbolized electromagnetic pulsations that – like lightning flashes – forever appear and disappear throughout the entire cosmos in countless times and places. The more simple interpretations of quantum theory predict that quantum fluctuations in the vacuum state of space give rise to energy fluctuations whose average expectation values could be 40-100 orders of magnitude greater than all of the energy contained in cosmic matter.

The thermal radiation pressures of air are "positive" exerting "inward-pushing" pressures over the entire surface of any accelerating vehicle – such as the upper vehicle shown in Figure 62. Airflow is both compressed and expanded as the speed of sound is approached, reached, and exceeded and an adverse pressure gradient forms about the vehicle. Here, higher than ambient pressures acting upon the front of the vehicle cause a resisting "push" and lower than ambient pressures acting on the rear of the vehicle cause a resisting "pull". And it may be noted that air pressures are resisted by a very thin repulsive field region that is caused by electrical repulsion between electrons of the outermost atoms of the vehicle's skin and the most nearby electrons of the atoms of passing air. In air-less vacuum, such a thin repulsive field vanishes, and Bernard Haisch in 1994 [11] showed that an electromagnetic interaction between the accelerating vehicle and its quantum vacuum medium causes an inward-pushing electromagnetic radiation pressure gradient that is somewhat similar to an aerodynamic pressure gradient. This flight resistance is equivalent to and can be viewed as vehicle inertia, thus opening the door to FTL travel in the near future as this discovery of Froning and Haisch are fully implemented in the aerospace industry.

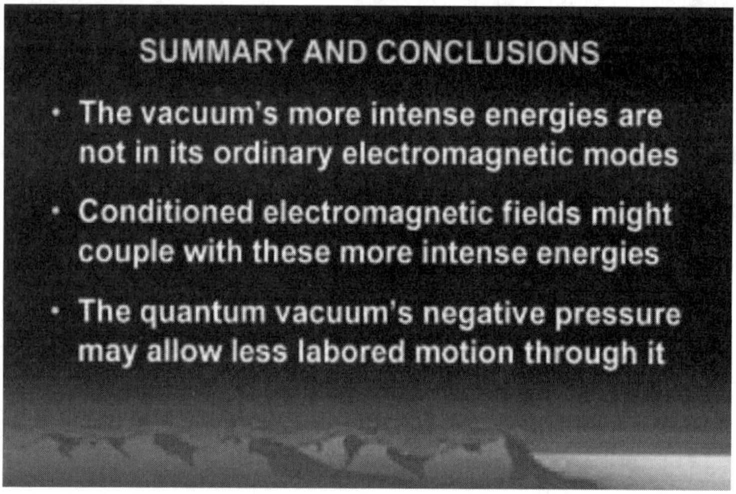
SUMMARY AND CONCLUSIONS
• The vacuum's more intense energies are not in its ordinary electromagnetic modes
• Conditioned electromagnetic fields might couple with these more intense energies
• The quantum vacuum's negative pressure may allow less labored motion through it

FIGURE 64 SUMMARY AND CONCLUSIONS

10 CAN HUMANITY EVER EXPAND BEYOND THE LIMITS OF EARTH?

A lot was happening between 1953 and 1983 when the most rapid advances in human flight occurred. But flight progress seemed to diminish shortly after that. However, aerospace conferences and talks did not diminish, and some of the more boring talks where planned government initiatives for advancing spaceflight. Some initiatives were expensive, going far into the future. But none seemed bold enough to revolutionize spaceflight – to reduce its cost enough to make spaceflight a commercial, economic opportunity instead of continual taxpayer burden. Then the government was also rather poor in controlling spaceflight costs. Shuttle costs were about 40 times more than promised and Space Station costs were exceeding estimates by factors of 10. So NASA estimates of $450 billion taxpayer dollars for a planned Mars Expedition with chemical rockets was being increased by senior cost experts to over a trillion.

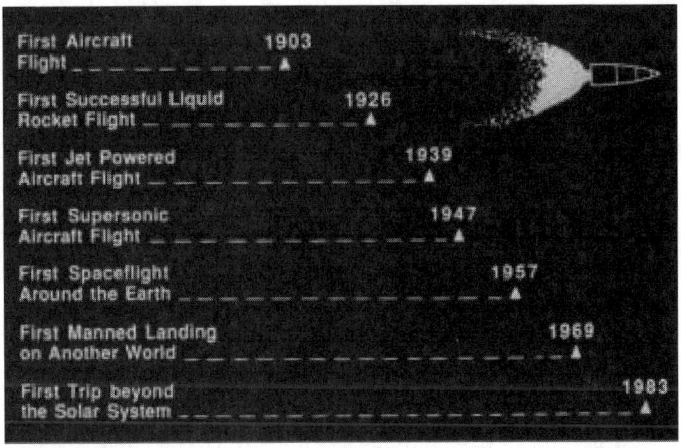

FIGURE 65 PAST ADVANCES OF FLIGHT IN FIRST CENTURY OF FLIGHT

Also, reasons for spaceflight initiatives seemed vague and mundane, though they would always claim good things like "increasing youth interest in science" and "maintaining U.S. pre-eminence in space". So, after hearing many plans for future spaceflight initiatives, I decided to try to plan my own. And I began by reviewing past advances during the first century of flight – as is shown in Fig. 66, it is seen that relatively rapid flight advancement during the first 80 years of flight had sometimes been enabled by a significant propulsion system advance.

The beginnings of rocket and air breathing jet propulsion were in 1926 and 1939, but, in about 55 years after that, jet propulsion superseded propeller propulsion to revolutionize airline travel by about 1980. And by then rocket propulsion had already allowed spaceflight to begin. Also, during those Halcyon years, when the Apollo program was going on, so the very successful "Rover" nuclear rocket program was going on. And by 1972 the Nerva NRX/XE nuclear fission rocket had accumulated 2 hours of ground firing time, including a 28-minute full-power run at 334 kN thrust, which was then believed sufficient for sending humans to Mars.

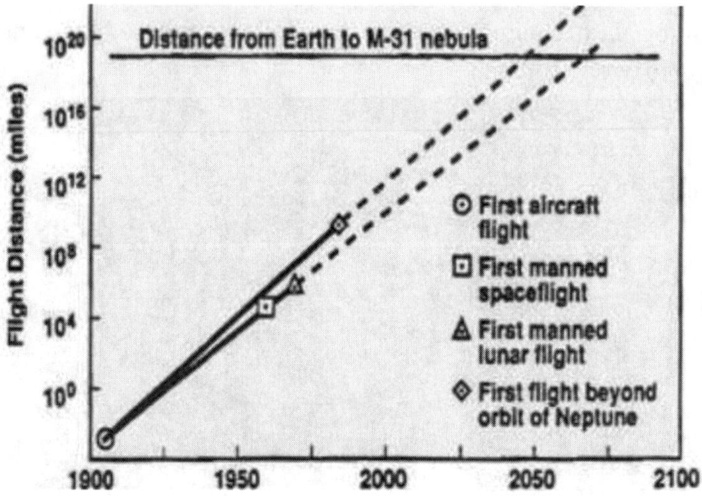

FIGURE 66 EXTRAPOLATING FLIGHT DISTANCES AND DATES TOWARD THE FUTURE

So, with this nuclear foundation, Werner von Braun lobbied strongly for departing on a human expedition to Mars by means of nuclear propulsion in 1982.

Next, I attempted extrapolations of past progress in increasing flight distance during the first 80 years of flight into the future, as shown above. And I was pleased that sustaining this past rate of progress into the future could result in humans reaching extremely distant star systems before this 20th century ends.

Extrapolations of flight distance progress of our past in Figure 66 shows we could be as far away as distant stars by 2075 if we can maintain the flight distance progress of our past. Straight line extrapolations of flight speed are not as encouraging as distance. Extrapolated times indicated that light-speed won't be reached until about 2120. But slightly non-linear extrapolations indicate it might be reached by about the end of this century. So, I plunged ahead with an initiative that strove for revolutionary spaceflight advances, to enable much safer and less expensive solar system exploration, possibly before the year 2100.

FIGURE 67 MCDONNELL DOUGLAS DC-10 JET POWERED AIRLINER

During the first 80 years of flight, invention of the jet engine (by Von Ohain and Whipple) and the transistor (by Shockley and others) led to computers, avionics, and turbofan engines that revolutionized air travel by enabling swifter, safer, more economical airliners like the

Boeing 747 and the McDonnell Douglas DC-10 shown in Figure 67. And this led to today's worldwide air travel industry of aircraft companies, airline companies, airports, hotels, and ground travel that has given careers and expansive flight experiences to millions of people on Earth.

A new air and space travel revolution would enable new opportunities like space tourism and high-speed, high-altitude airline travel, giving the exalting experience of spaceflight to thousands, not just a few. So, just as propeller propelled craft were replaced by jet propelled craft to revolutionize air flight; current jet and rocket propelled craft might be replaced by craft propelled by new modes of propulsion. And with help from other new technologies to dramatically reduce aerospace travel time, cost and hazard; these commercial craft would create exciting new careers for millions of more people on Earth.

Examples of such craft are air breathing and field-propelled aerospace planes discussed in previous sections and another one is shown below.

FIGURE 68 SWIFT REUSABLE AEROSPACE PLANES THAT CAN TAKE SPACE TOURISTS TO ORBIT OR FLY PASSENGERS THROUGH SPACE BETWEEN ANY TWO CITIES ON EARTH IN LESS THAN 2 HOURS.

Except for space organizations and steadfast space enthusiasts there was little passion in the 1990's for taxpayer-funded trillion-dollar chemical rocket expeditions to Mars. But if Mars expedition costs could be 100 times less than such costs with chemical rockets, they could be within investment resources of combined corporations. But such expeditions would require such drastic fuel reductions by its vehicles, that their engines would have to develop much of their power and thrust by the actions of fields, not by combustion of matter.

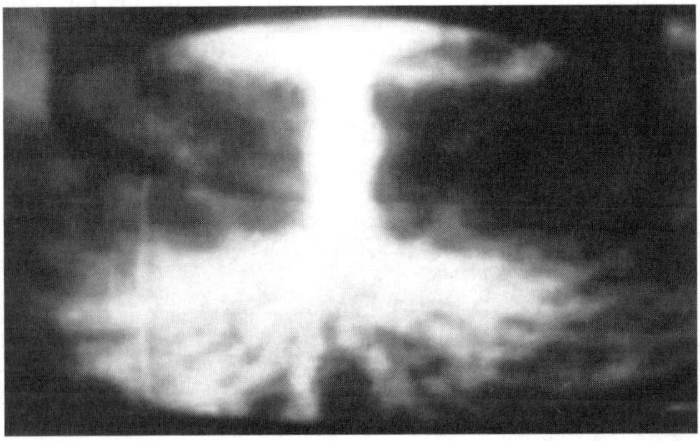

FIGURE 69 INTENSE ENERGY GENERATION FOR POWER AND THRUST BY ENORMOUS ACCELERATION AND COMPRESSION OF LIGHT PLASMAS BY STRONG ELECTRIC AND MAGNETIC FIELDS-BY PROFESSOR NARDI AT STEVENS INSTITUTE.

My initiative showed examples of possible propulsion advances initiatives during the next 80 years of flight, beginning with much more energetic rocket or ramjet systems and continuing with even more futuristic systems that would consume even less propellant by developing much of their thrust by actions of fields, not combustion of matter. The next page shows possible levels of field propulsion that could be striven for during the future years of this 21st century.

Studies by others indicated even much slower-than-light propulsion could enable colonization of our entire Milky Way Galaxy if advances in nano-technology and manufacturing would enable self-replicating ships and self-replicating crews that could multiply and spread to all this galaxy's habitable planets. Near light-speed travel would of course allow swifter settlement of this galaxy and settlement of even other galaxies as well. And faster-than-light travel could not only swiftly settle many galaxies, but also link their many worlds into a vast cosmic confederation by space commerce and communication.

But my pleasant vision, of faster propulsion allowing settlement of more worlds was somewhat spoiled by the unpleasant thought that faster propulsion for more settlement might not profit the universe as a whole if Earth's evil greed and belligerence was merely spread more swiftly through it. In fact, a successful interstellar initiative could do net harm if future Earth space-faring civilizations were just like all past Earth seafaring civilizations – who brought misery and misfortune to indigenous societies everywhere they went. This thought didn't stop my work but caused it to proceed with somewhat less zeal.

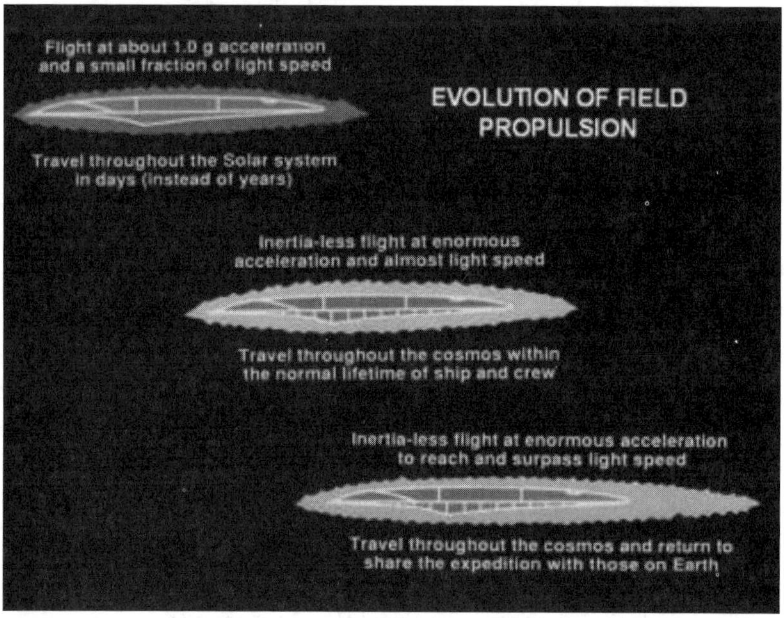

FIGURE 70 EVOLUTION OF FIELD PROPULSION IN THREE STAGES: (1) SUB-LIGHT SPEED; (2) INERTIA-FREE FLIGHT CLOSE TO SPEED C; (3) INERTIA-FREE FLIGHT AT FTL SPEEDS

I proposed an Interstellar Initiative [26] which would begin developments to help revolutionize space flight sooner in our solar system and to allow leaving this system before this Century ends. I made initial initiative events consistent with 1989 government advanced plans that called for nuclear fission and nuclear fusion rocket developments and human expeditions to the Moon and Mars in the 2020-2025 period by chemical rockets.

The Initiative selects its bold spaceflight developments (conceivably ones that might include things like zero-point energy and field-propulsion) and its activities from those of other initiatives. I have summarized my work in a paper "An Interstellar Initiative for Future Flight" and gave it at various forums. Below was a preliminary set of Initiative milestones and the preliminary completion dates it contained.

Preliminary Milestones of Interstellar Initiative

- **Develop clean fusion propulsion 2020**
- **Make first expedition to Mars 2030**
- **Develop field propulsion science 2040**
- **Make first human outer planet expedition 2060**
- **Develop relativistic field propulsion 2070**
- **Launch interstellar probe to nearby star 2080**
- **Embark on human expedition to distant star ...2099**

FIGURE 71 PRELIMINARY MILESTONES OF THE AUTHOR'S INTERSTELLAR INITIATIVE

My initiative was bolder than existing initiatives in attempting challenging science and technology programs and it had a challenging long-term goal. But it was much more modest than current major programs because it focused on science and technology advances that could soon feed into major programs if successful, as new programs were begun. And it would avoid becoming static and staid by being staffed with many part time people borrowed from academia, industry and government. Unfortunately, other company work allowed little time to delve into the details my initiative required, such as: how its research could favorably feed into work of other initiatives instead of competing with them, and how projects and key people doing its advanced work would most appropriately be selected and most appropriately be funded?

The reason for my initiative was probably too high-sounding, but it was simple: to bless more people by helping revolutionize spaceflight – in order to: provide new resources and careers for Earth and its people; open new frontiers (exciting endeavors) to stir more hearts and lift more spirits; and to overcome the flight barriers that prevent us from going where we need to be.

My paper was novel enough to stir up some interest from students, scientists, engineers, and even some government managers. I gave it at spaceflight forums and universities while at McDonnell Douglas and my own company. But nothing really changed. For with no national urgency or military pressure, government- industry-academic aerospace research had already seemed to relax into a more leisurely pace than during the Cold War. And unfortunately, something more like the urgent Cold War pace was what my Interstellar Initiative required. Air Force technical emphasis was already shifting from "speed" to "stealth" and government interest in nuclear power and propulsion was beginning to wane. So now, 30 years after the halcyon years ended, there has been no increase at all in flight speed or distance – with Voyager 1, launched during the halcyon years, is still our fastest and most distant probe. NASA's futuristic Breakthrough Propulsion Physics Program (BPP) eventually ended. (But, fortunately, some of the revolutionary research BPP began is now carried on by the NASA "Eagleworks" organization in at the Johnson Spaceflight Center).

Astronomer Michael Papagiannis of Boston University gave insight into the difficulty of evolving into a space-faring interstellar civilization. He described a civilization's industrial phase, beginning with unchecked growth, followed by growth limits: over-population, resource-depletion, environment-degradation. He suggested strength and procreation ability, important in early evolution, must finally be superseded by collective wisdom and ethical development to limit population growth and materialistic expansion. But this requires unselfish societies who conquer belligerence and love of wealth and power. And he argued that only such moral-ethical societies would survive long enough to Papagiannis's idea of intellectual and scientific prowess not being enough for going to the stars was discouraging. But his mention of ethics and morality resonated within me. I myself had once been confronted with an ethics-morality challenge and had seen the disastrous consequences of lack of it. A customer found our Douglas missile work superior to that of his incumbent contractor and wanted to replace this company with Douglas for forthcoming work. One day as I left his office, he gave me a thick manila envelope and said "read this – it should help you". Back at Douglas I opened the envelope. Inside was a report by the rival company, and I could see from the cover that it contained invaluable information to help defeat our rival. But the cover also contained a "Proprietary" designation. Immediately something within me ordered: "Do not open that report. Return it to the customer."

I did this, much to my customer's annoyance. From then on he dealt with others at Douglas and we never did win out over our rival. But many years later a similar thing happened in a similar competition between aerospace companies. One company had been improperly given proprietary information of the other. This was found out and the indicted company was severely fined and deprived of billions of dollars of future aerospace revenues – a very high price for lack of morality and ethics. Papagiannis's idea of ethical, moral space-faring civilizations were also comforting to me. For, if Earth itself, was ever visited by space-faring civilizations. The ethics and morality of these technically-superior beings would not allow the bad treatment past Earth seafaring civilizations imposed on Earth's native people.

Amazingly, near the end of his paper [27] Papagiannis suddenly brought God into his discussion, suggesting moral, ethical space-faring interstellar civilizations would be near the end of God's long, patient evolution of men and women out of the "dust of the earth" into perfect images and likenesses. I disagreed at once with Papagiannis's idea of interstellar civilizations being near the zenith of God's spiritual evolution of humanity beyond materialism and sin, towards perfection and holiness. It made no sense at all, for it did not correspond to today's reality of environment degradation-government corruption-corporate greed-human misery ever on the rise. Surely things weren't getting better – but worse. And there was no good news from religion. For Bible prophecy (Revelation Chapters: 4-19) described evils-plagues-quakes-fires ravaging most of the Earth.

But after I read past all of Revelation's years of tribulation – when angels with vials loosened calamities on Earth – I found there was more. I got to Revelation 20, when: "Satan is overcome and imprisoned in a bottomless pit. And, in the following Christ will reign with him on earth for 1,000 years". Almost nothing of detail is said about this 1,000-year reign, with most focus being on what happens after the millennium is over – when Satan is released and overcome, and the physical world is completely transformed into a spiritual world (in Revelation 21).

FIGURE 72 COMPOSITE IMAGE OF THE SOMBRERO GALAXY BY THE SPITZER AND HUBBLE SPACE TELESCOPES

An idea began growing in my mind that Papagiannis's religious view of interstellar flight was consistent with the Bible's "millennium" and if there was one, why could it not include interstellar flight? For, during this millennium, people who survived Satanic tribulation would be available for an Interstellar Initiative and surely they would be the kinds of people needed to organize into a moral-ethical space-faring civilization that Papaginnis requires for interstellar flight. So, even if millennium years are shorter than years today, trouble-making Satan is gone and the best possible Leader of any initiative is in charge. Therefore, from my limited Biblical scholarship, surely there will be ample time for an Interstellar initiative and final halcyon age of flight.

REFERENCES

[1] Report No. SM-35775, "Proposed Canard Control Nike Zeus Missile," Missiles and Space Engineering Department, Santa Monica Division, Douglas Aircraft Company, Inc., Santa Monica, California, June, 1959

[2] Douglas Report SM-41425, "Sprint Missile Preliminary Urban Defense Study, "Missiles and Space Engineering Department, Santa Monica Division, Douglas Aircraft Company, Inc., Santa Monica California, March, 1962

[3] Report No. SM-43268, 1962, "Medusa Interceptor Study Report," Missiles and Space Engineering Department, Santa Monica Division, Douglas Aircraft Company, Inc., Santa Monica, California, March, 1963

[4] Froning, H.D.,Jr., "Investigation of Antimatter Airbreathing and Rocket Propulsion for Single-Stage-to-Orbit Ships," MDC H2618A, 1987 JANAAF Propulsion Meeting, San Diego, California, 15-17 December, 1987

[5] Froning, H.D., Jr., Little, Jay," Fusion-Electric Propulsion for Aeropace Plane Flight," AIAA- 93-5126,AIAA/DLGR Fifth International Aerospaceplanes and Hypersonics Technologies Conference, Munich, Germany, 1993

[6] Bussard, R.W., "The Advent of Clean Nuclear Fusion: Superperformance Space Power and Propulsion," 57th International Astronautical Congress (IAC 2006), October, 2006

[7] Froning, H.D.Jr., Czysz, P., "Advanced Technology and Breakthrough Physics for 2025 and 2050 Military Vehicles," Space Technology and Applications International Forum (STAIF2006), AIP Conference Proceedings, Editor S. El-Genk, American Institute of Physics, Melville NY, 2006

[8] Barrett, T.W., "Topological Foundations of Electromagnetism, "World Scientific," 2008

[9] Barrett, T.W., "On the distinction between fields and their metric," Annales de la Louis de Broglie, 14, 1., 1989

[10} Froning, H. D. Jr., "Reducing Fusion Plasma Confinement Energy With Specially Conditioned Electromagnetic Fields, "Journal of Space Exploration, Vol.2 (3),2013, p126-185, Menta Press,2013

[11] Haisch, B., Rueda,A., Puthoff,H., Inertia as a zero-point field Lorentz Force," Phys. Rev. A, Vol. 49, p.678 (1994)

[12] Froning, H.D. Jr Roach, R.L.,"AIAA-2002-3925, Preliminary Simulations of Vehicle Interactions with the Quantum Vacuum by Fluid Dynamic Approximations" 38th AIAA/ASME/SAE/ASEE Joint Propulsion Conference, Indianapolis, Indiana, 2002

[13] Froning, H.D. Jr," AIAA 96-4329, Economic and Technical Challenges of Expanding Space Commerce by RLV Development",1996 AIAA Space Programs and Technologies Conference
[14] Sänger, E., "The Attainability of the Stars," 7th International Astronautical Congress, Rome, Italy, 1956
[15] Froning, H.D. Jr., "Some Preliminary Propulsion System Considerations and Requirements for Interstellar Flight," 18th International Astronautical Congress, Belgrade, Yugoslavia, 1967
[16] Froning, H.D. Jr., "Interstellar Flight – A Potential Space Vehicle Opportunity for
International Collaboration," 19th International Astronautical Congress, New York City, USA, 1968
[17] Bussard, R.W., "Galactic Matter and Interstellar Flight,"Astronautica Acta, Vol.6,171-193, 1960
[18] Wheeler, J.A., " Superspace and the Nature of Quantum Geometrodynamics." Topics in Nonlinear Physics, pp. 615-644; Proceedings of the Physics Section of the International School of Nonlinear Mathematics and Physics, Springer Verlag, 1968
[19] Froning, H.D. Jr., "Propulsion Requirements for a Quantum Interstellar Ramjet," Journal of the British Interplanetary Society, Volume 33, No.7, July, 1980
[20] Forward, R.L., "Extracting electrical energy from the vacuum by cohesion of charged foliated conductors", Physical Review B, Vol. 30, No. 4, 1984, 30, 1700
[21] Froning, H.D., Jr., "A Metaphysical Interpretation of Tachyons," Speculations in Science and Technology, Elsevier, Sequoia S.A., Lausanne 1, Switzerland, 1981-1982
[22] Stevens, W.C., UFO, Contact from the Pleiades, Volume 1, ,1983
[23] Froning, H.D. Jr., "Propulsion Requirements for Rapid Transport to Distant Stars," Journal of the British Interplanetary Society,
Volume 36, No.5, May, 1983
[24] Froning, H.D. Jr., "Reaching the Further Stars, Overcoming the Barriers of Time and Space", Spaceflight, Dec. 1983, Vol.25 No.12, Published by the British Interplanetary Society
[25] Alqubierre, M.," The Warp Drive, hyper-fast Travel within general relativity," Classical Quantum Gravity, 11(5)L73-L77, 1994
[26] Froning, H.D. Jr.," An Interstellar Exploration Initiative for Future Flight," MDC 91 H7041 Mc Donnell Douglas Space Systems Company, Presented at 28th Space Congress, Cocoa Beach, Florida, April, 1991
[27] Papagiannis, M.E., "Natural Selection of Stellar Civilizations by the Limits of Growth," Astronomy Department, Boston University,1984 <html:file:\1984QRAS-25-309>

[28] Froning, H. David, "Quantum Vacuum Engineering for Power and Propulsion from the Energetics of Space", Proceedings of the Third Conference on Future Energy, Washington DC, Integrity Research Institute publishers, 2009

Note: All Bible quotations are from the "Authorized King James Version" (Printed by Cambridge University Press)

AFTERWORD

by Thomas Valone

Upon completing this book, the reader may agree with me that David Froning is a genius. He collaborated with all of the right experts to give us the best theoretical foundation, experimental proof of principle, and computer simulations that one man can do in one lifetime. With that said, my function as editor was to add captions for all figures, pull together a final chapter which was partially written, and assemble an Appendix full of journal articles that were not too redundant. Since the page count of the book was not an issue with the publisher, we felt a little bit of overlap in the Appendix would not be bad, if a new perspective or explanation was presented in a later article to clarify what may have been difficult to understand in the chapters or in a previous article. After all, this book is really for future space-worthy generations who may be more familiar with advanced aerospace propulsion and will be eager to implement these concepts for governments or private industry ready to go to the stars and thereby need to understand everything that Froning has discovered. An important point to make is the comparison of **toroidal drag reduction** and **polarization-modulation (pol-mod)** drag reduction with a laser, either one or both aimed forward of the craft. (Two different fonts here help you keep track.) Dave clarified the facts for us by saying:

(1) The toroid experience is in the 400 kHz - 110 MHz frequency range and pol-mod is in the 1.0 -10 micron wavelength range;
(2) The toroid testing time being three weeks (Froning) and pol-mod testing time being three years (Froning); while the
(3) Toroid energy source is EMF alternating electric current; and
(4) Pol-mod energy source is EM microwaves or laser diodes.

He concludes by stating, "finally, that toroid superiority over pol-mod is that it combines transmitter and antenna functions in one system. Cheers, Dave"

This seemed to bring some resolution to the two competing methods but then a few weeks later on, as this book was being finalized for the printer, Dave sent the following message which shed a new light on the laser pol-mod approach (remember that Dr. Mead was the Director of the Air Force Research Lab and White Sands was his playground):

"Incidentally the laser and lightcraft shown in this addendum ppt night photo is Dr. Frank Mead's lightcraft and a laser at White Sands."

(See pulsed laser photo on the next page.)

Dave Froning went on further to say,

"This laser was modified and combined with another laser to test TERENCE BARRETT'S "polarization-modulation" laser idea (shown in a later slide [below]). The research was successful and written up but not yet published."

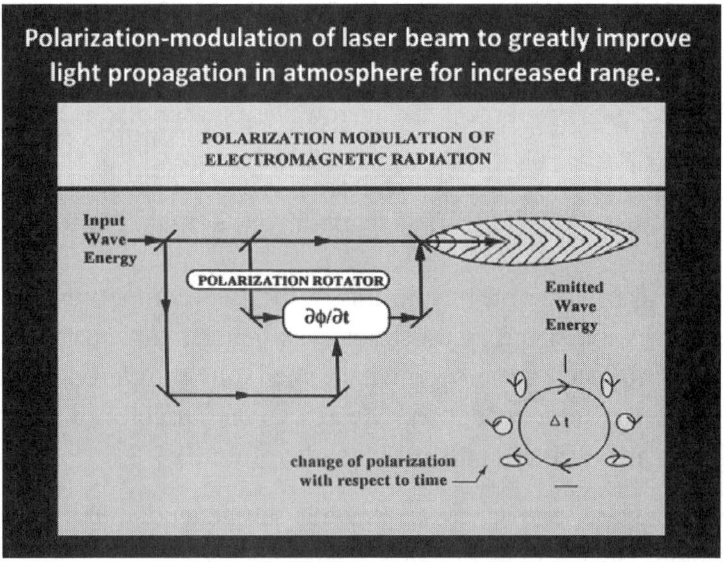

This White Sands missile test seemed so significant that this Afterword started to become a necessity. Even though we have seen the graphic diagram above many times in this book, here Dave indicates the pol-mod test with lasers was further developed most recently at his current institution, the University of Adelaide, for Beamed Energy Propulsion research:

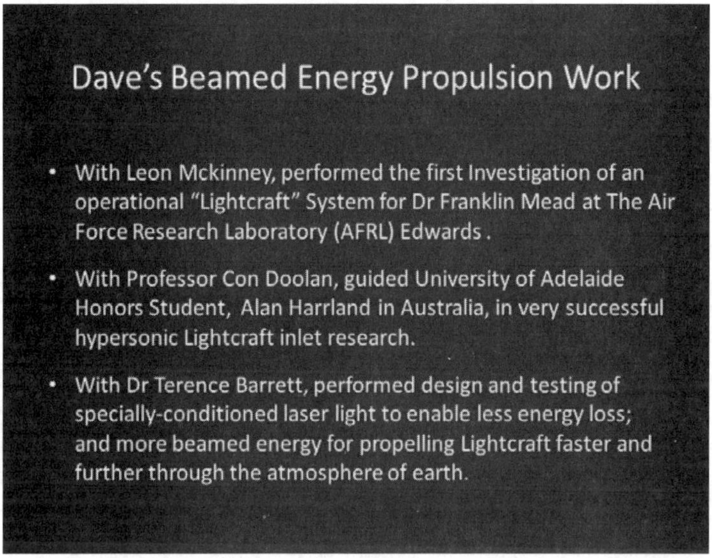

Note that Dr. Barrett is a theoretical physicist who proposed both the toroid and pol-mod designs. Below Dave's next two slides showing how this Beamed Energy Propulsion can be implemented in the future:

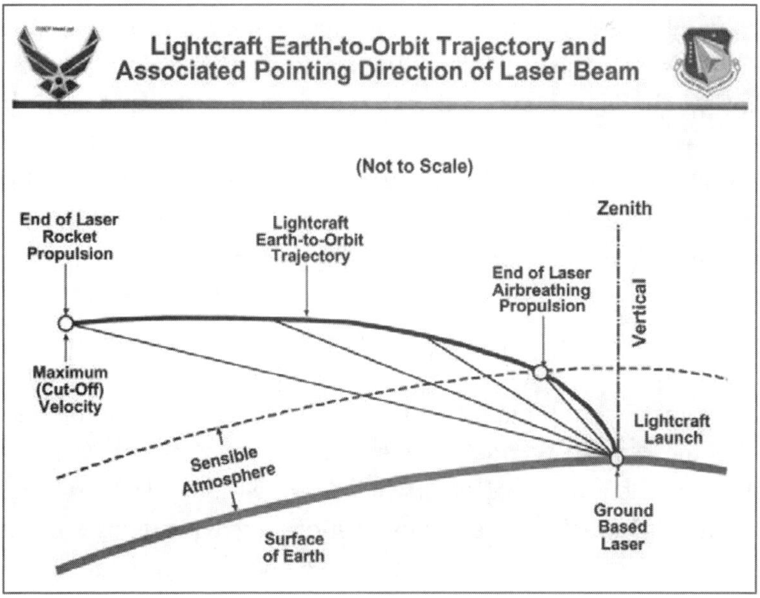

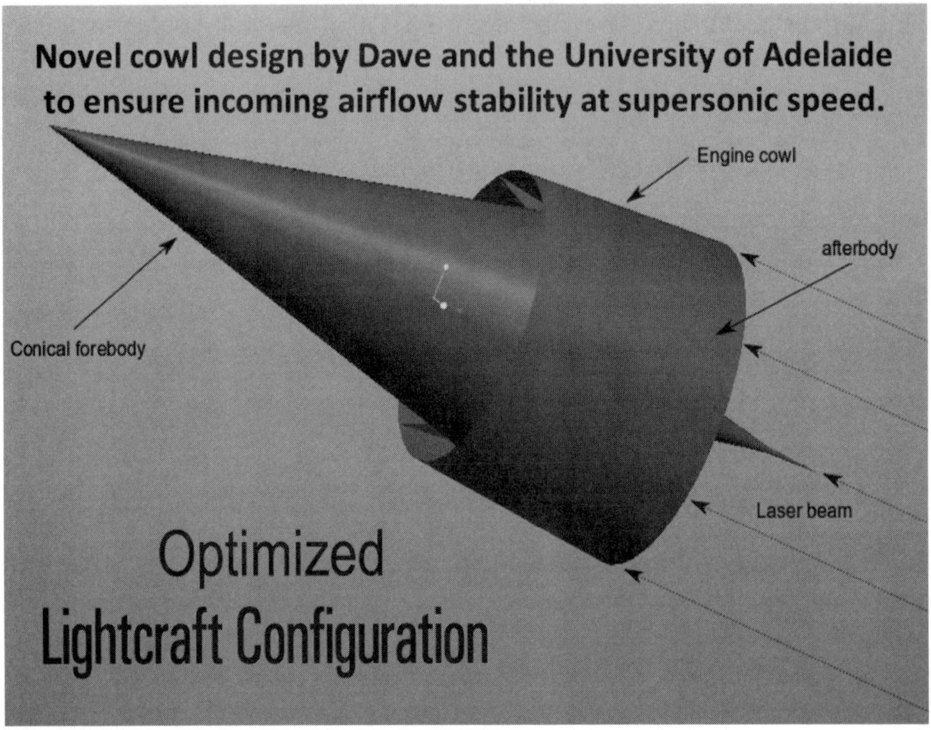

To finish up this section, Dave also offered the following to me for an Editor's Note regarding **pol-mod**:

Ordinary EM fields are believed to couple weakly with gravity. Those that might be able to couple are the higher SU2 and SU3 Lie symmetry of the Standard Model fields associated with the weak and strong nuclear forces of gravitating matter. For this reason, Froning became interested in Terence Barrett's development of SU2-SU3 EM radiation fields that have the same Lie symmetry as SU2 and SU3 matter and radiation fields. As such, these interacting EM fields might overcome or reverse gravity and inertia by EM discharge from swiftly accelerating, beam-propelled ships, for example, among other situations. Unfortunately, few engineers or scientists are capable of modeling and computing the complex tensor calculus needed to describe SU2 and SU3 EM radiation fields. Froning found that Derek Lineweber's Supercomputer Physics Group at the University of Adelaide are one of those that can. Lack of Computational Electromagnetics (CEM) has severely limited the amount of long, time-consuming non-abelian tensor calculations that Barrett has been able to do for detailed investigations of SU2 and SU3 polarization-modulations of EM radiation. Therefore, this book contains less accurate but more rapidly produced field pattern computations from Poynting vector and Lissajous figure plots.

As to three years of **toroid test results** done at the Hathaway Labs in Canada and at various universities, Dave's supplemental summary is as follows:

- Circumferential EM standing waves form on surfaces of toroids at resonant frequencies.
- Intense and narrowly focused magnetic fields form over all toroids at resonant frequencies.
- An asymmetric toroid focused magnetic field energy into a desirable, force-causing shape.
- More tests needed, but the emitted radiation seemed to maximize at resonant frequencies.

In the following two supplemental slides, Dave offers an insight as to the aerodynamic tests that were suggested for the implementation of small toroidal EM radiators placed strategically on a supersonic airplane.

EM discharges from the nose cone, wing and trailing edge can significantly reduce drag and sonic boom at the speed of sound and in this book the same symbol c is used for the speed of sound and the speed of light, since Froning and Barrett's thesis is that analogous results can be expected in each of the corresponding media, at fractions/multiples of c.

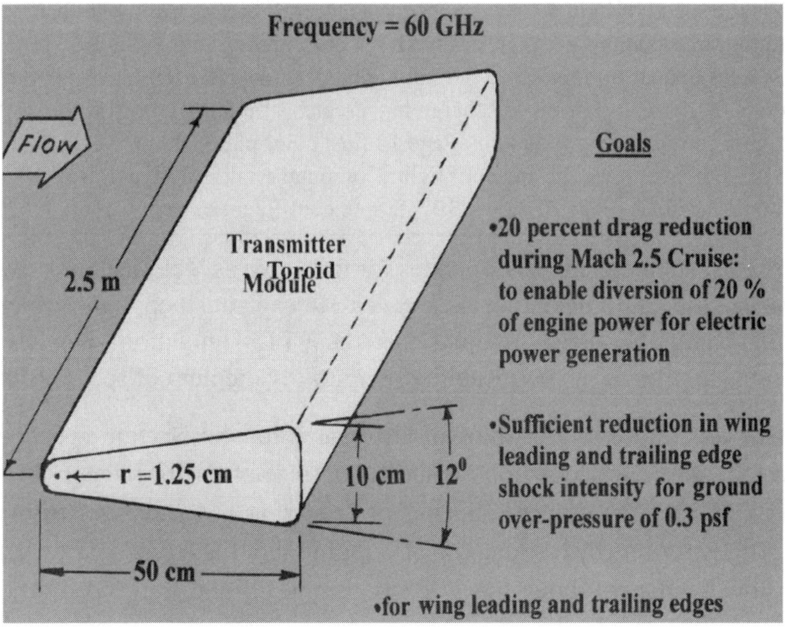

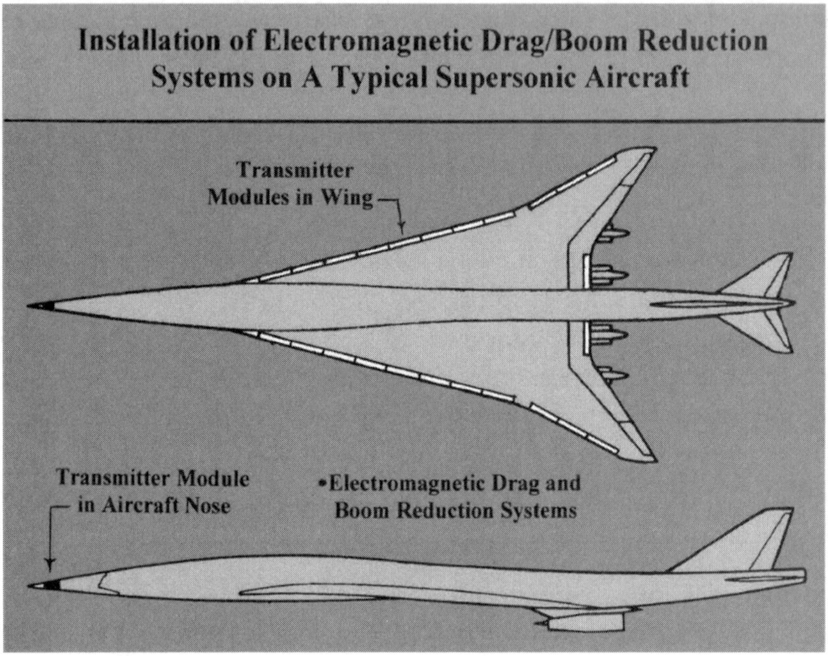

To bring this Afterword to a close, we finish up with the Roach computer simulations that used **Computational Fluid Dynamics (CFD)**, seen on the Color Plates in the center of the book and also on the back cover.

Key to this book's Color Plates: Page 3 (may not be numbered) shows M = 1.00 which is the speed of light c. This shows the spacecraft breaking the light barrier, as with the subsequent color plates as well. All craft are moving to the LEFT, so shock waves appear trailing off to the right. Page 9 shows M = 0.99, Cd=-.0486, gle=gte=4.0, gfs=1.4 with EM discharge that creates a bubble on the left (in front) of the ship, showing drag reduction. Page 12 (last color plate) shows M = 5.0 which is 5c or five times the speed of light. Computational fluid dynamics simulation variables gle = 10, gte = 10, gfs = 0, coefficient of drag Cd = 0.006295

Dave offers a supplemental set of slides for this topic as well, the last of which is also reprinted in COLOR on the back cover. He explains their significance in the following paragraph (keep in mind that the CFD simulation analyzes an air medium but applies to the permeability/permittivity medium of space as well):

In the event that rapid forming of EM beams for the aircraft or spaceship is vital, in the next four slides Bob Roach explores what has to happen after rapid discharge of such resistance-reducing or reversing EM radiation from a ship begins. This was done by tracking the development of the E beam pattern, power, and airflow exchange about the ship, as computed and iterated by the CFD's

Navier Stokes equations. First, the intensity gradient within the EM energy is deposited in surrounding still air at zero flight speed (Slide 1 below).

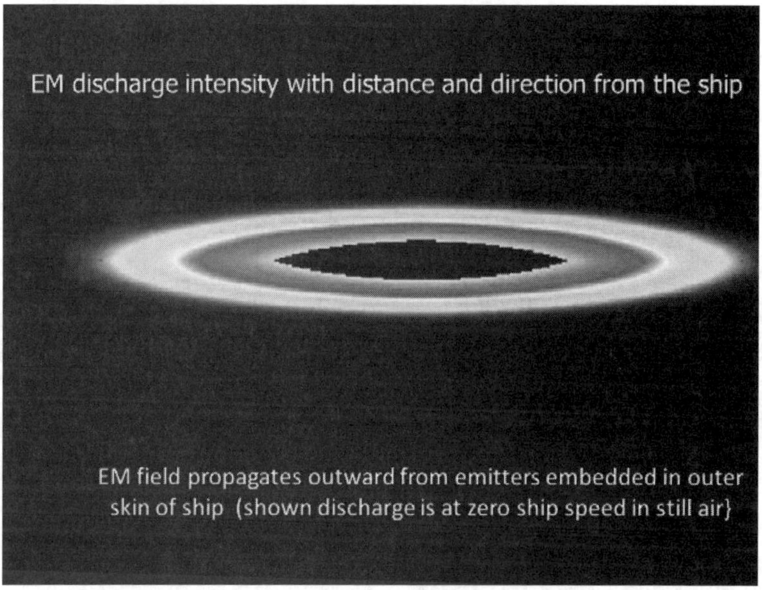

Second, the airflow pressures occurring at maximum drag is seen as the ship increases speed to Mach 0.99 for a non-radiating ship with no toroid turned on (Slide 2 below and also see similar version in the Color Plates section, M= 0.99).

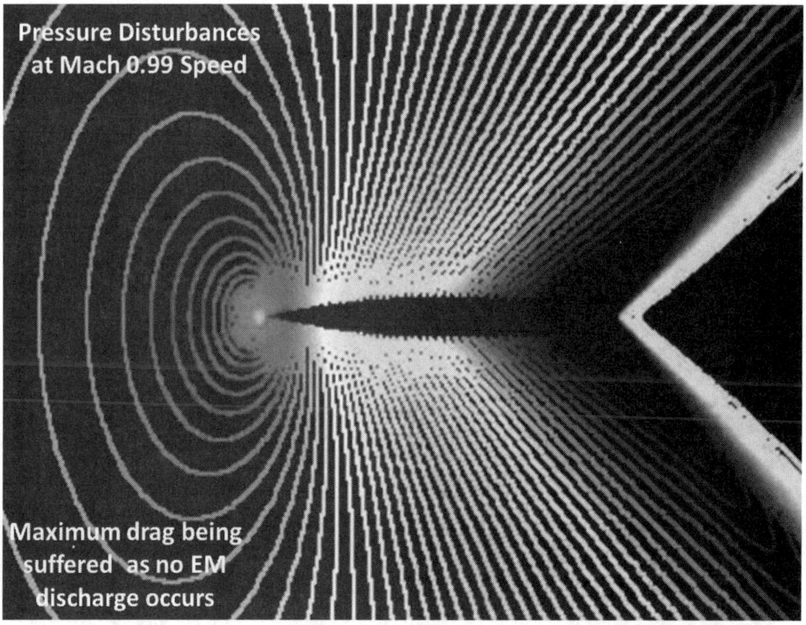

Thirdly, the EM discharge from our toroids or fore lasers is now developing and dimly seen, while the ship drag is already reducing (Slide 3 below).

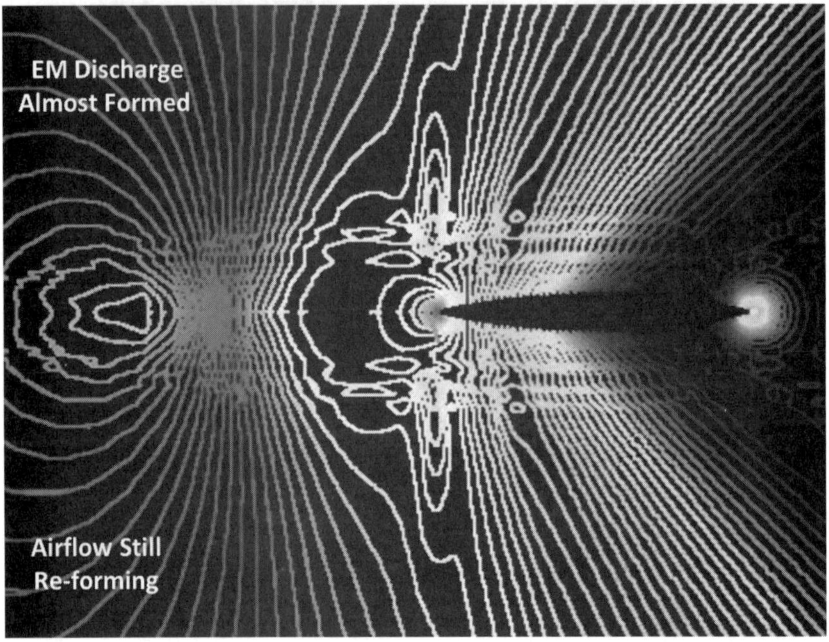

Finally, the CFD computation converges as the final drag value is reached as ship-emitted EM/medium disturbances coalesce into their correct and final patterns (Slide 4 below and fortunately, also in COLOR on the **back cover** of this book).

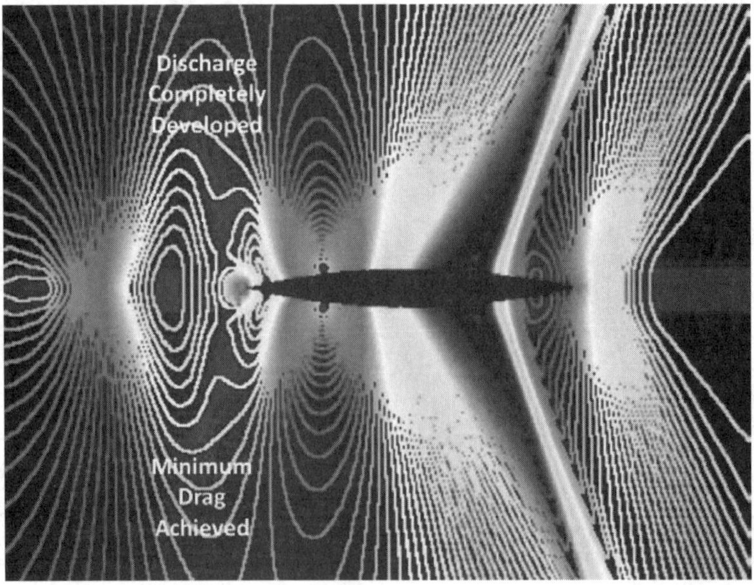

Hal Puthoff originated the great phrase "engineering the vacuum" which I believe is a good description of what we tried to do in this investigation. But much better simulations that ours could better model energy and propulsion interactions inside and outside a moving ship and allow engineering the vacuum for both impulsion and power. Here, one power need might be funneling energy into a ship from a favorable zero-point energy gradient outside it. This could be for needed EM energy that was emitted from the ship we studied. It's not known if this ship possesses a favorable zero-point gradient. But a favorable gradient might be in the central yellow and red region of the figure below. This region is formed adjacent to the ship by its fast speed and its fore and aft-pointing beams.

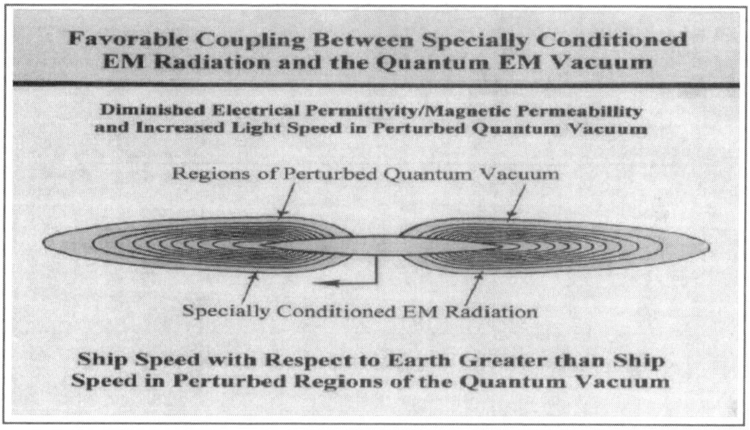

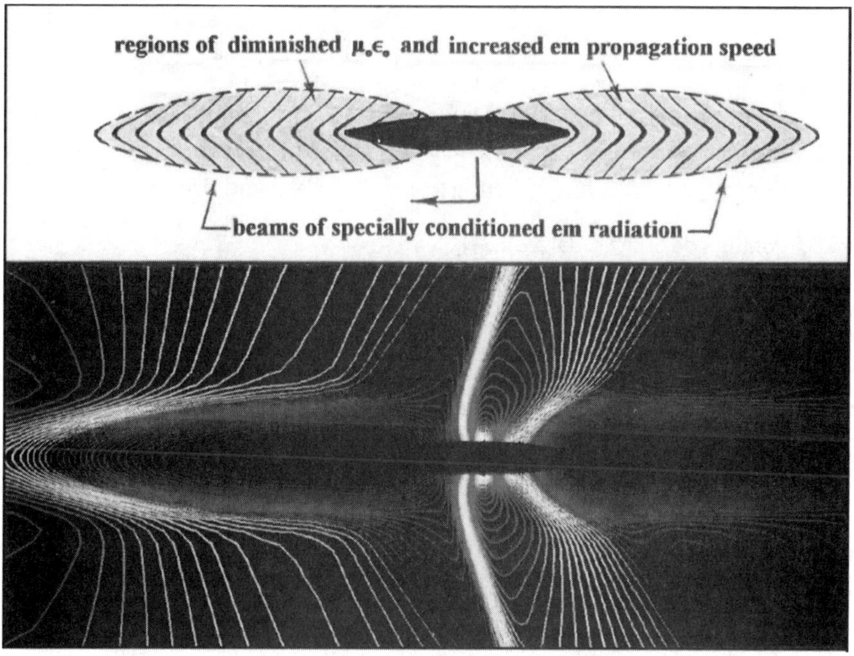

To answer the question that may arise concerning what the "minimum" drag may be for the chosen medium, Dave also presented a slide elsewhere for low, medium, and high EM radiation levels versus the percentage of drag reduction (below).

Zero-Lift Wing Drag Reduction

Discharge	Mach 0.99	Mach 2.50
Weak	31 %	16 %
Moderate	77 %	41 %
Significant	84 %	61 %

A last few revealing emails from Dave solidifies the details that this editor has wanted in order to represent the "two systems" faithfully, since each of them are so revolutionary and both have been lab or field tested and proven to work.

From: D. Froning <dfron4@gmail.com>
Sent: Monday, May 20, 2019 12:30 PM
To: Tom Valone <iri@erols.com>
Subject: Book Cover
Dear Tom. The existing Book cover is fine, for it indeed fits in with our "tear-drop" toroid design. Terence and I still like toroid systems even though Terence has not gotten as much customer interest in them as for polarization-modulated (polmod) systems that radiate EM fields. Thus, our toroid EM data base is much skimpier than our Polmod one.

From: D. Froning
To: Tom Valone
Subject: Re: Book Cover and Tale of Two Systems
Date: Tuesday, May 21, 2019 9:55:14 AM

Dear Tom I like your idea of a toroid - polmod comparison. I filled out your table here and will make some more toroid-polmod comparisons tomorrow. Dave

Characteristics	TOROID	POLMOD
Radiates energy forward	YES	Yes
Radiates energy backwards	Yes	Yes by mechcanical or electronic scanning.
Warps spacetime	Yes	Yes
Uses (needs) lots of Power	YES	Yes
Can get craft to FTL	Yes	Yes
Needs secondary propulsion	Yes for 1st generations	Yes for 1st gemerations
Burns rocket fuel	No	No
Needs Coolant	Maybe	Maybe

All further updates, such as this table, will be posted on David Froning's website listed below. It is my sincere hope that all of the extra time and effort that has been put into this book will help the reader appreciate the true value of the extraordinary work that David Froning's contribution represents for the future of mankind. While there are others that claim to have achieved nuclear fusion propulsion (e.g., David Adair), Froning offers the details with his prestigious coauthors so that the construction of such a rocket can be understood and someday may be implemented.

Below is Dave receiving the Integrity in Research Award, which he is holding, at the IRI COFE4 in 2011. I'm presenting the award and shaking his hand.

We also have a number of Dave Froning articles and PowerPoint slideshows, as well as a wonderful "Final Addendum" slideshow that he sent in too late to reformat for the book. All of the extra ones not used in the book are being posted to our Integrity Research Institute website, on David Froning's webpage, for free download. The easy to remember webpage is https://tinyurl.com/DavidFroning while Dave's email is dfron4@gmail.com .

Onward and upwards…ad astra!

Thomas Valone, PhD, PE
October 21, 2019

APPENDIX

Landscape slideshow images of the author are presented first on the following pages, which are easy to read by turning the book sideways.

The rest of the Appendix has portrait style reprints of the author's journal articles and supporting documents.

Be sure to visit www.tinyurl.com/DavidFroning for more.

MHD Power from Air Breathing Propulsion

- In the 1990's, the Russian Leninetz Company claimed that significant energy could be extracted from hypersonic air breathing engine airflow by MHD interaction and be re-deposited in the engine's nozzle flow to greatly increase the hypersonic air breathing engine's thrust or Isp.
- Investigations by many groups confirmed possibility for significant MHD power extraction from an air breathing engine airflows at Mach 7 speed or higher but they did not confirm that re-deposition of this energy in air-breathing engine flow would significantly increase engine thrust or Isp.
- Our MHD implementation doesn't re-deposit extracted MHD energy into engine nozzle flow. Instead, MHD energy flows to systems like a high energy laser or aneutronic fusion system as Mach 7 to 10 is reached.
- Extraction of power from engine airflow diminishes its enthalpy. So loss in engine thrust or ISP occurs. Therefore, extraction would only occur if there is significant power need. But, above about Mach10, the extracted MHD power is enormous and airflow enthalpy loss is less than 10 %.

Air Breathing Aerospace Plane Powered by MHD and Fusion Energy and Propulsion

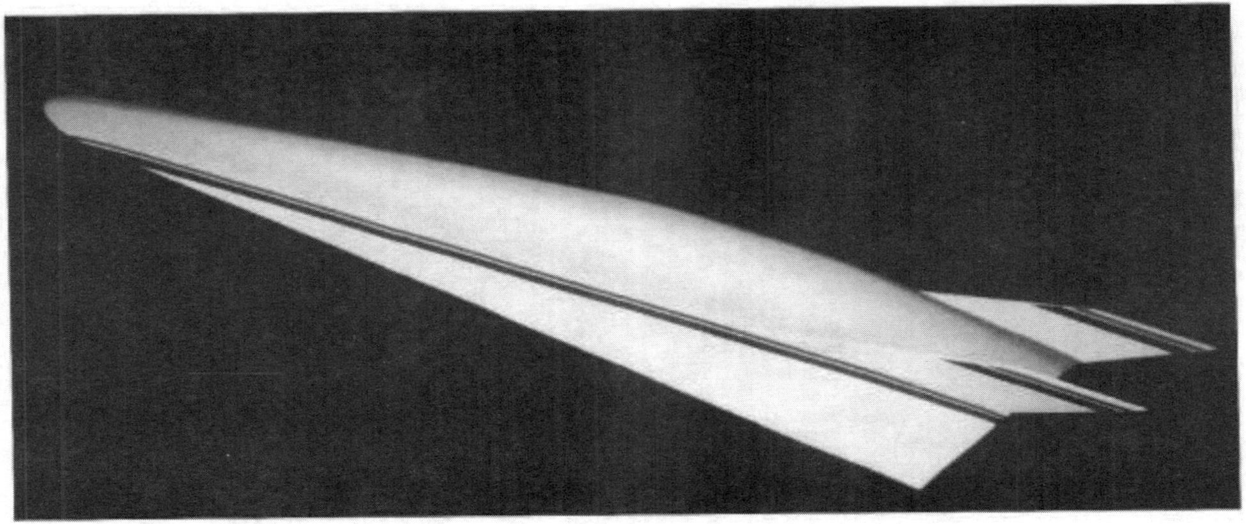

© STAIF 2001, 2005 by H. D. Froning and George Miley

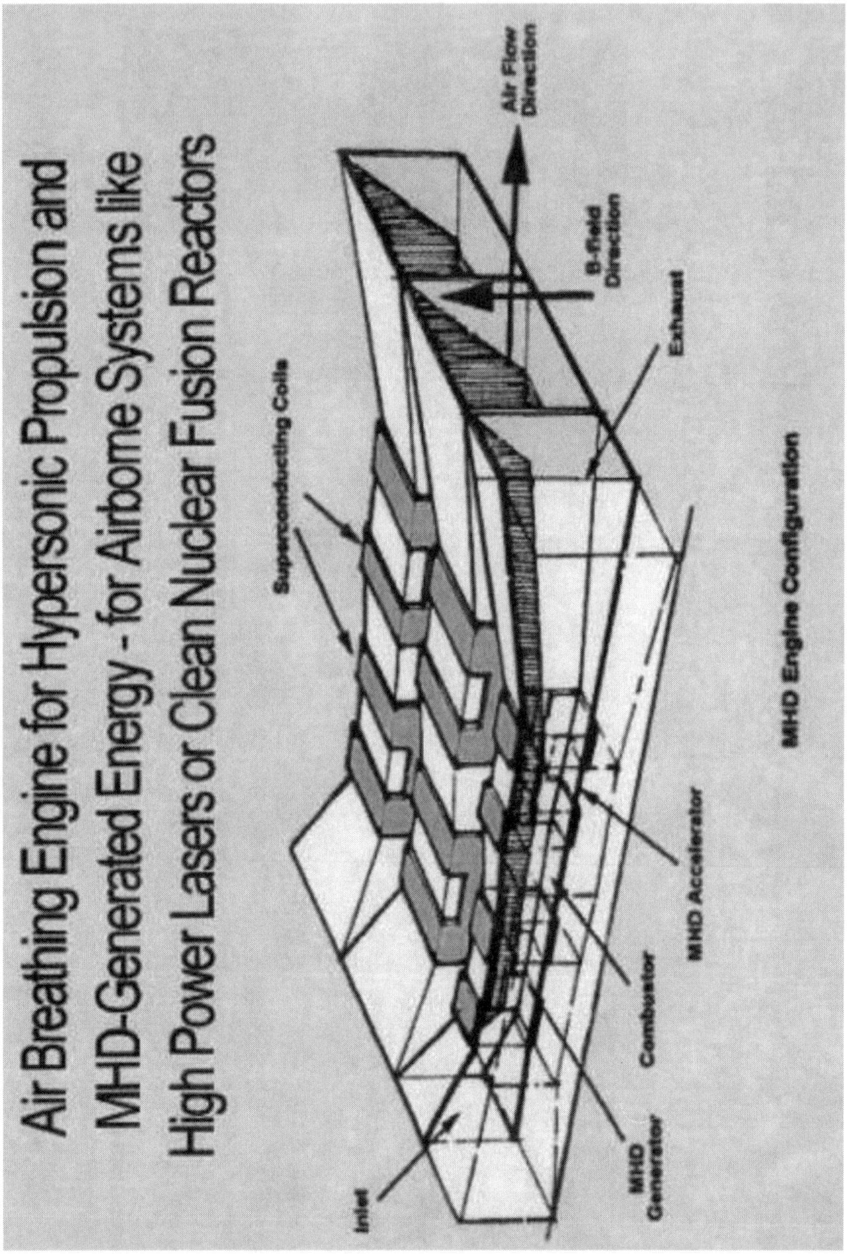
Air Breathing Engine for Hypersonic Propulsion and
MHD-Generated Energy - for Airborne Systems like
High Power Lasers or Clean Nuclear Fusion Reactors
Superconducting Coils
Air Flow Direction
B-field Direction
Exhaust
Inlet
MHD Generator
Combustor
MHD Accelerator
MHD Engine Configuration

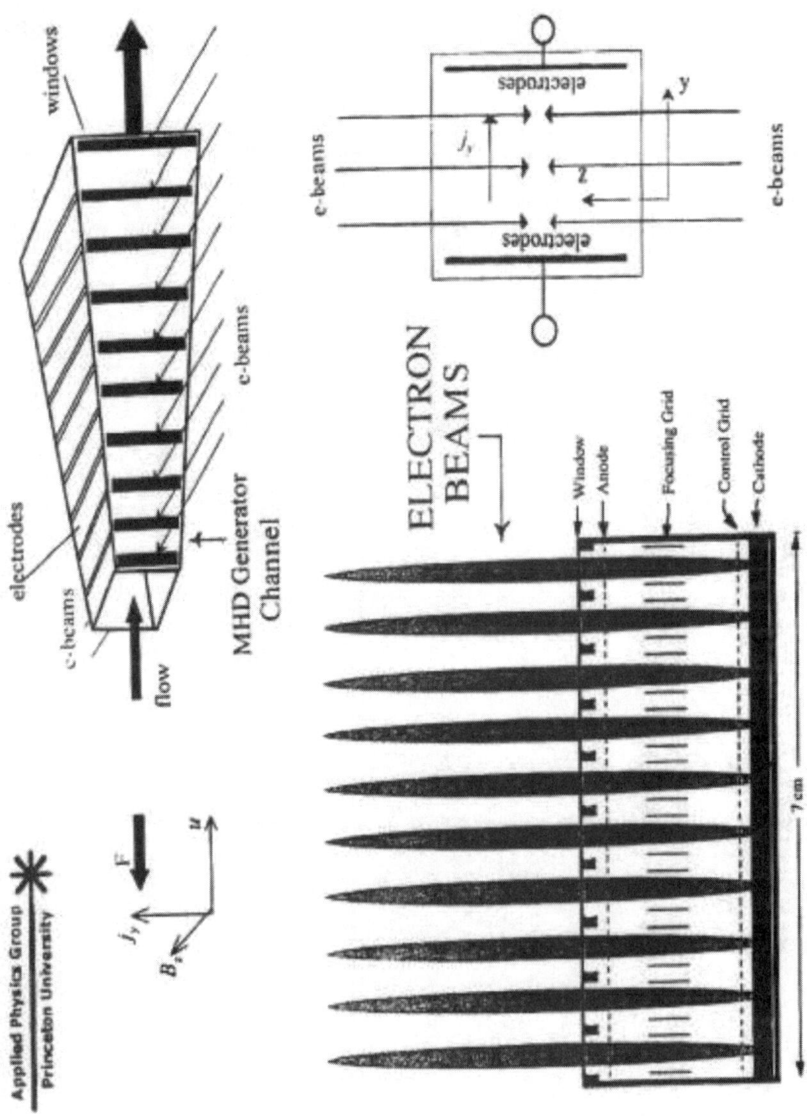
Applied Physics Group
Princeton University
windows
electrodes
e-beams
flow
MHD Generator
Channel
e-beams
electrodes
electrodes
e-beams
e-beams
ELECTRON
BEAMS
Window
Anode
Focusing Grid
Control Grid
Cathode
7 cm

Results of 1D Modeling of Hall MHD Generator

M_{free}=8; wedge angle =10^0 , B=7 T
input parameters: T=432 K, p=7.9e-2 Atm, u=2401 m/s; M=5.76.
geometry: A_{in}=0.15x0.15 m^2, A_{out}=0.45x0.45 m^2. E-beam slots: 0.5 cm; intervals - 2 cm

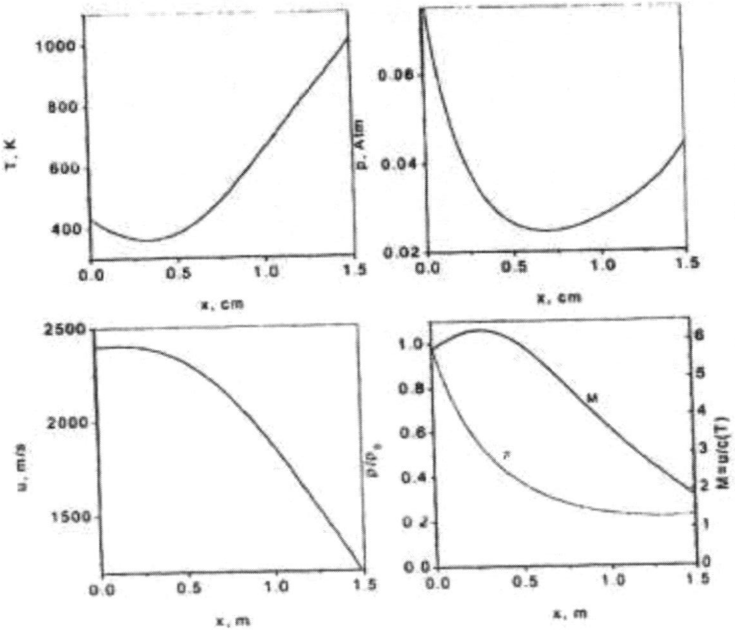

B=7 T; j_b=3.5 mA/cm^2; k=0.45.

W_{output}=4160 kW;
(η =36 %)

$W_{beab\ input}$=32.3 kW

T_{stag}=1545 K, M_{exit}=1.85

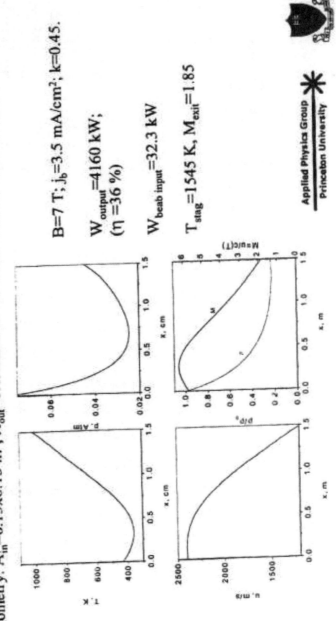

Applied Physics Group
Princeton University

Flight Testing MHD Power Systems

- An air-launched hypersonic X-Plane in the 25 t mass range, with a reusable rocket as its basic propulsion, enables the sustained, intensive flight testing of advanced hypersonic systems, to rapidly advance and mature their development.
- This plane may be able to flight-test systems that develop both air breathing propulsion and MHD power – with power directed to other vehicle systems – like high power lasers.
- This requires useful thrust and MHD power from air breating and MHD components of flight-weight mass and size - like superconducting magnets and electron beams embodied in engine structure - and which tend to be bulky and heavy.

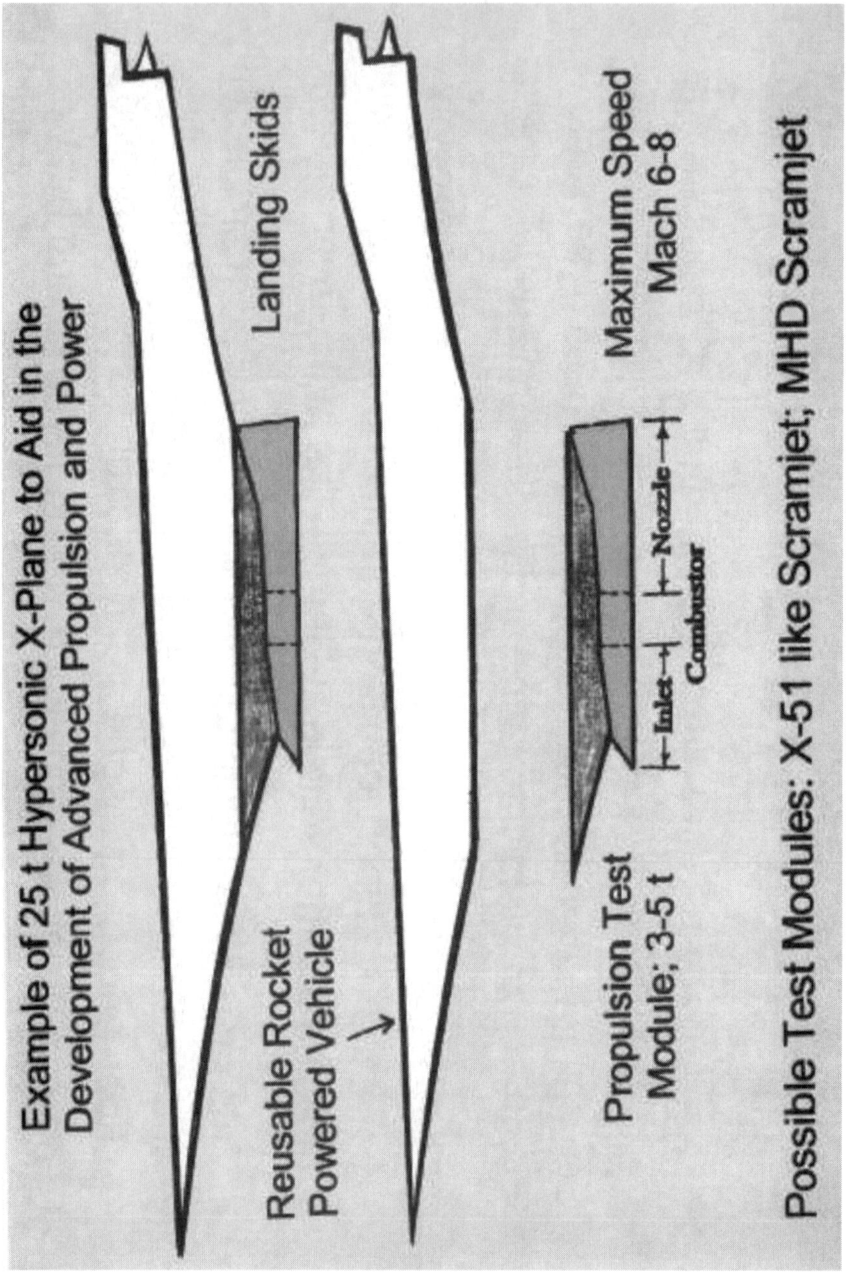
Example of 25 t Hypersonic X-Plane to Aid in the
Development of Advanced Propulsion and Power
Reusable Rocket
Powered Vehicle
Landing Skids
Propulsion Test
Module; 3-5 t
Inlet
Combustor
Nozzle
Maximum Speed
Mach 6-8
Possible Test Modules: X-51 like Scramjet; MHD Scramjet

Fraction of Enthalpy Extracted from Engine Airflow and Net MHD Power Generation Increase with Speed

Altitude	Mach 4	Mach 6	Mach 8	Mach 10
15 km	8.0 % 3.8 MW	12.3% 5.6 MW		
20 km	13.2 % 1.3 MW	15.8 % 5.3 MW	12.0 % 11.8 MW	9.2 % 20.9 MW
30 km	32.5 % 0.6 MW	33.0 % 2.7 MW	23.0 % 5.4 MW	23.1 % 22.1 MW

B=7T, Inlet=25x25 cm, Exit=50x50 cm, Engine L=3m

Data from Applied Physics Group, Princeton University

Air Breathing MHD + Fusion for Aerospace Planes

- Energy extracted from engine airflow by MHD interactions is not immediately deposited in engine nozzle flow. Instead, the MHD energy is used to initiate nuclear fusion – which results in enormous energy generation.
- It is this enormous fusion-generated energy that is deposited into engine exhaust flow, to provide more than 2-fold increase in air breathing Isp. In space, fusion power is deposited in the combustor to heat propellants like LH2 to provide more than twice the Isp of oxygen-hydrogen rockets.
- Pure air breathing is used until ~Mach 10 is reached. MHD operates from Mach 10 to12 - where MHD-generated power starts and maintains fusion.
- Studies[1] of combining air breathing MHD and clean fusion systems in a single-stage-to-orbit aerospace plane show more than a doubling of plane air breathing and rocket ISP and more than a halving of its takeoff mass.

1 Papers by Froning and Miley at STAIF 2005 and STAIF 2006 Forums

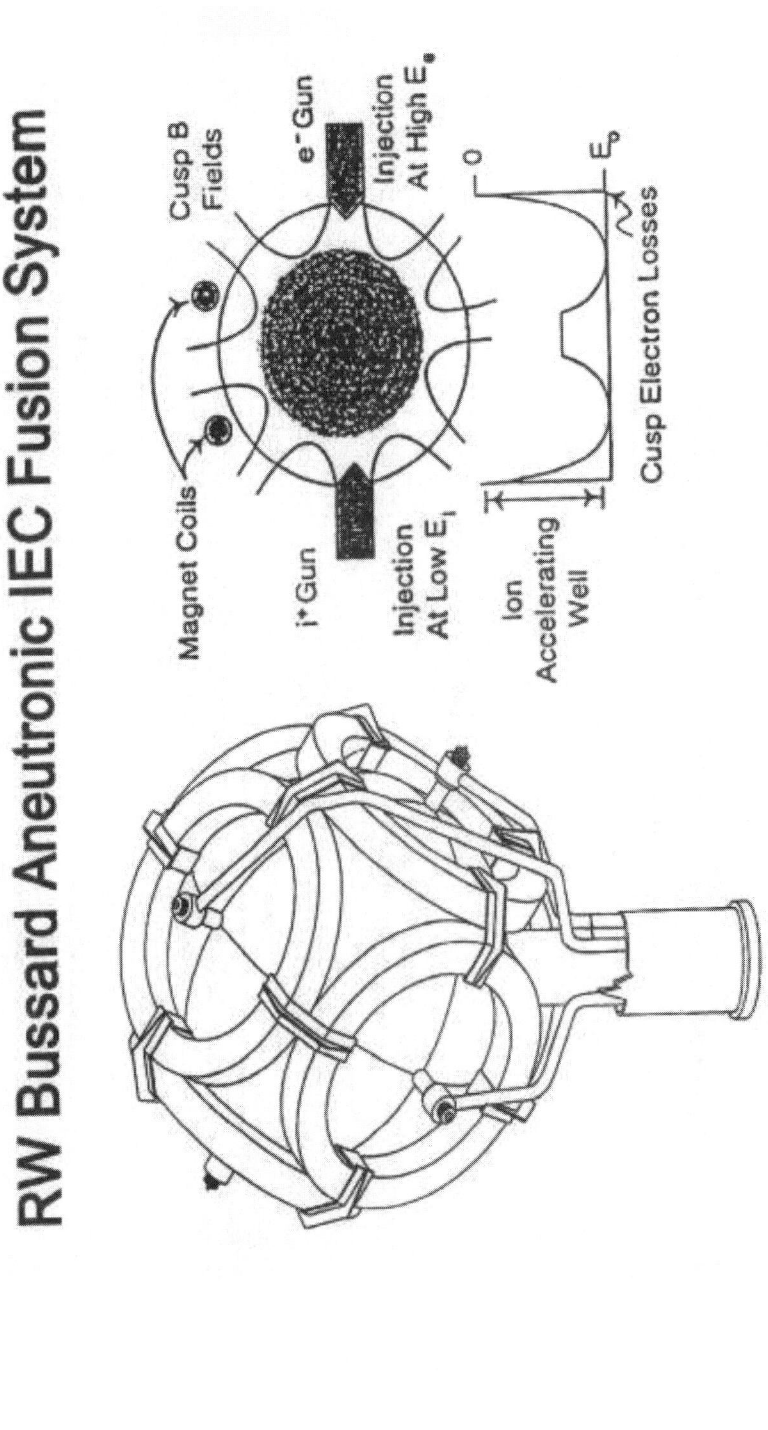
RW Bussard Aneutronic IEC Fusion System
Magnet Coils
Cusp B Fields
i+ Gun
e− Gun
Injection At Low E_i
Injection At High E_e
Ion Accelerating Well
0
E_o
Cusp Electron Losses

MHD Airbreathing and IEC Fusion within SSTO Vehicle

(Schematic Representation)

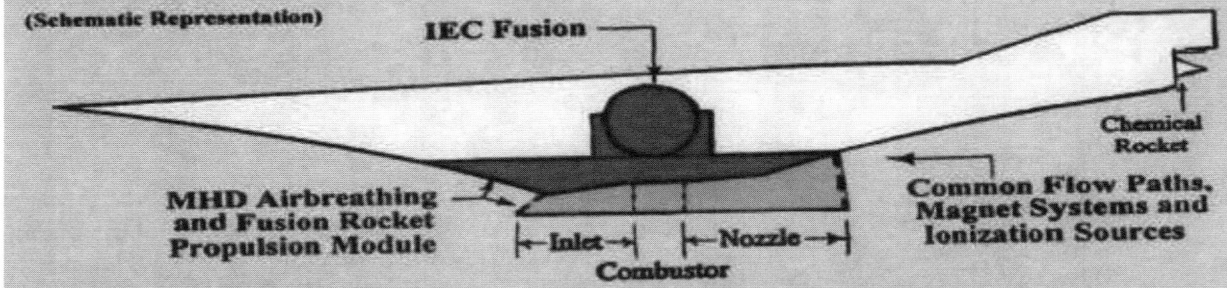

Earth-to-Orbit Flight Phases	Propulsion Mode	"Combustion" Energy	Ionization Energy
Zero to Mach 12 Speed within Sensible Atmosphere	Ducted Rocket and MHD Airbreathing	Chemical and IEC Fusion Reactions	MHD and IEC Fusion Reactions
Mach 12 to Orbital Speed above Sensible Atmosphere	Rocket	Chemical and IEC Fusion Reactions	IEC Fusion Reactions

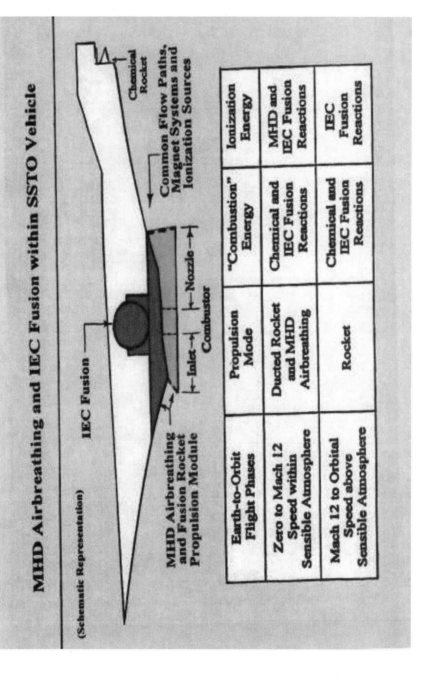

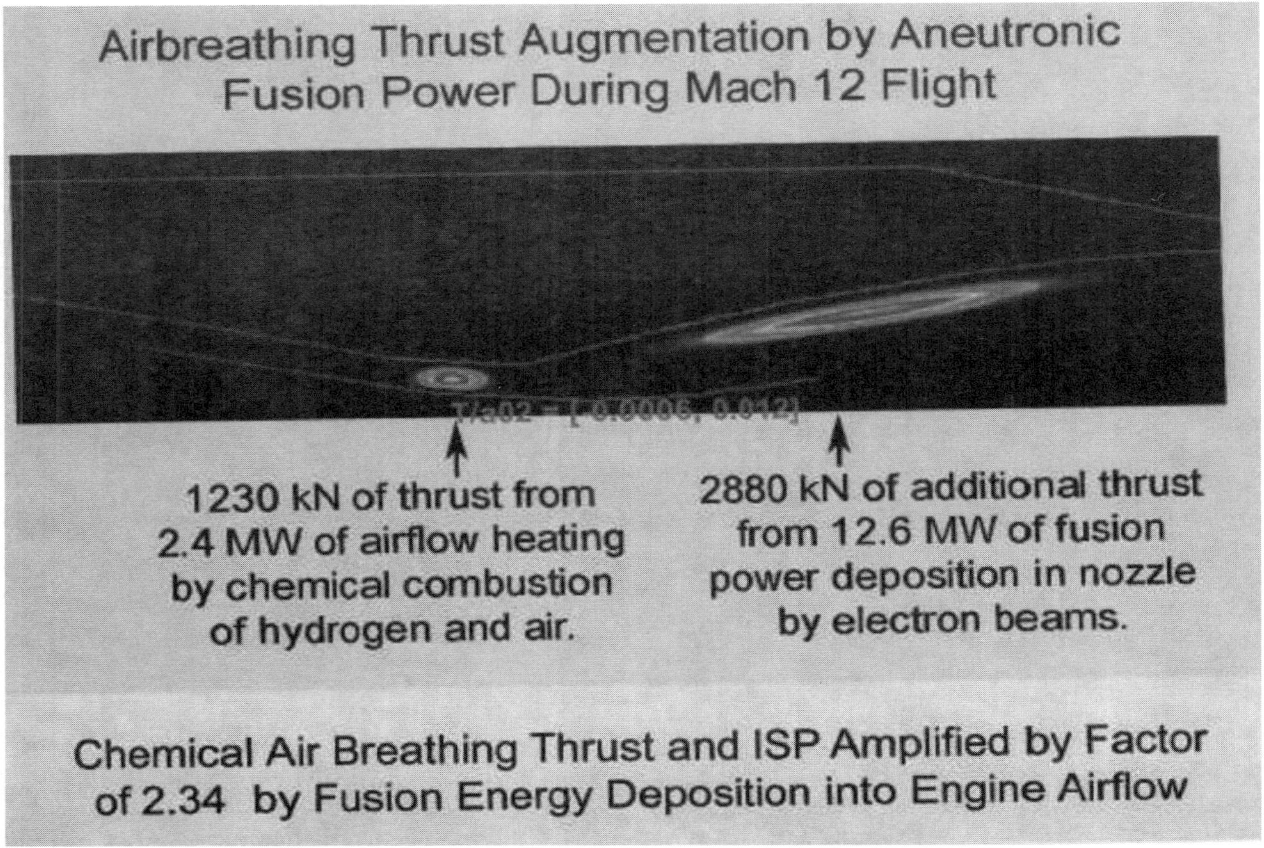
Airbreathing Thrust Augmentation by Aneutronic Fusion Power During Mach 12 Flight
1230 kN of thrust from 2.4 MW of airflow heating by chemical combustion of hydrogen and air.
2880 kN of additional thrust from 12.6 MW of fusion power deposition in nozzle by electron beams.
Chemical Air Breathing Thrust and ISP Amplified by Factor of 2.34 by Fusion Energy Deposition into Engine Airflow

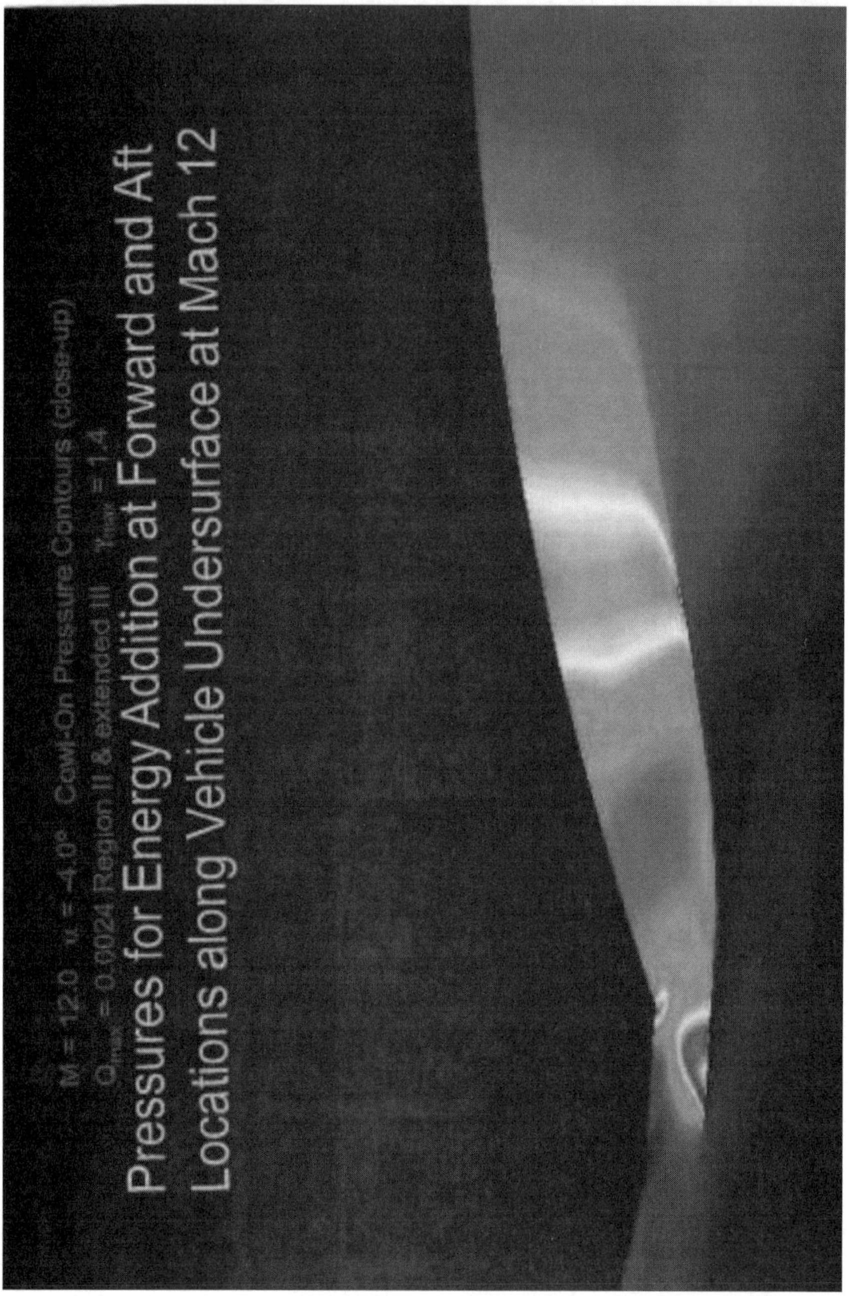
M = 12.0 α = -4.0° Cowl-On Pressure Contours (close-up)
Q_max = 0.0024 Region II & extended III γ_max = 1.4
Pressures for Energy Addition at Forward and Aft Locations along Vehicle Undersurface at Mach 12

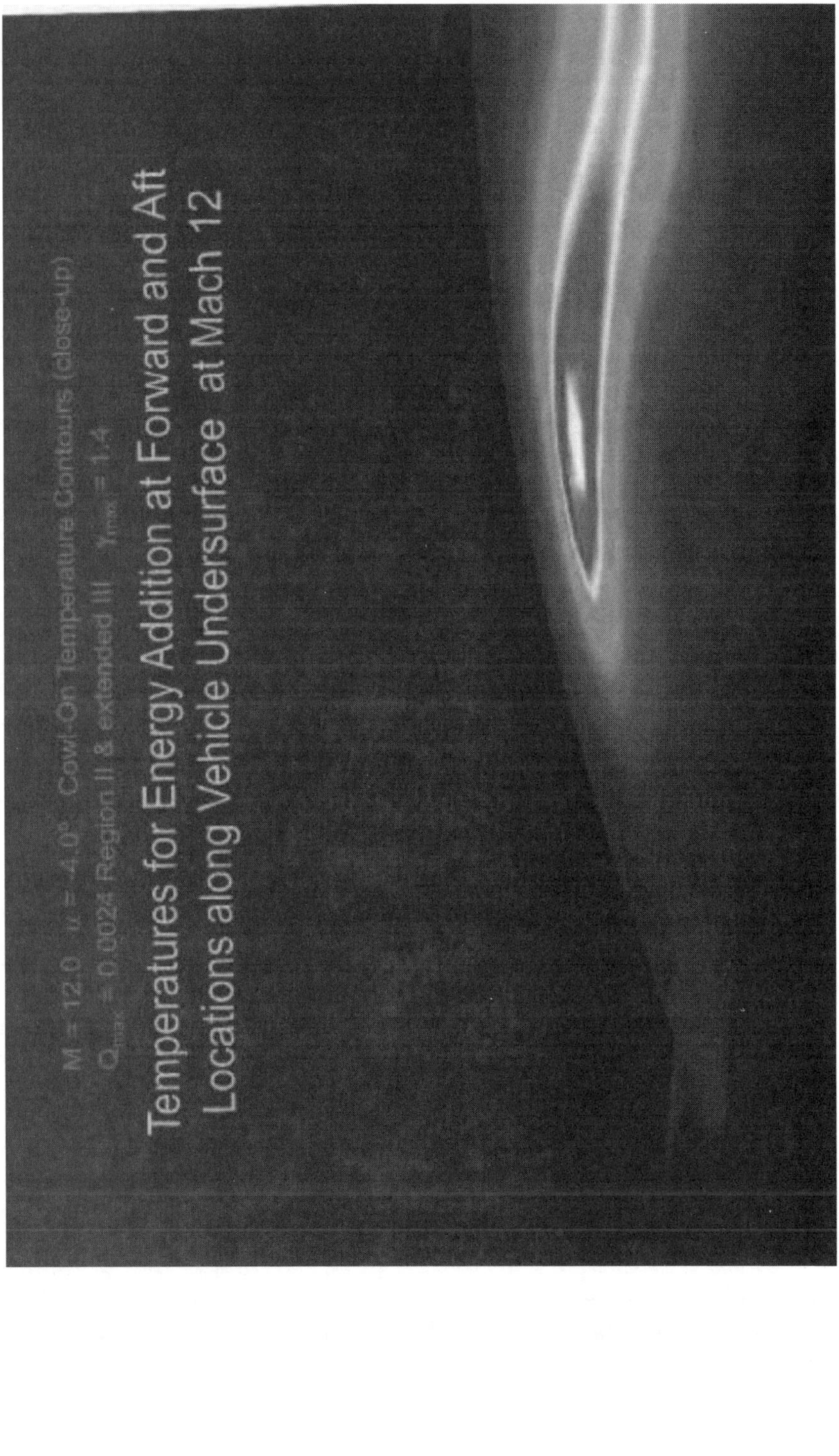
M = 12.0 α = 4.0° Cowl-On Temperature Contours (close-up)
Q_{max} = 0.0024 Region II & extended III γ_{max} = 1.4
Temperatures for Energy Addition at Forward and Aft Locations along Vehicle Undersurface at Mach 12

Single-Stage HTHL Spaceplane Requirements for Orbital and Sub-Orbital Hypersonic Space Tourism Flight

H. D. FRONING*

PO Box 180,
Gumeracha, South
Australia, 5233,
Australia

If hypersonic vehicles are first developed and operated for government needs, additional vehicles for space tourism might be procurable by space tourism developers and providers. However, profitable earth-to-orbit space tourism would require such vehicles to possess about a 2-fold increase in both propulsive efficiency (thrust per propellant flow rate) and structural efficiency (surface area per unit weight) compared to current hypersonic rocket plane art. But, profitable sub-orbital space tourism planes would require much less hypersonic art increase to fly space tourists 8,000 nmi distances between major international airports in 60 minutes; or, to fly space tourists, in 90 minutes, once around the entire Earth.

I. Introduction

* Visiting Lecturer, Department of Mechanical Engineering, University of Adelaide, AIAA Associate Fellow

Space tourism officially began in April 2001, when billionaire Dennis Tito paid $20 million dollars for a flight to and from Space Station Freedom on a Soyuz launch vehicle. But it will more truly begin in several years when commercial space tourism vehicles operated by commercial companies such as Virgin Galactic begin flying tourists to space at about $200,000 per passenger. These first space tourism planes (Spaceship2 and Rocketplane) will be, essentially, supersonic (Mach 3.0+) rocket planes that climb higher than 100 km above Earth.

And if these supersonic tourism planes prove popular and safe, one can envision future possibility for hypersonic tourism planes that, by flying faster, higher, further and longer, would provide even more exciting and inspiring rides in space. But it is well known that exponential increases in structural temperature, propulsive energy and vehicle mass occur as speed increases from supersonic into the hypersonic range. And this, in the opinion of many, would make hypersonic vehicle development risks and costs far above what private space system developers could afford. It has thus been assumed that: hypersonic space tourism vehicles may have to be derivatives of government developed vehicles that meet crucial civil or military needs; and that space tourism providers would, of course, still face the challenge of acquiring, modifying, and operating these relatively large and complex aerospace vehicles economically and safely.

In this respect, Chase[1] describes a military, rocket-powered, hypersonic aerospace plane whose "global reach" flight profile would also be attractive for space tourism flight. This plane would: take-off from military runways; fly suborbitally around the entire earth on its longest missions: and then land on the same runway from which it took off. Froning[2] shows that such a hypersonic plane must have high hypersonic lift/drag ratios (L / D in the 3 .5 to 4. 0 range at Mach 10) for long, high-speed glide trajectories around the entire Earth. The very compressed distance scale of Figure 1 shows such a vehicle skipping in and out of Earth's upper atmosphere many times during its 90-minute around-the-world flight, providing – in some respects – much more spectacular views of Earth and space than seen from ordinary earth orbits. Figure 2 shows lateral acceleration increasing to almost 1.0 earth gravity for many seconds at the bottom of each of the plane's many swoops. It also shows lateral acceleration decreasing to near zero gravity for many seconds at the top of each of the plane's many zooms. Therefore, since skip-glide flight profiles provide space tourists with extreme exhilaration in addition to spectacular views, such flight is flown by all single stage horizontal takeoff and horizontal landing (HTHL) vehicles that were configured and examined in the following analysis.

II. All-Rocket, Single-Stage, Suborbital Space Tourism Plane

Chase[3] briefly describes military missions that could be performed by a sub orbital, rocket-powered, aerospace plane, which could launch payloads that reach geosynchronous orbit and carry up to 11 tons of payload around the circumference of Earth. The rocket plane used two modified SSME Shuttle rocket engines and the metal matrix

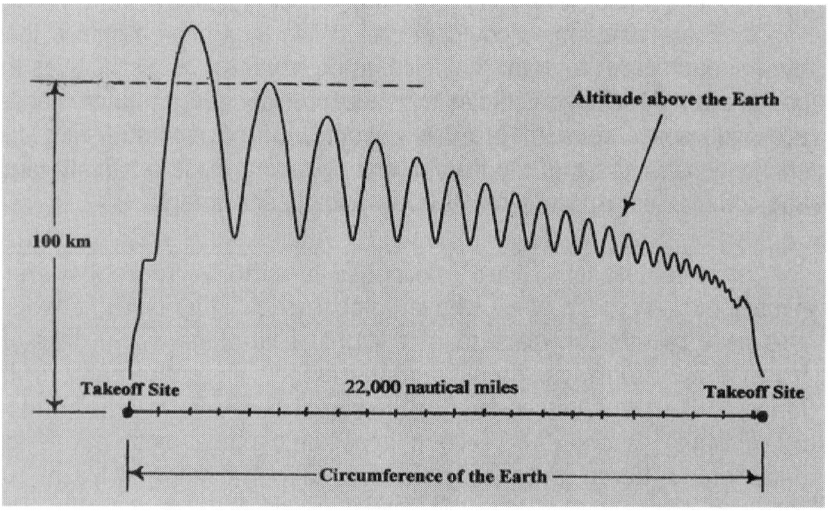

FIGURE 1. SUBORBITAL AROUND THE WORLD FLIGHT

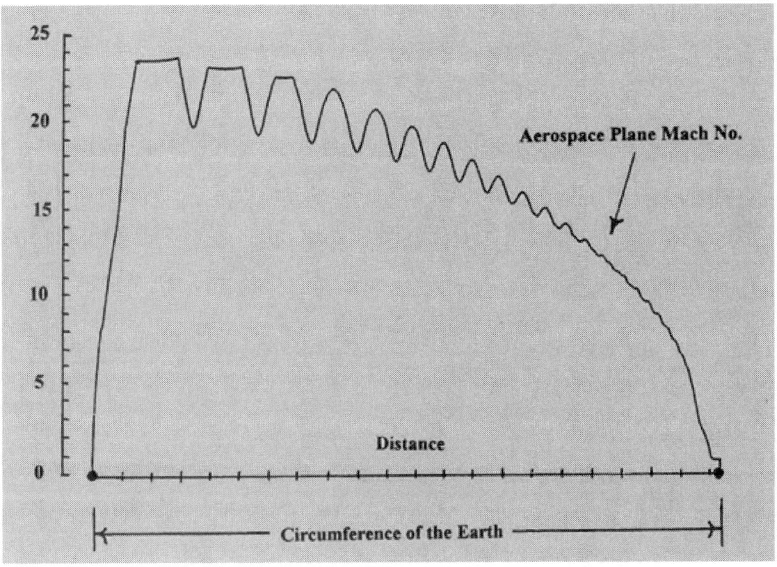

FIGURE 2A. SUBORBITAL CURING AROUND THE WORLD FLIGHT

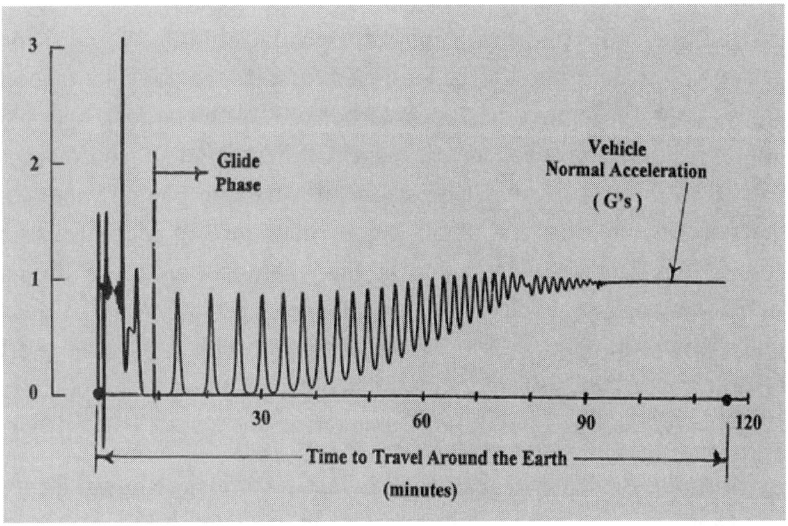

FIGURE 2B. NORMAL ACCELERATION AROUND THE WORLD FLIGHT

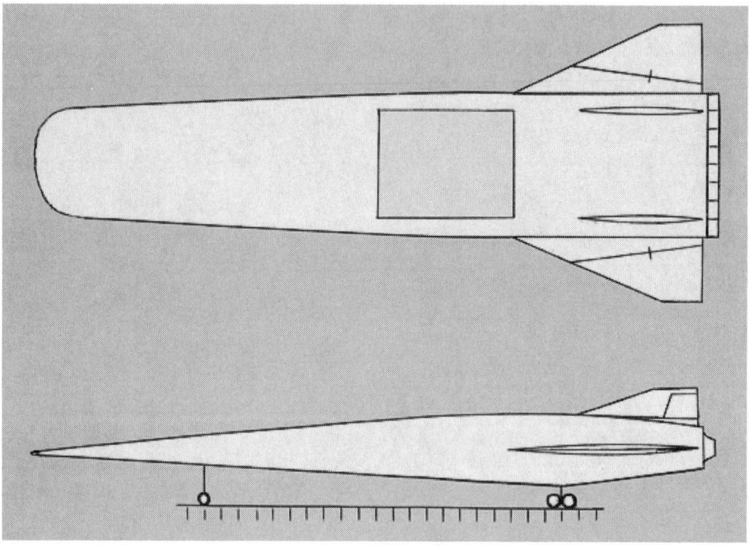

FIGURE 3. SPACE TOURISM ROCKET PLANE

"hot structure" was used over most of the vehicle and carbon-carbon refractory TPS structural material was used in hotter regions. Its takeoff weight is about the same as a 570 ton Airbus A380 – the largest, heaviest jet airliner in operation today. In this respect, two students, Richard Jones and Artur Menton, in the University of Adelaide Department of Mechanical Engineering are beginning to configure a space tourism version of the Chase[3] military rocket plane as an aerospace design project. Their work will explore the possibility of a 570 ton vehicle – such as that shown in Figure 3 – being achieved with state-of-the-art hypersonic materials, systems and satisfactory volume, size and aero shape.

Although vehicle design work is just beginning, the significant landing gear mass required for unassisted runway takeoff (about 3 % of takeoff mass) is significantly limiting the allowable amount of payload mass. Work will therefore include, if required, investigation of vehicle take-off assist concepts that reduce landing gear mass. These will include larger, heavier and smaller, lighter trolleys, which support all and some of the vehicle takeoff weight. Student work will also determine space tourism module shape, volume and mass, together with passenger seating and passenger viewing arrangements (such as synthetic vision from sensors embedded within vehicle skin, vs. actual vision through high temperature, transparent glass). As indicated in Figure 3, Linear Aerospike Engine arrays will be considered in addition to modified SSME engines for sub orbital, space tourism rocket plane flight.

III. Possible Advances in Hypersonic Space Tourism Vehicle State-of-the-Art

Three levels of hypersonic vehicle propulsion and structural efficiency were considered in addition to that associated with the current propulsion and materials state-of-art (SOA) of the Chase[3] rocket vehicle. One was a factor of two increase in thrust per fuel flow rate by use of air breathing propulsion. Another was a factor of two decrease in dry mass per surface area by use of nanostructured airframe and propulsion materials. The third was a factor of two improvement in both propulsion and structural materials efficiency. Figure 4 shows the 4 different vehicles associated with the 4 different levels of hypersonic SOA, together with their takeoff and dry masses for around-the-world flight. And Table1shows takeoff and payload masses for the different hypersonic SOA levels for different missions.

Here, it is well known that use of air breathing propulsion during a spaceplane's upward ascent through the atmosphere of Earth enables significant reduction in vehicle propellant and takeoff mass. As an example, Dissel4 shows that air breathing propulsion from Mach 2.5 to Mach 15 (with 2-D inlets) results in a 2-fold increase in mission-averaged ISP (thrust per fuel flow rate) compared to that of the all-rocket vehicle of Chase[3]. And, as shown in Table 1, this enabled a space tourism payload increase from 11 ton to 31 ton for around-the-world suborbital flight.

Just as a 2-fold improvement in propulsive efficiency by use of air breathing propulsion was investigated, so a 2-fold improvement in structural efficiency by use of carbon nanotube materials within vehicle airframe, propulsion and system subsystems of the vehicle was investigated. Here, carbon nanotube-structured materials have been estimated to be 7 to 10 times lighter than steel and to possess 70-250 times more tensile strength. For the purposes of vehicle sizing, use of nanotube materials was assumed to result in a 4-fold reduction in dry mass per surface area for 80% of the airframe structure and for 40% of the propulsion and systems structure. Table 1 shows this enabling in a 2-fold rocket vehicle dry weight decrease for around-the-world flight; and a 4-fold increase in payload.

Finally, a 2-fold improvement in propulsive efficiency by air breathing propulsion from Mach 2.5 to Mach 15 in conjunction with a 2-fold improvement in structural efficiency by much lighter nanotube-structured materials was examined. Table 1 shows this increasing air breathing vehicle payload from 44 ton to 78 ton for around-the-world flight.

IV. Influence of Space Tourism Mission

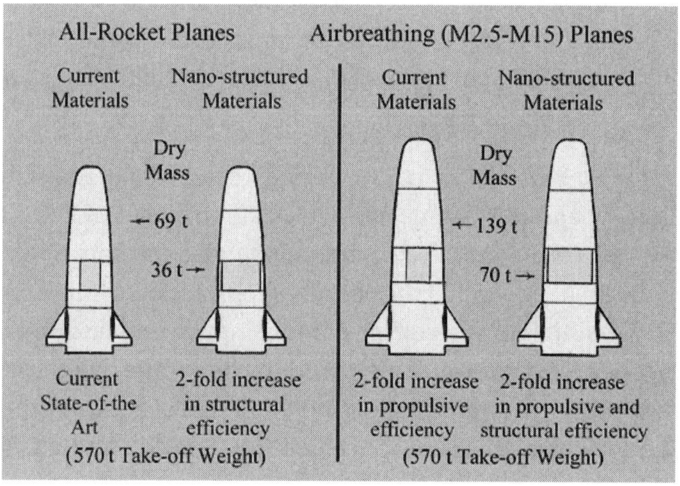

FIGURE 4. HYPERSONIC ROCKET PLANES

Three space tourism missions were considered for each of the 4 vehicles. The first was earth-to-low equatorial earth orbit, which

Mission / Plane	Earth to Orbit	Around the World	8,000 nm Range
Rocket with SOA Structure		570 t 10.5t	570 t 30.9t
Rocket with Light Structure	570 t 21.4t	570 t 43.9t	570 t 64.3t
Airbreather with SOA Structure	570 t 9.1 t	570 t 35.1t	570 t 78.2t
Airbreather with Light Structure	570 t 68.2t	456 t 78.2t	343 t 78.2t

TABLE 1. VEHICLE PAYLOAD CARRYING CAPABILITY (T = TON)

required a cut-off velocity of 7.93 km/s. The second was flight around the circumference of Earth, which required a cut-off velocity of 7.22 km/s for a vehicle hypersonic L/D of 4.0. The third was flight over an 8,000 nmi distance (the same distance as A380 aircraft maximum range), which required a cut-off speed of 6.22 km/s. Since take-off mass was the same for each of the different missions, the reduced propellant load required for the shorter-range missions, enabled more payload to be carried. The rocket plane of Chase[3] could not achieve earth-orbit, but Table 1 shows the payload carrying capability of each of the vehicles for each mission. It is seen that payload-carrying capability of the different vehicles is extremely sensitive to their space tourism mission. As an example, the air breathing vehicle with SOA structure can carry only about 9 ton of payload (space tourism module+passengers) to low earth orbit. But with skip-glide flight, it can fly 44 ton of payload around the Earth; and 78 ton of payload can be flown over distances such as those separating Spaceports in the United States and South Australia.

It should be mentioned that investigations such as Dissel[4] show some benefits in SSTO vehicle mass by use of hydrocarbon propellants during the initial portions of flight. But, in the spirit of not adding more global-warming greenhouse gases to Earth's atmosphere, only environmentally favorable liquid oxygen and liquid hydrogen have been used. This study and student work has not yet considered two-stage space tourism vehicles, - which have many advantages despite the additional complexity of an additional flight vehicle system and its staging. It is therefore planned to also study two-stage space tourism vehicles in future space tourism vehicle design and analysis work.

V. Return-on-Investment by Hypersonic Space Tourism Planes

It was far beyond the scope of this preliminary study to attempt a properly comprehensive estimate of costs and Return on Investment (ROI) for hypersonic space tourism planes. However, very top level cost and ROI estimates were attempted to obtain an order of magnitude indication of the impact of hypersonics advancements on possible attractiveness of hypersonic spaceplanes to space tourism providers and investors. In this respect, Chase[3] includes a compilation of some fairly comprehensive "bottoms up" life cycle cost estimates by major American aerospace contractors of the hypersonic military space plane that he describes. These development, acquisition, and operational cost estimates were converted into 2006 dollars and used to estimate the cost to: acquire 6 additional vehicles; modify these already-developed vehicles for space tourism over a 3 year period; and operate the fleet of 6 vehicles over a period of 5 years. Figure 5 shows the resulting cost-per-passenger for a space tourism provider as a function of the number of space tourism passengers carried over its 5 years of operation. Also shown from Ellingsfield[5] is the ticket price space tourism passengers would be willing to pay as a function of passengers carried – based on market surveys of how many people would pay given amounts for spaceflight. It is seen that investment breakeven by a space tourism provider of around-the-world space travel with a hypersonic rocket plane similar to that of Chase[3] would be about 750,000 passengers carried at an average ticket price of $15,000 to $20,000 (in US 2006 dollars) Czysz[6] and Froning[7] show life cycle cost estimates for all-rocket and rocket + air breathing SSTO vehicles that embody various levels of hypersonics technology and which therefore provide various levels of payload delivery efficiency in terms of payload to takeoff mass. Vehicle descriptions and costs from these references were used to determine vehicle acquisition and operations cost as a function of payload delivery efficiency, and these costs were normalized with respect to the costs of the all-rocket vehicle in Chase[3] as shown in Figure 6. The payload to takeoff mass ratios of the 2 different rocket planes and 2 different rocket + air breathing planes are also indicated on the Figure 6 trend curve for the around-the-world space tourism mission. Curves such as Figure 6 were then used to calcu- late ROI for a 5 year investment period for the two rocket and two rocket + air breathing space tourism planes. Al- though theses preliminary cost estimates only indicate trends, they seem consistent (with the exception of future hydrogen cost estimates) with spaceplane costs and trends by Penn[8, 9] for high volume spaceflight and space tourism.

Figure 9 shows ROI as a function of passengers flown over 8,000 nmi distance for the all-rocket and the air breathing plane with SOA structure. It is seen that flights per plane over their 5 year operational life is very impor- tant – with the rocket plane program loosing money if flights for each plane of the fleet is 200. Flights needed for breakeven by the air breathing plane are only 50 because it carries more passengers per flight. But neither rocket nor air breathing plane achieve their full ROI potential unless every plane of each 6-vehicle fleet can fly 1,000 times to space. And this would require that each plane fly, on the average, about 4 times per week during its 5 years of life.

Table 2 summarizes ROI potential for each level of hypersonics technology for each space tourism mission. It is seen that profitable earth-to-orbit space tourism (ROIs greater than about 20 percent) requires effective air breathing propulsion to about Mach 15, together with a 50 percent dry weight reduction by use of nano-structured materials). And this is clearly far beyond currently achievable hypersonic art. But either of these air breathing propulsion or ma-terials advancements (or lesser advancement in both) would enable profitable around-the-world space tourism; -and even more profitable space tourism over the shorter distances between possible commercial spaceports on Earth

VI. .Airport and Spaceport Noise Considerations

Space tourism flight over distances of the order of 8,000 nmi would allow travel between spaceports at remote locations such as White Sands, New Mexico and Woomera, South Australia. But it would be even more desirable if such travel could also be between major international airports of the world.

Here, airport noise becomes important. And needed suppression of takeoff noise for the rocket and rocket-based combined cycle (RBCC) engines assumed in this study may be difficult to achieve, compared to the lower noise achievable by turbine engines of similar thrust.

One possibility for the rocket planes would be two-stage configurations, with the first stage being powered by air breathing jet engines, such as the 4 Trent 90 or GP turbofan engines used on the Airbus A380 aircraft (which already meet all international airport noise requirements). a very preliminary concept and mass estimate for a 570 t two- stage configuration has been made. Like the Scaled Composites "White Knight" a lower air breathing stage accelerates, a rocket plane to about 10 km launching altitude and Mach 0.8 launching speed. A possibility for the air breathing planes would be turbine-based combined cycle (TBCC) engines with less noise (but possibly more mass) than RBCC engines. Figure 8 shows a NASA aerospace plane, based on TBCC work at the Glenn Research Center.

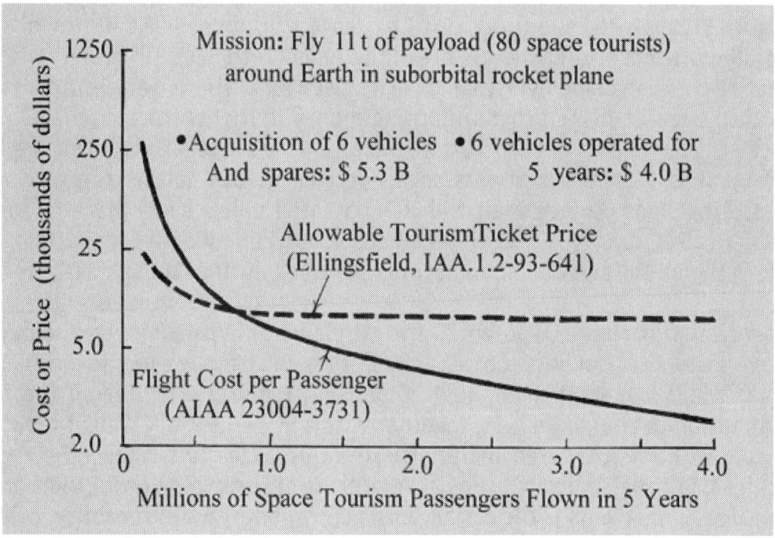

Figure 5. Influence on Passengers carried on cost

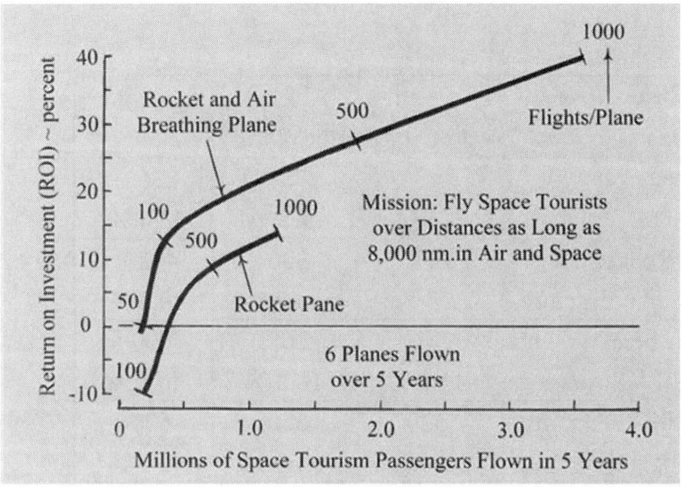

FIGURE 6. INFLUENCE OF TOURISTS FLOWN (FIG. 7 OMITTED)

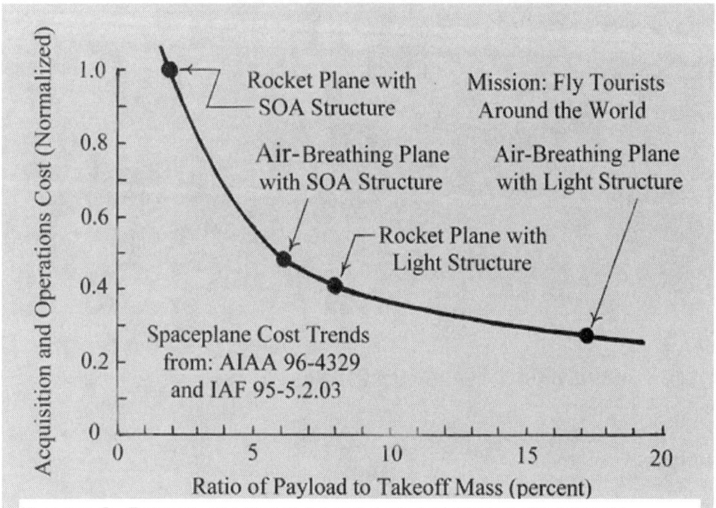

FIGURE 8. EFFECT OF VEHICLE PAYLOAD DELIVERY EFFICIENCY

Mission / Plane	Earth to Orbit	Around theWorld	8,000 nm Range
Rocket with SOA Structure		78 - 9.5 %	236 13.4 %
Rocket with Light Structure	160 7.7 %	336 23.7 %	492 30.6 %
Airbreather with SOA Structure	70 -11.0 %	268 18.3 %	600 36.1 %
Airbreather with Light Structure	520 32.0 %	600 38.0 %	600 42.2 %

Figure 9. Table of Tourists/Flight and Return on Investment for Space tourism missions and Planes

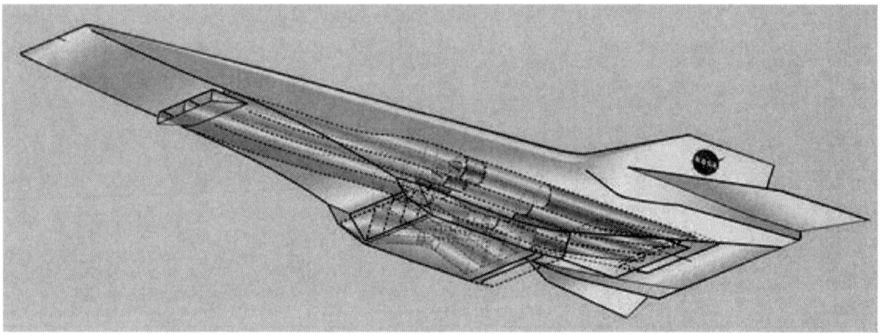

FIGURE 10. TURBINE BASED AEROSPACE PLANE

Conclusions

If government bodies absorb risks and costs of hypersonic single-stage sub-orbital vehicle developments to meet civil or military needs, there is the possibility of acquiring, modifying and operating more of such vehicles for space tourism. But this study shows that profitable earth-to-orbit space tourism by such vehicles would require them to embody advancements beyond current rocket propulsion and hypersonic structural art. Such required advancements would include a 2-fold increase in propulsive efficiency (thrust/ fuel flow) by effective air breathing propulsion to about Mach 15, combined with a 2-fold increase in structural efficiency (surface area/ dry mass) by use of nano-structured materials, such as carbon nanotubes, in airframe, propulsion, and systems components. But profitable sub- orbital space tourism requires much less hypersonics advancement: a 2-fold increase in either propulsion or structural efficiency; or more modest improvement in both. And even more profitable space tourism between the major airports of the world appears possible if: acceptable takeoff noise is achievable by single stage air breathing vehicles powered by RBCC or TBCC engines; or by two-stage vehicles whose upper rocket stages are boosted by lower stages with low noise, high thrust turbofan engines like the Trent 90 or GP 7000. A final, formidable challenge is vehicle lifetimes of the order of 1,000 flights. This is 20 times greater than that expected for Space Shuttles, but it may be possible with new materials/ structural designs, and the milder environment of sub-orbital skip-glide flight.

References

1 Chase, R.L., "The Design Evolution of a Combined Cycle Air-Breathing Powered SSTO Design Concept", AIAA 95-6010, 6th International Aerospace Planes Conference, Chattanooga, TN, USA, April, 1995

2 Froning, H.D., McKinney, L.E., Chase, R.L., Aerospace Plane Trajectory Optimization for Sub-Orbital Boost-Glide Flight", AIAAA 96-4519, 7th International Aerospace Planes Conference, Norfolk, VA, November, 1996

3 Chase, R.L., "Formulation of a Near Term Stepping Stone to a Low Cost Earth-to-Orbit Transportation System Based on Legacy Technology", AIAA/ASME/SAE/ASEE Joint Propulsion Conference and Exhibit, Fort Lauderdale, FL, July, 2004

4 Dissel, Kothari, A.P., Raghavan, V., Lewis, M.J., "Comparison of HTHL and VTHL Air-Breathing and Rocket Systems for Access to Space", AIAA-2004-3988, 40th AIAA/ASME/SAE/ASEE Joint Propulsion Conference and Exhibit, Fort Lauderdale, FL, July, 2004

5 Ellingsfield, F., et al. “Space Tourism in Europe”, IAA.1.2-93-65A, 44th International Astronautical Federation Congress, Gratz, Austria, October, 1993

6 Czysz, P., etal. “A Propulsion Technology Challenge-Abortable, Continuous Use Vehicles”, IAF-95-5.2.03, 46th International Astronautical Federation Congress, Oslo, Norway, October, 1995

7 Froning, H.D., “Economical and Technical Challenges of Expanding Space Commerce by RLV Development”, 1996 AIAA Space Programs and Technologies Conference, Huntsville, AL, September, 1996

8,9 Penn, J.P., Lindley, C.A.,”Spaceplane Economic Tradeoffs over a large Projected Travel Range” and ”Space Tourism Optimized Reusable Spaceplane Design”, Space Technology and Applications Forum, (STAIF1997), Albuquerque, NM, 1997.

Combining MHD Air Breathing and Aneutronic Fusion for Aerospace Plane Power and Propulsion

H. David Froning

PO Box 180, Gumeracha, South Australia, 5233, Australia

Visiting Lecturer, University of Adelaide, AIAA Associate Fellow

An aerospace plane concept that combines the extraordinary propulsion and power generating capabilities of MHD air breathing engines and aneutronic (neutron-free) fusion rocket engines is summarized. This vehicle, using liquid hydrogen as coolant, fuel, and working fluid, for power and propulsion, generated: air breathing propulsion from takeoff to Mach 7; MHD air breathing propulsion and power from Mach 7 to Mach 14; combined MHD airbreathing and aneutronic fusion power and propulsion for Mach 14 cruise; or for fusion power and propulsion from Mach 14 to orbital speed. This resulted in a reusable air breathing SSTO launch vehicle with 1/4th the propellant consumption of chemical airbreathing designs. And this enabled space access with aircraft-like operations and takeoff weights.

I. Introduction

Recent Air Force and NASA air breathing and rocket propulsion developments have focused on two-stage-to-orbit (TSTO) vehicles to improve military and civil access to space because propulsion and airframe technology developments for more revolutionary single-stage-to-orbit (SSTO) vehicles are not deemed achievable in the foreseeable future. In this respect, studies by Murthy and Froning[1] indicated that SSTO air breathing vehicles powered by "clean" aneutronic fusion rocket reactions would have about ½ the takeoff mass and 1/3 the propellant consumption of chemical air breathing SSTO vehicles. And Froning and Bussard, in a study for NASA2, estimated about 17 years to develop aneutronic fusion to a NASA Technical Readiness Level (TRL) of 5. Since this 17-year span measured from the current epoch would enable start of full-scale vehicle engineering development in about 2025, the Air Force Research Laboratory (AFRL) initiated a study performed in 2003 and 2004 by Flight Unlimited to determine what a revolutionary military SSTO aerospace vehicle might be like if it could be powered and propelled by aneutronic fusion and commence development in about 2025. This study was performed with the assistance of the University of Illinois Fusion Physics Laboratory. The military SSTO vehicle work is described in some detail in Froning [3], and the fusion work in Thomas [4, 5]. Both works are summarized in less detail in this paper.

II. MHD Air breathing and Fusion Rocket Vehicle Concept

The general requirement for the vehicle that will be described, is to :(a) provide the US Air Force with global- reach and rapid space-access with military aircraft-like vehicles that are no heavier than Air Force combat and transport aircraft such as the B1-B, B-2 and C-17; (b) deliver up to 18t of payload to space; and (c) be capable of commencing full-scale development in about 2025. Aircraft-like operations required aneutronic fusion power and pro- pulsion to eliminate destructive neutron emissions and residual radioactivity. "Dense Plasma Focus" (DPF) fusion reactors were deemed by AFRL to show the most promise of achieving the plasma dynamics needed for aneutronic fusion and of having the lightness and compactness to achieve needed power and thrust-to-weight. Rocket and tur- bine based combined cycle air breathing engines were considered for accelerating the vehicle to Mach 14 – the highest possible fusion system ignition speed and altitude (which also minimizes fusion rocket thrust and mass). MHD power generation is used during Mach 7-14 air breathing flight because it enables the hundreds of megawatts of needed electrical power for DPF fusion system ignition. The DPF fusion system provides propulsion and power at speeds of Mach 12 or greater. And a thrust vectoring chemical rocket system provides additional thrust and control any time during vehicle flight. The total consumption of liquid oxygen oxidizer depends upon whether Rocket or turbine-based air breathing propulsion is eventually selected. But liquid hydrogen, used as: regenerative coolant; working fluid; and propellant; constitute most of the consumed fluid during air breathing and fusion rocket flight.

III. MHD Air Breathing Power and Propulsion

Enormous amounts of electrical current are created within air breathing engines by very strong MHD (**J** x **B** = **F** which is the current crossing a magnetic field) vector force interactions within ionized and magnetized airflow at hypersonic (Mach 7 to Mach 14) flight speeds. Such current is a consequence of flow-slowing J x B interactions within airflow that is ionized to about 10^{13} electrons/cm^3 and subjected to magnetic fields of about 7 tesla. This current is: extracted from airflow by electrodes; conditioned to needed power and voltage within an MHD generator; and distributed to appropriate vehicle subsystems, which, of course, include the MHD flow ionizing and magnetizing components. MHD components are located around and embedded within engine walls, and significantly increase air breathing engine mass. They include: superconducting coils-to create magnetic fields; electron beams for flow ionization; and electrodes to extract MHD-created current.

Figure 1, taken from an MHD air breathing aerospace plane study for NASA by Chase6, shows a typical embodiment of MHD elements, such as "saddle magnets" (built up from superconducting coils) wrapped around the air- breathing engine walls. Also shown is flow path direction and the direction of the B-field generated by the magnets.

The enormous amount of MHD-generated electrical power extracted from hypersonic airflow (of the order of 1.0 gigawatt) is vital for providing the large amount of electrical power needed for high altitude ignition of the DPF fusion system at Mach 12. After fusion system ignition, fusion energy is immediately injected into the air breathing engine flow by electron beams – where intense ohmic heating of nozzle airflow increases air breathing thrust with negligible addition of mass. CFD computations indicate that this energizing of airflow also cause higher pressures to act along the vehicle aft-expansion surface to significantly decrease vehicle drag. And this increased thrust and de- creased drag results in more than a 2-fold increase in effective air breathing ISP between Mach 12 and Mach 14.

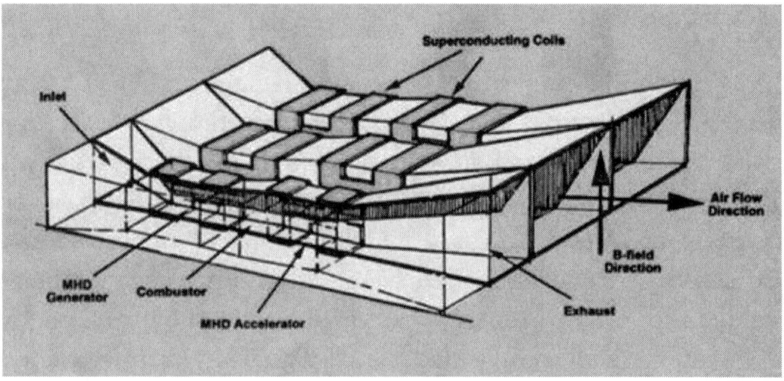

Figure 1. Typical MHD Airbreathing Engine Systems

IV. Aneutronic DPF Fusion Power and Propulsion

Figure 1 shows the DPF propulsion and power system selected for the 2025 vehicle. It fuses boron 11 and hydrogen nuclei to yield 3 helium 4 ions and 5.68 MeV of energy during each fusion. Very light shielding is needed to shield against soft x-rays, but elimination of neutron emissions eliminates neutron impacts upon vehicle structure and resulting radioactivity. And this enables aircraft-like maintenance and servicing of vehicle engines and structure.

The DPF consists of a coaxial electrode configuration in which a capacitor bank is discharged across the electrodes, ionizing injected nuclear fuels into a gas that forms into a plasma sheath. Current flows radially between electrodes, inducing an azmuthal magnetic field. "Rundown" then occurs wherein the plasma sheath is accelerated by **J x B** MHD interactions down the length of the anode as it proceeds rearward within the annular volume inside the concentric cylinder of the DPF. And finally, "Focus", wherein the accelerated plasma sheath collapses towards the central axis of the anode, compressing to fusion temperatures and pressures by enormously strong magnetic fields within a small pinch region immediately downstream of the DPF device. The released fusion energy then accelerates and expands injected hydrogen to high exhaust velocity and specific energy within a magnetic nozzle.

This aneutronic DPF fusion cycle is repeated about 10 times every second – with injected charges of nuclear fuel being vaporized by electrical pulses that are approximately: 400 kV in voltage; 20 MJ in energy; 200 MW in power.

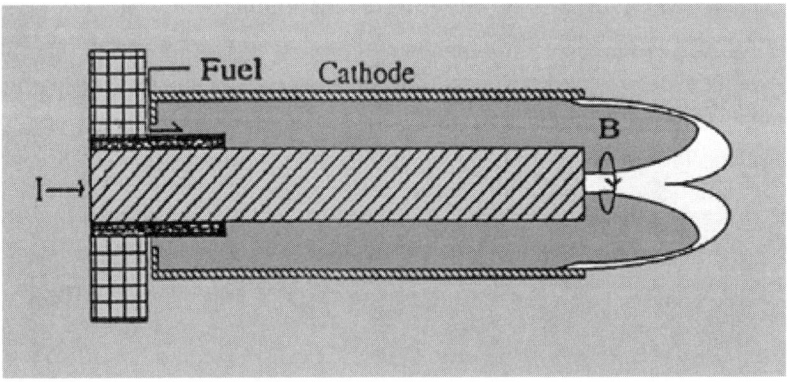

Figure 2. Typical Plasma Focus Device

An attractive fusion vehicle design required fusion rocket ISP (thrust/fuel flow rate) of 1500s to 2000s for thrust levels in the 750 kN to 500 kN range, enabling vehicle acceleration from Mach 14 to Mach 25 speed in about 15 minutes of time. Thomas indicate that the energies and power levels needed to achieve these thrust levels with satisfactory fusion power and propulsion system mass require capacitor energy storage/discharge capabilities would have to be about 15 kJ/kg. This is beyond current capacitor state of the art, but sources such as Sargeant[7] describe metalized capacitor developments that should yield capacitor efficiencies of 15-20 kJ/kg during the next 10 years.

Above the atmosphere, fusion must, supply needed power to vehicle subsystems, and Thomas4 has identified an attractive MHD power generation approach. Here, nozzle expansion of the DPF plasma exhaust causes magnetic flux trapping between the plasma jet and a stator pick-up coil, and flux diffusion into both. Then rapid field compression transforms plasma kinetic energy into magnetic pressure, forcing field lines out of the stator ring, and inducing high voltage pulses propagated through appropriate transduction equipment for electric power distribution.

V. Vehicle Configuration Definition

MHD air breathing integrated reasonably well with DPF fusion, sharing some propulsive flow paths and having similar components, such as electron beams and superconducting magnets. The resulting vehicle, shown in Figure 3, had less than 40% the takeoff mass and less than 20 % the propellant mass of an ordinary SSTO airbreathing plane.

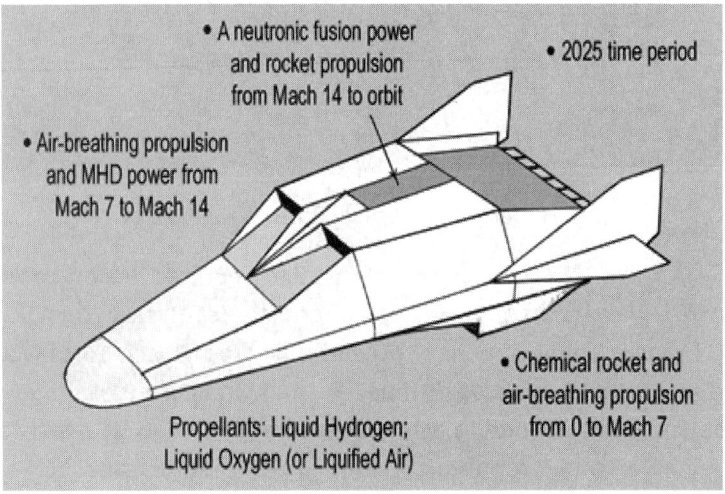

Figure 3. MHD Air-breathing Rocket Fusion Aerospace Plane

Since the vehicle's takeoff mass is only 175 tons – less than current Air Force bombers and long range jet airliners – it could use available runways and much of the other infrastructure at many military and/or civil airports in the USA.

Significant improvement in structural efficiency (vehicle surface area / structural+TPS mass) was assumed to be achievable by 2025 with use of high temperature (~950 deg C) metal matrix materials – such as rapidly solidified titanium and silicon carbide – over most of the vehicle. This resulted in about 25 percent improvement in structural efficiency over that of air breathing vehicles studied and documented in reports such as "NASA Access to Space Final Report Volume 1, July, 1993". A 10 percent improvement in superconducting magnet mass over that used in a previous NASA MHD study by Chase[5] was assumed. But recent YBCO and carbon nanotube materials work by Putnam[8] and Chapman[9] indicate that even lighter superconducting magnet weights than those which were assumed in this study may be possible. Shown in Table 1 are estimated masses of the various subsystems of the 2025 vehicle

Table 1. Estimated Masses of the 2025 Vehicle (ton = t)

Takeoff Weight...174 t
Payload.. .... 18 t
Dry mass.. ....82 t

Structure..27 t
Systems.................................... ....12 t
Propulsion (chemical 22t, fusion 21t)...43 t

Propellant (airbreathing 47t, rocket 27t)..............74 t

VI Critical Issues for DPF Fusion Power and Propulsion

Although many formidable challenges are associated with MHD air breathing power and propulsion, most technical uncertainties are associated with development of the aneutronic DPF fusion power and propulsion system. One uncertainty is achievement of the high ignition temperatures needed for fusion of protons and Boron[11] ions with acceptable input energy and power. Bremsstrahlung radiation from high temperature excitation of the electrons within grated plasma can result in excessive energy dissipation during fusion power build-up unless electron temperatures are lower than ion temperatures. This appears possible for the non-equilibrium dynamics of plasmas created within DPF devices, and for the narrow non-Maxwellian energy distributions within the DPF plasmas - which minimize wasteful heating of particles that have a fairly low probability of participating in the fusion reactions.

An associated challenge is retaining a significant amount (at least half) of the bremsstrahlung radiation inevitably emitted during aneutronic fusion - by reflecting and re absorbing it within the inner flow. Single-film Hohraum-like reflecting cavities are under study by Murkami[10] and thin film multi-layer reflectors by Joenson[11]. Neither concept has yet to be found entirely satisfactory, but they are believed capable of about 10 reflections of photons before being lost. If so, power and flux densities of about 10^5 J/cm^2 and 10^6 W/cm^2 can be reabsorbed in flow.

Critical DPF engineering issues include: conversion of very high fractions of fusion power into electrical power; and conversion of a high fraction of generated electrical power into useful jet power in the rocket exhaust.

References

1.Murthy, S.N, Froning, H.D, "Combining Chemical-Electric-Nuclear Propulsion for High-Speed Flight", Proceedings of the International Society of Airbreathing Engines (ISABE), Nottingham, England, 1991

2.Froning, H.D, Bussard, R.W, "Roadmap for QED Fusion Engine Research and Development", NASA Purchase Order H28027D for the George C. Marshall Space Flight Center, Huntsville, AL, USA, 1997

3.Froning, H.D, Czysz, P.A, "Advanced Technology and Breakthrough Propulsion Physics for 2025 and 2050 Aerospace Vehicles", Proceedings of the Space Technology and Applications Forum (STAIF2005), Albuquerque, NM. USA, 2006

4.Thomas, R, Yang, Y, Miley, G.H, Mead, F.B, "Advancements in Dense Plasma Focus (DPF) for Space Propulsion", Proceedings of the Space Technologies and Applications Forum (STAIF 2005), Albuquerque, NM, 2005

5.Thomas, R., Miley, G.H., Mead, F.B., "An Investigation of Bremsstrahlung Radiation in a Dense Plasma Focus (DPF) Pro- pulsion Device", Proceedings of the Space Technologies and Applications Forum (STAIF2007), Albuquerque, NM, 2006

6.Chase, R.L, et al, "An Advanced Highly Reusable Space Transportation System", NASA Cooperative Agreement No. NCC8-104 Final Report, ANSER, Arlington, VA 1997

7.Sargeant, W.J, et al, IEE Transactions on Plasma Science, 26, p 1368, 19998

8.Putman, P, et al, "Superconducting Permanent Magnets for Advanced Propulsion Applications", Proceedings of the Space Technologies and Applications Forum (STAIF2005), Albuquerque, NM, 2005

9.Chapman, J.N, et al, "Flightweight Magnets for Space Applications Using Carbon Nanotubes", AIAA 2003-330, 41st Aerospace Sciences Meeting and Exhibition, Reno NV, 2003

10.Murakami, M., et al, "Indirectly Driven Targets for Inertial Confinement Fusion", Nuclear Fusion, 31, p 1315,1991

11.Joenson, K.D, "Broad-band Hard X-Ray Reflectors", Nuc. Inst. and Methods in Phys. Research B, 132, p 221,1997.

Reducing Fusion Plasma Confinement Energy with Specially Conditioned Electromagnetic Fields

H. David Froning, JOURNAL OF SPACE EXPLORATION, 2013 Vol: 2 (3)

University of Adelaide, Adelaide, South Australia, 8004, Australia (Retired) dfron4@gmail.com

ABSTRACT

A central fusion physics problem is strong repulsion between fusion fuel ions that must be overcome by strong confining fields that must drive ions close enough so their nuclear fusion can occur. The ions finally experience attraction when their separation becomes shorter than the short ranges of ion-attracting SU(3) strong nuclear fields, and ion fusion then occurs. In this respect, Barrett shows the possibility of conditioning ordinary U(1) electromagnetic (EM) fields with the same SU(2) and SU(3) Lie Symmetry as the SU(2) and SU(3) nuclear fields that accomplish hydrogen fusion in the Sun with less pressure and temperature than is required in fusion reactors on Earth. This has suggested the possibility of SU(2) and SU(3) EM fields causing terrestrial fusion less confinement energy than ordinary U(1) EM fields currently require for fusion. And, this possibility of SU(2) or SU(3) EM fields enabling terrestrial fusion with less confinement energy than U(1) EM fields currently require, is briefly explored for some promising nuclear fusion reactor designs.

KEYWORDS: COULOMB REPULSION; IEC FUSION; SU(2) SYMMETRY FIELDS; SU(3) SYMMETRY FIELDS

INTRODUCTION

Nuclear fusion of hydrogen occurs in our Sun because of charge-changing actions by SU(2) weak nuclear fields and ion-attracting actions by SU(3) strong nuclear fields. And accomplishing such relatively clean, safe fusion on Earth and in space was once believed Earth's best chance of meeting meet its future terrestrial and spaceflight needs. But, in space and on Earth a formidable fusion barrier must be overcome. This is enormous coulomb repulsion between fusion ions, which must be overcome by enormous levels of ion-compressing force by strong electrical or magnetic fields in fusion reactors. And so far, this formidable barrier has prevented any real success by any of the many nuclear fusion systems that have been proposed and built in the past 50 years.

The general historical trend in fusion energy development has been continual striving for more powerful fields for higher force for higher compression of fusion plasmas. Unfortunately, higher ion compressions tend to be accompanied by unexpected plasma behaviors or instabilities that negate some benefits of the increased input field power and force. So, progress is slow as unpleasant surprises usually accompany each increase in fusion power. Thus, national or international fusion research reactors (which receive most fusion funding) are now so large and expensive that many doubt that fusion can compete with renewable energies like solar and wind.

Short range, ion-attracting SU(3) strong nuclear fields can't cause ion fusions until the ions are driven close together by very strong electric (E), or magnetic (B) or electromagnetic (EM) fields. Such strong E or B fields are not needed for hydrogen fusion in the Sun, even though pressures and temperatures in the Sun's core are too nuclear fields. But quark-mutating actions in SU(2) weak nuclear fields convert some hydrogen protons into neutrons. So, strongly repelling proton-pairs transform into almost neutral proton-neutron pairs that are easily fused by ion-attracting SU(3) nuclear fields into deuterium ions. And these ions start a fusion chain of reactions that convert hydrogen into Helium 3 in the Sun and emit charged particles and solar radiation from its corona.

In this respect, Barrett [1] shows the possibility of transforming ordinary U(1) EM fields into EM fields with the same Lie Group symmetry as the SU(2) and SU(3) nuclear fields that enable successful nuclear fusion in the Sun. This led the Author, T. Barrett, and G. Miley[†] to explore the possibility of fusion ions possessing less charge differential or experiencing less repulsion if they are confined by higher symmetry SU(2) or SU(3) EM fields in fusion reactors [2]. And this would contrast with today's situation where ions possess more charge differential and experience more repulsion when confined by ordinary, lower symmetry U(1) EM fields in fusion reactors.

If SU(2) or SU(3) EM fields could reduce ion repulsion, less fusion reactor input power, mass, size and cost would be needed. In this respect, [2] indicated the possibility of 10-15 fold reduction in input field power for fusion if SU(2) or SU(3) EM fields could significantly reduce fusion ion repulsion at separation distances 2-3 orders of magnitude longer than the 10-15 m distances needed for final ion fusion by SU(3) strong nuclear force. This follow-on paper continues exploring use of SU(2) and SU(3) EM fields to reduce fusion ion repulsion.

RESULTS AND DISCUSSION

SU(2) AND SU(3) EM RADIATION FIELDS

Ordinary electric (E) and magnetic (B) fields are usually described by vectors and Abelian algebra and they develop in space and time in accordance with the four Maxwell Equations formulated by Clerk Maxwell about 140 years ago. In the 1950's Feynman and others made Maxwell's EM theory compatible with quantum theory and special relativity. Quantum Electrodynamics was the result. It precisely predicts interactions between matter and radiation. But, Maxwell's classical theory and equations have been used, essentially unchanged, in the

---

[†] Dr. George Miley is the former Director of the Fusion Research Lab, University of Illinois – Ed. Note

design of all electromagnetic devices since its inception. Maxwell theory describes electromagnetism in terms of: electric field strength **(E)**, magnetic flux density (**B**), and total current density (**J**). Electric and magnetic fields develop and propagate in accordance with the 4 Maxwell Equations shown in Table 1. Its fields embody U(1) Lie Group symmetry, and E an B vector fields are sometimes defined mathematically in terms of non-physical potentials. They are usually called the magnetic vector potential (A) and electric scalar potential (ϕ).

Table 1. Maxwell's Equations for U(1) Symmetry Electromagnetic Vector Field

Gauss' Law	$\nabla \bullet \boldsymbol{E} = \boldsymbol{J}_0$
Ampere's Law	$\frac{\partial \boldsymbol{E}}{\partial t} - \nabla \times \boldsymbol{B} - \boldsymbol{J} = 0$
Coulomb's Law	$\nabla \bullet \boldsymbol{B} = 0$
Faraday's Law	$\nabla \times \boldsymbol{E} + \frac{\partial \boldsymbol{B}}{\partial t} = 0$

$$E = -\frac{\partial A}{\partial t} - \nabla\phi, \quad B = \nabla \times A$$

Barrett [1] has used group and gauge theory and topology to develop SU(2) EM field theory, and Table 2 shows the Expanded Maxwell Equations associated with SU(2) EM fields. Ordinary U(1): J, E and B vector fields are transformed into SU(2): J, E and B tensor fields that are described by Non-abelian algebra, while the magnetic vector potential (which is non-physical in U(1) electromagnetism) is transformed into a physical A tensor field. And SU(2) EM, Maxwell Equations have additional terms that involve A tensor fields interacting with E and B tensor fields in various ways. SU(2) EM fields are not primordial and act over much longer distances than SU(2) nuclear fields do. Also, SU(2) EM fields are mediated by a single boson (a photon) while SU(2) nuclear fields are mediated by 3 bosons. On the other hand, mediating actions associated with the 3 different couplings of A, E, and B fields with each other (as is shown in the SU(2) Maxwell Equations of Table 2) may be somewhat analogous to mediating actions associated with the W +, W -, Z bosons of SU(2) nuclear fields.

Table 2. Extended Maxwell's Equations for SU(2) Symmetry Electromagnetic Tensor Fields

$\nabla \bullet \boldsymbol{E} = \boldsymbol{J}_0 - iq(\boldsymbol{A} \bullet \boldsymbol{E} - \boldsymbol{E} \bullet \boldsymbol{A})$
$\frac{\partial \boldsymbol{E}}{\partial t} - \nabla \times \boldsymbol{B} - \boldsymbol{J} + iq[\boldsymbol{A}_0, \boldsymbol{E}] - iq(\boldsymbol{A} \times \boldsymbol{B} - \boldsymbol{B} \times \boldsymbol{A}) = 0$
$\nabla \bullet \boldsymbol{B} + iq(\boldsymbol{A} \bullet \boldsymbol{B} - \boldsymbol{B} \bullet \boldsymbol{A}) = 0$
$\nabla \times \boldsymbol{E} + \frac{\partial \boldsymbol{B}}{\partial t} + iq[\boldsymbol{A}_0, \boldsymbol{B}] = iq(\boldsymbol{A} \times \boldsymbol{E} - \boldsymbol{E} \times \boldsymbol{A}) = 0$

Barrett has made a small start towards SU(3) electromagnetic field theory. SU(3) electromagnetic theory is, of course, more complex than SU(2) EM theory – with SU(3) EM fields embodying higher order tensors and higher order A, E, B couplings than SU(2) EM. But since SU(3) EM theory requires much more development, this paper focuses mainly on descriptions and hardware experiment possibilities for SU(2) EM radiation fields.

Table 2 shows SU(2) EM couplings between A and E and A and B fields that do not occur in ordinary U(1) EM. Thus, forces exerted on moving charged particles in an U(2) EM field can be different than forces on the particles when moving in an ordinary U(1) EM field. This is shown in Table 3 which is taken from [1]. Shown is Lorentz force exerted on a moving charged particle in an SU(2) E and SU(2) B tensor field – as compared to Lorentz force exerted on the particle in an ordinary U(1) E and B vector field. Here, U(1) Lorentz Force is described by E and B fields and A vector potentials; while SU(2) Lorentz force is described by E and B and A tensor fields. Differences in numbers of Lorentz Force terms and their different vector and tensor natures imply that an SU(2) EM field can exert a Lorentz force of different intensity and direction on moving fusion ions, compared to Lorentz Force exerted on the fusion ions by an ordinary U(1) EM field of equal strength.

Table 3 Lorentz Forces Acting on Moving Charged Particles

U(1) Lorentz Force	$\mathscr{F} = e\mathbf{E} + ev \times \mathbf{B} = e\left(-\frac{\partial \mathbf{A}}{\partial t} - \nabla\phi\right) + ev \times \left((\nabla \times \mathbf{A})\right)$
SU(2) Lorentz Force	$\mathscr{F} = e\mathbf{E} + ev \times \mathbf{B} = e\left(-(\nabla \times \mathbf{A}) - \frac{\partial \mathbf{A}}{\partial t} - \nabla\phi\right) + ev \times \left((\nabla \times \mathbf{A}) - \frac{\partial \mathbf{A}}{\partial t} - \nabla\phi\right)$

Barrett [3] has identified one way of radiating SU(2) EM field energy. It is by flowing alternating current at radio frequencies through a toroid coil at one of the possible resonant frequencies that are possible for a given toroid and coil configuration. Figure 1 shows the two A field patterns that form about a transmitting toroid. They are viewed as two U(1) A vector potential patterns (ϕ1 and ϕ2) which overlap in polarity. A resonant frequency occurs when the difference (ϕ1-ϕ2) in overlapping vector potential amplitude maximizes for the toroid at one of its alternating current frequencies. A single SU(2) tensor field forms about the toroid at this frequency. And, as indicated, many resonant frequencies are possible for the toroid and its coils in Figure 1.

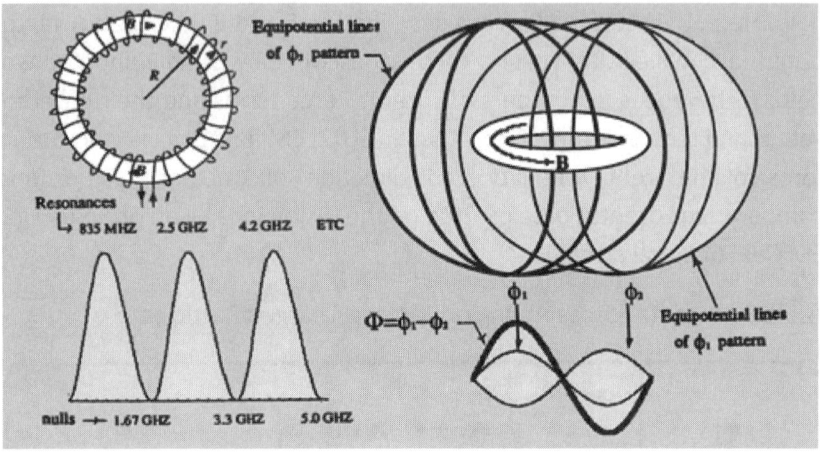

Figure 1. A vector potential patterns generated by alternating current flow in toroid coils. Phase difference (ϕ1-ϕ2) of the patterns maximize when a "resonant" frequency is reached. Shown are 3 different resonant frequencies for a given toroid and coil geometry, and A slightly different SU(2) EM field will be emitted at each different resonant frequency.

Toroid testing at radio frequencies [4] was not exhaustive enough to conclusively prove that SU(2) EM fields were emitted at resonant frequencies But the measured resonant frequencies were in good agreement with those predicted from Barrett's SU(2) theory work. And, more intense and highly focused magnetic fields were always measured above and below surfaces of tested toroids when the toroids were radiating at a resonant frequency.

Barrett (1) has identified another way of generating SU(2) EM field energy. Description of this way is taken from pages 46 and 61 of [5]. It is shown in Figure 2, which uses a waveguide paradigm to show oscillating U(1) EM wave energy being transformed into SU(2) EM wave energy by phase and polarization modulation. Here, input wave energy enters from left, is polarization-modulated, and is emitted as SU(2) wave energy to the right.

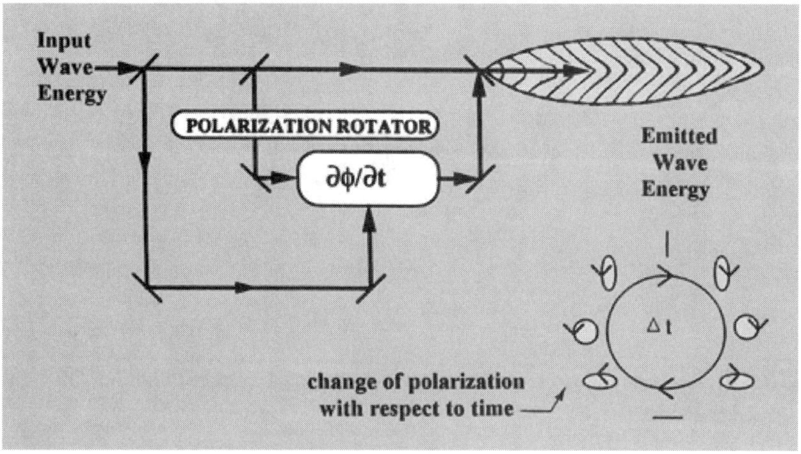

Figure. 2 Generation of Polarization-Modulated EM Radiation

Figure 1 shows: part of the input wave energy unchanged; another part phase modulated (∂A/∂t) and combined with a part that passes through a "polarization rotator". This results in two orthogonally polarized waveforms (with one being the unchanged fraction of input wave energy. These phase modulated and polarization modulated waveforms are combined and emitted as a single beam of SU(2) EM radiation of continually varying polarization. Shown in the lower right of Figure 2 is polarization-modulated EM radiation swiftly sweeping through many polarizations (linear, elliptical, circular) during a cycle of polarization modulation.

Figure 3 from [1] shows swift, variegated change in E field amplitude and direction during a cycle of polarization-modulation. Similar B field change occurs 90 degrees to E field. Such rapid field variation cannot be approached with fixed linear, circular or elliptical polarization. The higher angular dynamics of such a polarization-modulated EM beam will exert different forces and moments on particles like electrons or fusion ions (as compared to lower angular dynamics of ordinary polarized beams). But ability of higher dynamics of polarization-modulated EM beams to modify ion charge distributions or ion repulsions has yet to be determined.

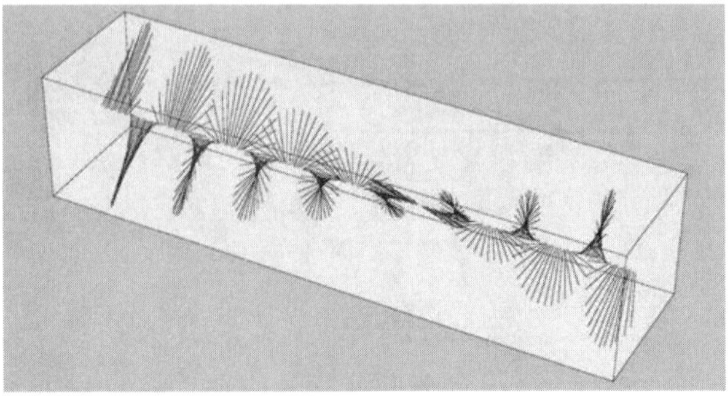

Figure 3. Rapid change in E field configuration during 1 polarization-modulation cycle by an SU(2) Beam.

Since ion-attracting SU(3) nuclear fields consummate every fusion reaction, ion-attracting SU(3) EM fields might be possible and reduce ion repulsion more than SU(2) EM fields. If so, Figure 4 from [5] shows an added modulation of polarization-modulated wave energy that may result in the emission of SU(3) EM wave energy. Developing SU(2) EM field theory to its present state has required much labor by a single individual (Barrett) and there are still SU(2) EM issues to be resolved. So, development of even more complex SU(3) EM field theory will surely require even greater effort.

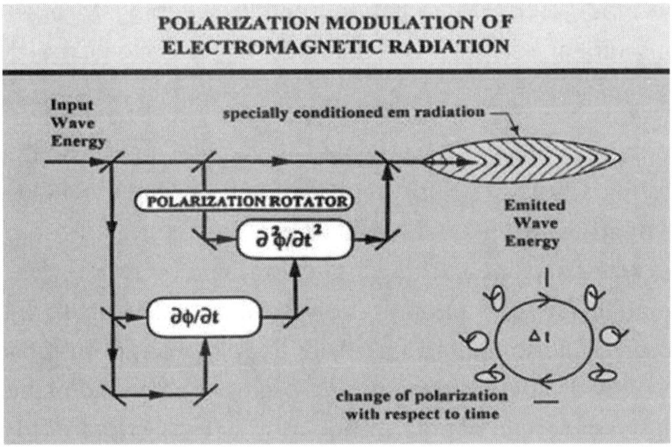

Figure 4. Added phase modulation of EM wave energy that may result in emission of SU(3) EM wave energy (compare to Figure 2)

SU(3) EM Experimental challenges will also be encountered. For example, emission of the SU(2) polarization-modulated wave energy shown in Figure 1 requires a level of phase modulation that is believed achievable with current undulator or oscillator bandwidth state-of-the-art.

But the additional wave modulation shown in Figure 4 indicates that a higher level of phase modulation would be required for emission of SU(3) polarization-modulated wave energy. Barrett [6] believes this higher level of phase modulation would require undulator or oscillator bandwidths in the 20-200 THz range. Such performance has not yet been achieved by any high-bandwidth device – even those in the highest performing FELs of today.

USE OF SU(2) OR SU(3) EM RADIATION FIELDS IN TYPICAL NUCLEAR FUSION SYSTEMS

Fig 5. Fusion Reactor at Univ. of Illinois

Reducing input power for fusion by use of SU(2) EM fields has been explored in most depth for "Inertial Electrostatic Confinement" (IEC) systems. The IEC system considered was pioneered by George Miley at the University of Illinois. Fig 5. shows an operating IEC system at the University of Illinois. It includes multiple ion beams (swiftly-moving streams of ions) converging toward the center of the IEC reactor. Ion beams are emitted from "ion guns" mounted on the reactor periphery. Ion inward motion is accelerated by a negatively-charged electrode (a spherical wire-woven grid) which allows positive-charge ions to freely pass through. And inside this grid, ions converge and crisscross as thousands of ion fusions per second cause the bright central glow. Figure 6 shows one University of Illinois ion beam chamber (ion gun) connected to the IEC reactor. The ion gun contains an RF antenna which heats (ionizes) flowing fusion fuels with EM wave energy. For SU(2) EM testing, the ion-gun's fixed-polarization RF antenna would be replaced by a polarization-modulated RF antenna. Like the current RF antenna does, this RF antenna would create ions by ionizing fusion fuels with EM energy.

Figure 6. Ion gun embodied in IEC Reactor at University of Illinois

Fig. 7 from [7] shows the currently used IEC ion gun. It contains: gaseous deuterium fuel; an electric power system; a "helicon" radio-frequency (RF) antenna whose EM energy intensely heats flowing fusion fuel to form fusion ions. The major ion gun change would be emission of polarization-modulated SU(2) RF wave energy instead of fixed-polarization U(1) RF wave energy for fuel heating and ionization. Fixed-polarization antenna elements would be replaced with orthogonally-polarized antenna elements. And the antenna elements would be driven with state-of-the-art phase modulators that Barrett [8] believes would be in the 10-12 to 10-15 Hz range.

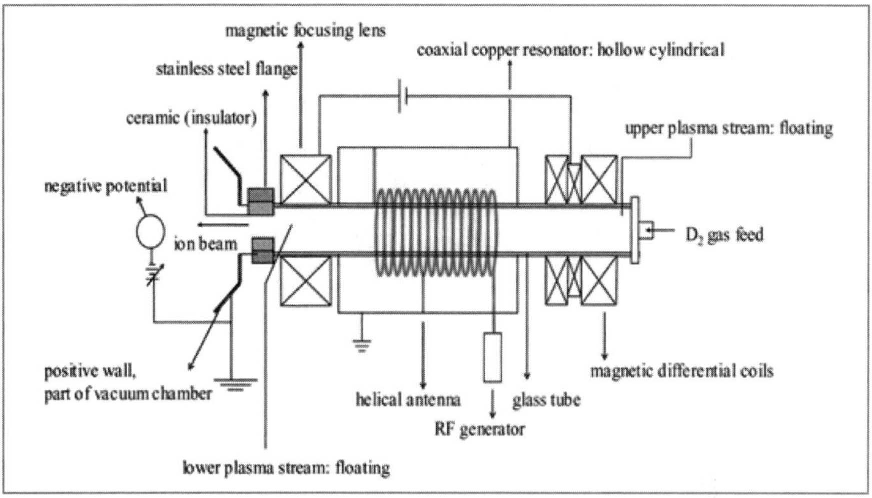

Figure 7. Elements embodied in one of the ion guns used in the University of Illinois IEC reactor

It is expected that emitted SU(2) or SU(3) wave energy would ionize fusion gases in ion guns with about the same efficiency as currently emitted U(1) wave energy does. But energy intensity falls-off with distance, and it is not known how much this fall-off would affect SU(2) or SU(3) EM field ability for reducing fusion ion repulsion at the reactor center. So, more SU(2) EM energy than needed for gas ionization might be required.

A good location for deposition of more SU(2) EM wave energy may be inside the negatively-charged IEC grid – where ions are forced close together for fusion. Effective ion Irradiation in this region might require as many as 4 EM beams – with widths comparable to ion beam widths. Like ion beams, EM beams would mount on the reactor periphery and operate at high frequency to be narrow enough to avoid enlarging the IEC grid so both ion and EM beams could pass through. Many SU(2) EM beam concepts for irradiating fusion ions in central regions would have to be examined. One such EM beam concept could be like the undulator of a "Free Electron Laser (FEL) as shown in Fig, 8. An undulator includes dipole magnets straddling a narrow beam of ions or electrons -- a beam which is then transformed into a narrow beam of SU(2) or SU(3) EM radiation.

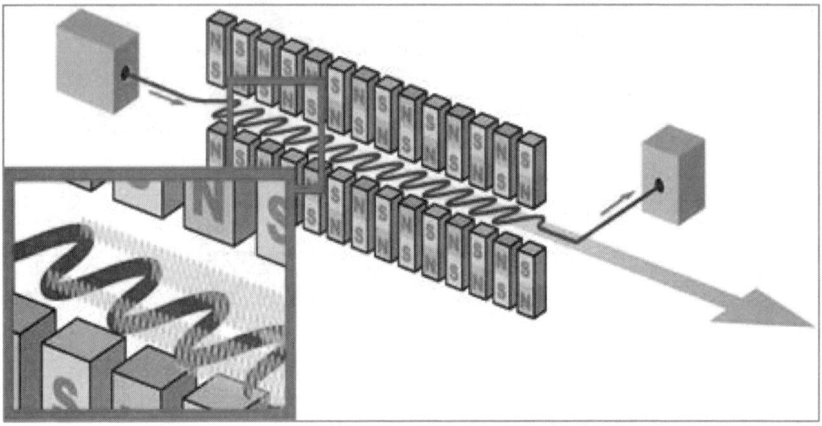

Figure 8. Precise, highly modulatable electromagnetic beam formed by electron beam passing through dipole magnets.

Undulator systems can radiate EM energy from either electron or ion beams. So, one possibility would be modifying an IEC ion gun so its emitted ions would enter a forward undulator section and be transformed into an SU(2) electromagnetic beam. Adding an undulator section to an ion gun may, of course, not be the best possible way to generate an SU(2) EM beam. But this might be the least expensive for early SU(2) testing.

APPLICATION OF SU(2) OR SU(3) EM FIELDS TO OTHER NUCLEAR FUSION SYSTEMS

Reducing input power for several other fusion systems have been very briefly explored. One is the EXL system by Robert Bussard [9]. It is shown on the right of Figure 9 and compared with Miley's IEC system (called IXL) on the left of Figure 9. In EXL, fusion ions are strongly-compressed together by a spherical distribution of electrons. The electron distribution itself is confined by inward-pushing magnetic fields formed from current flowing through the specially configured toroid coil shown in Figure 10. Therefore, highly-confined electrons strongly push fusion ions together in the reactor, where 10^9 fusions per second can occur.

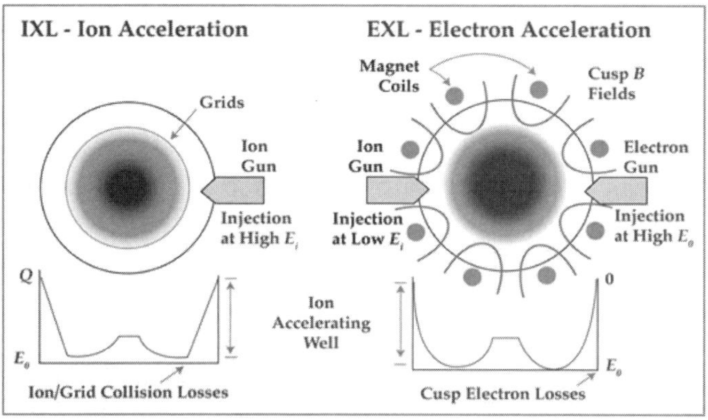

Figure 9. Comparison of George Miley's IX Fusion Reactor with Robert Bussard's EXL Reactor

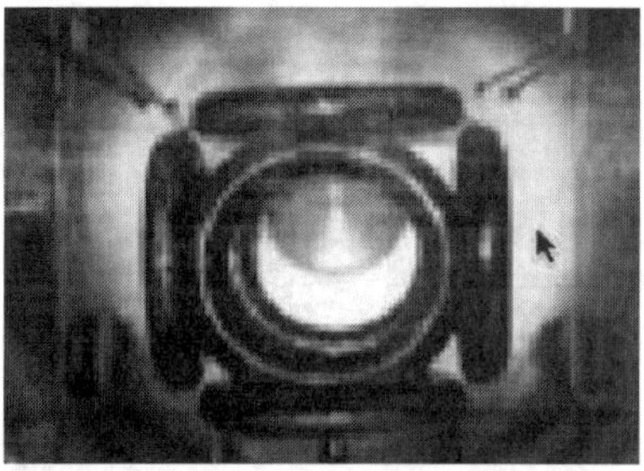

Figure 10. Toroidal coil geometry for EXL electron-confining magnetic field

Details of Bussard's toroid coil design and field pattern aren't known. But it is conceivable that Barrett's idea of emitting SU(2) magnetic field energy from a toroid at resonant frequencies (Figure 1) might enhance the efficiency of Bussard's electron-confining toroidal magnetic field. Also, Bussard's EXL uses an ion gun and accomplishes fusion in a reactor's center like Miley's IEC. But fusion ions tend to be surrounded by confining electrons in an EXL. So, more knowledge on how electron and ion patterns and distributions evolve in EXL reactors is needed to identify an effective possibility for depositing SU(2) or SU(3) EM energy within them.

Reducing input power for a "Focus Fusion" (FF) system by Lawrenceville Plasma Physics Inc.[‡] has also been considered. It is a variant of a fusion device called "Dense Plasma Focus" (DPF) studied by the US Air Force for

---

[‡] Eric Lerner, Dir. Of Lawrenceville Plasma Physics, was a COFE3 speaker – Ed.

fusion propulsion in the 1980's. Figure 11 from [10] shows one DPF system developed by the Air Force in the 2002-2004 time period. Here, gaseous fusion fuel is ionized by strong electrical discharges and a formed plasma sheath is accelerated between cylindrical electrodes. At the cylinder's open end, the sheath reverses direction and enormously intense, ion-confining magnetic fields form in a narrow pinch region, where fusion occurs. And, Figure 12 shows a 6-nanosecond exposure of a DPF test by Professor Nardi at Stevens Institute.

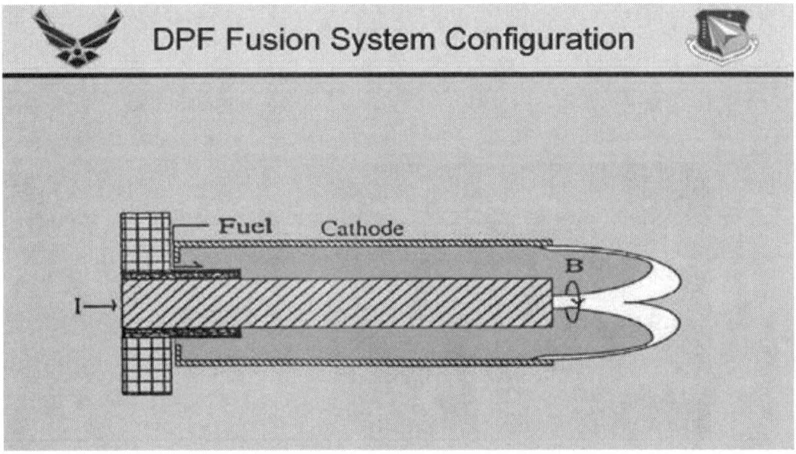

Figure 11. Dense Plasma Focus fusion propulsion for flight

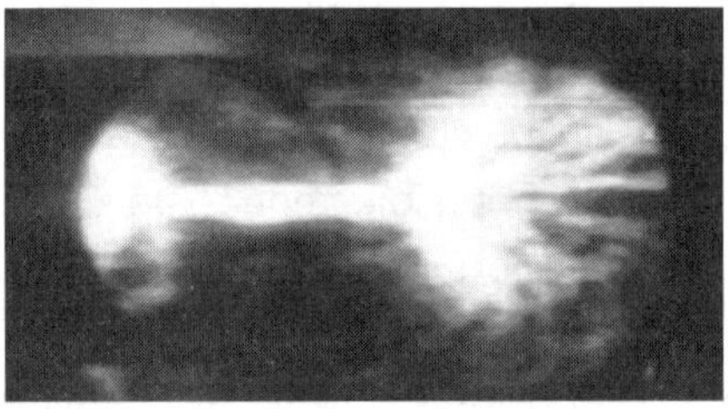

Figure 12. Dense Plasma Focus (DPF) testing at Stevens Institute

The Lawrenceville Plasma Physics Inc. website describes significant advancement in DPF-FF state-of-art with their Focus Fusion designs – with plasmoid densities of 8 x 10^{19} ions per cm^3 quoted. As with Bussard's EXL system, detailed FF information is needed before any

SU(2); SU(3) EM benefits could be claimed. But, it is known that accelerating-turning-compressing of DPF or FF fusion plasmas are strongly influenced by Lorentz Forces, which would be different for SU(2) A, E and B fields. SU(2) Lorentz force calculations could ascertain if SU(2) EM Lorentz force could considerably influence the accelerating-turning-compressing of FF plasmas.

SOME IDEAS ON THEORETICAL AND THEORETICAL RESEARCH WORK

As previously mentioned, Dr. Terence Barrett (though his consulting company "BSEI" has developed SU(2) EM radiation field theory sufficiently for numerical studies to be performed and SU(2) EM field generators to be designed, and this has enabled preliminary experiments to be performed with encouraging results. But some unresolved SU(2) EM field issues remain. Furthermore, theory development for SU(3)EM fields, which may be even more promising then SU(2) EM fields for reducing fusion input energy, will require much more effort than a lone person can accomplish in spare time. Thus, it is believed important that academic institutions be involved in advancing SU(2) electromagnetism and starting SU(3) electromagnetism development .

SU(2) and SU(3) theoretical work would require an estimated 3 year level of effort at fairly modest cost with university faculty, graduate students, computing facilities and some consulting help. Work would enable code development for numerical modeling of SU(2) EM processes associated with fluid and plasma dynamics involved in typical power and propulsion systems. Research would also begin to lay down foundations of SU(3) EM field theory, with consulting help from people like Dr. Barrett.

Two experimental options are possible for relatively quick and inexpensive testing to confirm or refute the possibility of SU(2) EM fields modifying coulomb repulsions. One possibility is: (a) use of an available IEC reactor and ion gun system at one of several universities that possess IEC fusion systems; and (b) modification of the helicon RF antenna of an IEC gun to emit SU(2) EM wave energy. Then, measured IEC fusion intensity with U(1) EM discharges from existing fixed-polarization RF antennas can be compared with measured IEC fusion intensity with SU(2) EM discharges from a modified ion gun with polarization-modulated RF antenna.

The other possibility is use of less expensive non-nuclear plasmas formed by RF discharges in gases like Argon or Xenon. Such tests can be done in plasma chambers with helicon RF antennas to heat (ionize) their gases. These RF antennas are very similar to RF antennas in University of Illinois IEC ion guns. Figure 13 shows such a chamber. It has a helicon RF antenna and is located at the Australian National University (ANU).

Figure 13. Chamber with Helicon RF Plasma Generating System - Australian National University

Just as small differences in fusions for U(1) and SU(2) RF discharges must be detected in IEC reactors, so small differences in ion and electron behaviors would have to be detected in plasma chambers for U(1) and SU(2) RF discharges. This is possible in the ANU chamber. For it incorporates an innovative "movable energy analyzer" that allows plasma properties to be accurately and rapidly mapped throughout the entire chamber.

As with theory, a concurrent 3 year level of effort is recommended for experiment work. It is admitted that this theory and experiment work is exploratory. Hence, it is somewhat "high risk" in that specific accomplishments or success cannot be assured. But, at the very least, advances in EM field theory will be made, new plasma physics will be learned, and the experiments will significantly extend plasma physics state-of-the-art.

CONCLUSIONS

T.W. Barrett, has derived EM fields with the same SU(2) and SU(3) Lie Group symmetry as the SU(2) weak and SU(3) strong nuclear fields that bring about hydrogen fusion in the Sun with less confinement power and temperature than is required in terrestrial fusion reactors. And it is suggested that the A, E, and B tensor field couplings embodied in SU(2) and SU(3) electromagnetism could conceivably modify charge distributions in fusion ions and repulsive forces between them. This reconstituting of fusion ions into higher symmetry form – which removes stronger ion repulsion present at lower U(1) symmetry state - would reduce confining power and temperature needed for terrestrial fusion. And this would be somewhat like what SU(2) and SU(3) nuclear fields do in modifying charges and repulsions of hydrogen ions in the Sun's core to bring about efficient solar fusion.

Reducing needed input power for fusion with SU(2) or SU(3) EM energy has been explored. One possibility is depositing such EM energy into gaseous fusion fuels - to transform them into ions whose lessened repulsion would require less confining force for their fusion. Another is irradiating ions in fusion regions with charge-changing or repulsion-reducing SU(2) or SU(3) EM field energy, to lessen needed confining force for their fusion. Finally, SU(3) EM may be more promising than SU(2) EM, but it needs more development.

Initial theoretical SU(2) and SU(3) EM research would require: about three years at relatively modest cost at an interested university. A concurrent three year SU(2) experimental effort at a university reactor or plasma facility is also recommended. This theoretical and experimental research would be high-payoff if successful. But, at the very least, it would greatly advance electromagnetic field theory and plasma physics state-of-art.

ACKNOWLEDGEMENTS

I would like to acknowledge the extensive, pioneering work of Dr. Terence Barrett, who is the originator and developer of the advanced electromagnetic field theory described in this paper. I am also grateful to Dr. Barrett for many stimulating discussions on theoretical and experimental issues associated with his work. I also want to acknowledge all the nuclear fusion knowledge I have had the privilege of gaining from Professor George H. Miley on our many collaborations on Inertial Electrostatic Confinement and Dense Plasma Focus fusion systems – collaborations based on Professor Miley's very innovative and ground-breaking fusion energy work.

REFERENCES

[1] Barrett, T.W. "Topological Foundations of Electromagnetism", World Scientific, 2008

[2] Froning, H.D., Barrett, T.W., Miley, G.H., "Specially conditioned EM fields to reduce nuclear fusion input energy needs", Space Propulsion and Energy Sciences International Forum – 2012, Elsevier Physics Procedia, 2012

[3] Barrett, T.W., "The impact of topology and group theory on future progress in electromagnetic research", Progress in Electromagnetic Research Symposium (PIERS 1998) Nantes, France, 1998

[4] Froning, H.D., Hathaway, G.W. "Specially Conditioned EM Radiation Research with Transmitting Toroid Antenna",

2001-3658, 37th AIAA / ASME/SAE/ASEE Joint Propulsion Conference and Exhibit , Salt Lake City Utah, 2001

[5] Barrett, T.W. " Topological foundations of electromagnetism" Annales ds la Foundation Lious de Broglie – special issue on George Lochak pour non 70eme anniversaire , 26 ,55-79, 2001

[6] Barrett, T.W., Personal Communication

[7] Miley, G.H., Shaban, Y., Yang,Y., "RF Gun Injector in Support of Fusion Ship II Research and Development",

Proceedings of Space Technology and Applications Forum – STAIF 2005 , Edited by E.L. Genk, American Institute of Physics, Melville, New York, 2005

[8] Barrett, T.W., Personal Communication

[9] Duncan M., “Should Google Go Nuclear”, askmar.com, video of talk at <htpp://video.google.com/>

[10] Froning, H.D., Czysz, P., “Advanced Technology and Breakthrough Physics for 2025 and 2050 Military Space Vehicles”, Proceedings of Space Technology and Applications Forum -STAIF 2006, Edited by E.L. Genk, American Institute of Physics, Melville, New York, 2006

[11] Froning, H.D., Little, Jay. “Fusion-Electric Propulsion for Aerospaceplane Flight”, AIAA 93-5126, AIAA/DLGR 5th International Aerospaceplanes and Hypersonics Technologies Conference, Munich Germany, 3 December, 1993.

Quantum Vacuum Engineering for Power and Propulsion from the Energetics of Space

H. David Froning*

PO Box1211, CA, 90262, USA

Phone: 310-459-5291; Emailfroning@infomagic.net

 Presented at Third International Conference on Future Energy Washington Hilton Hotel, Washington DC, October 9-10, 2009

Abstract. This talk describes challenges and possibilities for exploiting the stupendous energetics that may be contained within the zero-point ground state of seemingly "empty" space for meeting Earth's future power and propulsion needs. In particular, it's suggested that intense vacuum energetics, in addition to those that reside within the vacuum's more well-known electromagnetic modes, and research on specially conditioned electromagnetic radiation to interact with these energetic modes is described. And, in addition to its energy, it is shown that quantum vacuum negative pressure could conceivably enable less labored vehicle motion in it.

INTRODUCTION

The most bizarre possibility for meeting Earth's power and propulsion needs may be the harvesting the stupendous energetics that possibly exist within the quantum mechanical ground states of the vacuum of "empty" space. These energetics are symbolized in Figure 1 as electromagnetic pulsations that – like lightning flashes – forever appear and disappear throughout the entire cosmos in countless times and places. The more simple interpretations of quantum theory predict that quantum fluctuations in the vacuum state of space give rise to energy fluctuations whose average expectation values could be 40-100 orders of magnitude greater than all of the energy contained in cosmic matter. Wheeler, 1968 speculates that the enormous vigor and vitality within the submicroscopic intensity of individual vacuum energy fluctuations is diminished by some sort of wave interference-like effect that almost totally diminishes their intensity over the much larger scales of time and distance that unaided senses can perceive. So, space is inert and empty to ordinary macroscopic observation. But Figure 1 shows that zero-point energy densities of 1.0 J/cm3 would be revealed to hypothetical probes that could sense and resolve physical activity occurring over 10-6 cm scales of distance. And at 10-7 cm scales of distance, much more violent 104 J/ cm3 activity would be perceived.

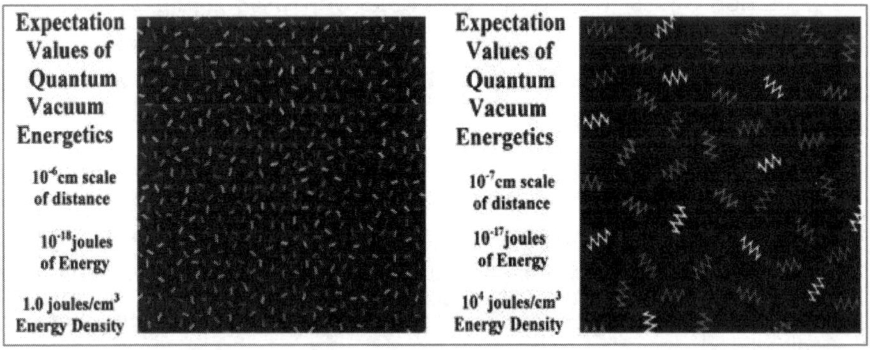

Figure 1 Expectation Values of Zero-Point Energy in 10-6and 10-7cm. Regions within the Quantum Vacuum of Space

In his book "The Lightness of Being" Frank Wilczek defines the quantum vacuum as 'the grid', with its various constituents being: (1) quantum fields associated with radiation, matter and gravitation; (2) virtual particle pairs and weak superconducting condensates; and (3) dark energy. Unfortunately, many quantum vacuum unknowns remain, with no scientific consensus as to its essence and extreme doubt over possibility of exploiting its energetics for power or propulsion. Typical questions: Is the quantum vacuum an enormous sea of stupendous zero-point energy, or does it have a much more meager energy allocation? Is the quantum vacuum the supreme source from where matter gets all its energy; or does it only exchange energy with matter? And, of course, the critical engineering and science issue: Can more zero-point energy be extracted from quantum vacuum than the required input of energy into vacuum for its extraction? These questions will be touched on more from the standpoint of engineering than science.

QUANTUM VACUUM ENGINEERING TO DEFINE ENERGY EXTRACTION NEEDS

Froning, 1980 is the first known peer-reviewed article to suggest the possibility and problems of extracting zero-point energy from the quantum vacuum for power and propulsion. Figure 2, from this article, indicates spaceflight as the chosen application. Zero-point energy expectation values given in Wheeler, 1968 were used, and processes that could: (a) materialize electrons and positrons out of the vacuum for mutual annihilation; and (b) gather the resulting photons into a thrusting beam of light were assumed. The investigation identified the scales of time and distance that zero-point energy must be interacted with and extracted from the quantum vacuum, to swiftly accelerate starships to almost light-speed. It was found that interaction with and extraction of zero-point energy from the vacuum must occur over 10-6 to 10-7cm scales of distance – distance 4 to 5 orders of magnitude larger than the diameters of atoms.

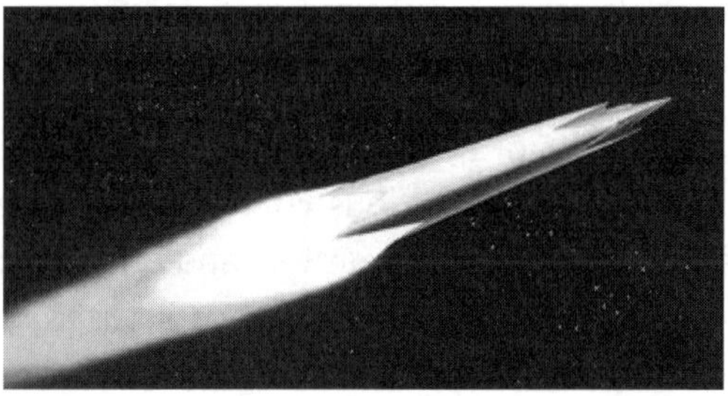

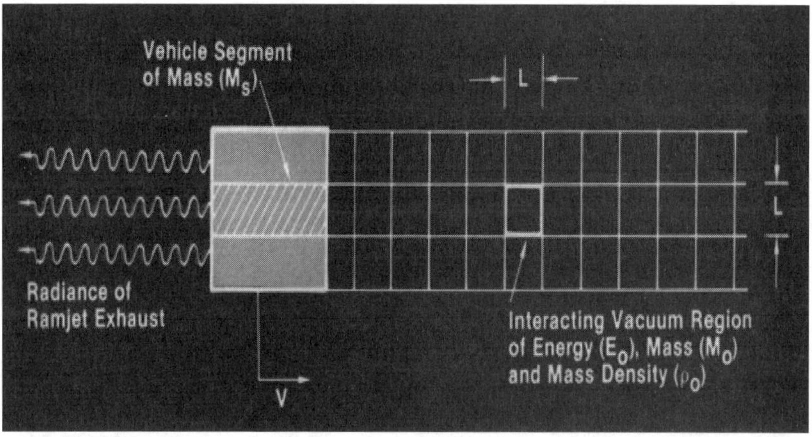

Figure 2. Schematic Representations of a "Quantum Interstellar Ramjet" Extracting Zero-Point Energy from Space

Energy extraction within 10^{-6} cm of distance requires times as short as 10^{-16} s. At the time of this investigation, the shortest achievable times were electrical switching times of about 10^{-11}s. But femtosecond lasers operating in the extreme ultraviolet now have beam widths of the order of 10^{-6} cm and pulse widths of the order of 10^{-16} s. Thus, electromagnetic devices now operate at times and distances that may be of interest for zero-point energy interaction.

INTERACTING WITH THE QUANTUM VACUUM BY SPECIAL EM RADIATION

It's generally agreed that quantum vacuum fields are associated with the matter fields that comprise weak and strong force interactions in nuclei and with the radiation fields that are associated with electromagnetism and gravitation. In this respect, zero-point energy modes associated with the quantum fields associated with electro-magnetism have been fairly extensively studied the most and have revealed extremely interesting quantum field phenomenon such as the Casmir Effect. But more robust zero-point energy modes may be associated with some of the other quantum vacuum fields. One example is the intense zero-point vacuum field associated with the strong force that guides and controls quark motions within protons. As shown in Figure 3, This intense vacuum field action is manifested in the mass-less, ever-changing gluon fields that materialize-dematerialize-rematerialize in proton-occupied space in less than a trillionth of a trillionth of a trillionth of a second, and which comprise about 97 % of a proton's total energy.

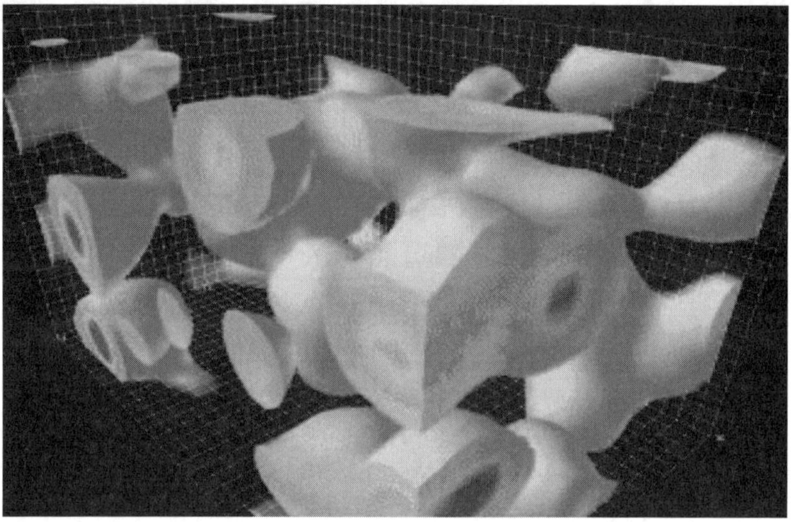

Figure 3. Manifestation of the Quantum Vacuum as Gluon Field Activity in Hadron-Occupied Space – Image Computed by Dr. Derek Leinweber , University of Adelaide, ARC Special Research Center for the Subatomic Structure of Matter

Robust quantum vacuum field activity may also be associated with the weak-interactions that change neutrons into protons by mutating quarks into different quarks within hadrons. Here, it is generally believed that a Higgs vacuum field is associated with the "Higgs Boson" that gives rise to the Z and W bosons that mediate such weak interactions. The matter fields associated with weak and strong interactions in nuclei possess a high SU(2) and SU(3) symmetry. Similarly their associated zero-point quantum vacuum fields are of SU(2) and SU(3) form. In this respect, classical electromagnetic radiation possesses a lower U(1) field symmetry; and, presumably, a U(1) vacuum field as well. But Barrett, 2007 has used group and gauge theory and topolological analysis to show the possibility of conditioning ordinary U(1) electromagnetic radiation to SU(2) or even higher field symmetry form.. Such electromagnetic (EM) radiation fields are considered classical. But they have (in addition to electric and magnetic field content) A-vector fields whose mathematical structure closely resembles that of the quantum vacuum fields associated with the strong and weak force. Figure 4 shows expanded Maxwell Equations that describe actions of specially conditioned SU(2) EM fields in accordance with Gauss's and Ampere's and Faraday's Laws. It is seen that Maxwell Equations for specially conditioned SU (2) EM fields contain electric and magnetic field terms - as ordinary U(1) EM fields do – plus added terms that involve coupling of electric and magnetic fields with A-vector and A-scalar potential fields. Like quantum fields, A-field terms involve **non-abelian algebra** (wherein vector and dot products such as (A E - E A) and (A x E – E x A) are not zero. This quantum mechanical similarity of classical SU(2) and SU(3) EM fields and SU(2) and SU(3) quantum vacuum fields is encouraging us to investigate the possibility of the A-vector field energetics in SU(2) or SU(3) EM beams coupling with the zero-point energetics in SU(2) or SU(3) quantum fields.

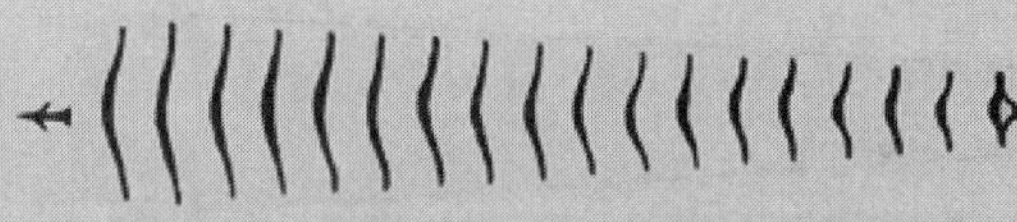

SU(2) Maxwell Equations

$\nabla \cdot E = J_0 - iq(A \cdot E - E \cdot A)$
$\frac{\partial E}{\partial t} - \nabla \times B + J + iq(A_0 E - E A_0) - iq(A \times B - B \times A) = 0$
$\nabla \cdot B + iq(A \cdot B - B \cdot A) = 0$
$\nabla \times E + \frac{\partial B}{\partial t} + iq(A_0 B - B A_0) + iq(A \times E - E \times A) = 0$

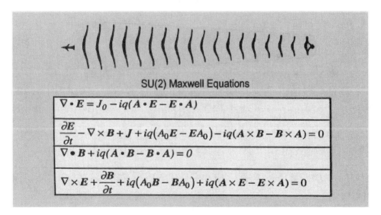

Figure 4. Expanded Maxwell Equations for Specially Conditioned SU(2) Electromagnetic Radiation

Up to now, two ways of generating specially conditioned EM fields have been developed. One generates SU(2) and possibly SU(3) fields by the polarization modulation of radio-frequency or microwave or laser beams (by orthogonal polarization and phase modulation of input waveforms). The other generates SU(2) fields by toroids of appropriate geometry and coil winding; and resonant frequency – whereby mainly A-vector field energy is emitted. Figure 5 shows several of the successfully tested toroid transmitters. The tests are described in Froning and Hathaway, 2002.

Figure 5. Asymmetric, Caduceus-Wound, Toroidal RF Field Generators of SU(2) EM Radiation

Although not designed or tested for zero-point energy interaction, both ordinary and symmetrical toroid transmitters were swept through radio frequencies between 0.4 and 110 MHz and enormous increase in near field intensity and signal strength occurred when the toroids operated at their different resonant frequencies within this frequency band.

QUANTUM VACUUM AND CONDITIONED EM FIELDS IN HIGHER DIMENSIONS

Wheeler, 1967 deemed the seeming inertness of the dynamic quantum vacuum to be due to a wave interference-like processes that diminish the intensity of localized vacuum energy fluctuations over the much larger scales of distance and time that can be resolved by the senses. But more recently Wheeler, as have others, speculated that the seeming inertness of the quantum vacuum may be due to its stupendous vigor and vitality being distributed within the much greater vastness of a higher-dimensional space – as symbolized in the left portion of Figure 6 below. Wheeler's view is consistent with current supercomputer computations of gluon-quark dynamics occurring in atomic nuclei at places such as the University of Adelaide. Here, simulations of the strong force interactions occurring in hadron-occupied spaces involve computation of 3-D, time-varying gluon fields that control quark motions in nuclei. And, the gluon fields are found to be 3-D manifestations of higher dimensional 4-D structures of the quantum vacuum that move in and out of 3-D space. This action is described by Yang-Mills theories in terms of "instantons"- mathematical entities that minimize 'energy functionals' within 4-D Riemannian manifolds. Shown, in the left potion of Figure 6, is a simplified representation of a hadron-occupied region of spacetime - where quark-gluon interactions are taking place trillions upon trillions of times per second in nuclei. And above this region is higher dimensional quantum vacuum that gives rise to quark-confining gluon fields and 97 % of nuclear matter's energy in the spacetime region below it.

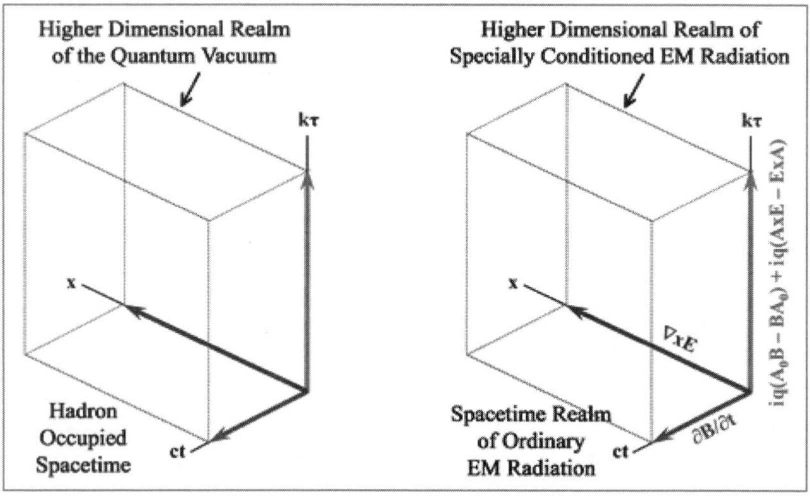

Figure 6. Similarities in the Higher Dimensional Fields of Quantum Vacuum and Specially Conditioned EM Radiation

The specially conditioned SU(2) electromagnetic fields that have been previously described are based on the same Yang-Mills theories that involve the same instanton propagation as that involved in actions of the quantum vacuum fields that give rise to the gluon field energies that confine quarks and comprise 97 percent of a proton's mass. And, as shown in the right hand portion of Figure 6, A-vector field actions of specially conditioned electromagnetic fields extend in the very same orthogonal direction relative to 4-D spacetime as do the quantum vacuum fields that give rise to weak and strong interactions inside matter. As mentioned, the 4 Maxwell Equations for specially conditioned electromagnetic fields contain the same electric and magnetic field terms as ordinary electromagnetic fields do. But additional terms, that involve vector and dot products of A-fields and electric and magnetic fields, are multiplied by an operator (i) that denotes action that is orthogonal to 4-D spacetime. It can, thus, be shown that A-vector field structures of specially conditioned electromagnetic fields closely resembles quantum field structures of the vacuum.

It has therefore been concluded that specially conditioned SU(2) and SU(3) electromagnetic fields have compelling similarities with the quantum vacuum fields that are associated with SU(2) weak and SU(3) strong nuclear forces. Both have the same overall field structure. Both obey nonabelian mathematics and commutation rules. Both evolve (propagate) by "instanton" action. And both exist in a deeper realm than 4D spacetime. One major differences, of course, are the different scales of distance over which they interact - 10-16 cm and 10-13 cm scales of distance for the SU(2) weak and SU(3) strong force; and much larger 10-6 to 10-7 cm scales of distance that may be required for significant coupling between SU(2) or SU(3) electromagnetic fields and quantum vacuum. It is, of course, premature to claim the feasibility of such interactions. But the needed types and levels of analysis to do so have been identified.

PRELIMINARY SIMULATIONS OF FLIGHT THROUGH ZERO-POINT VACUUM

Haisch etal.1994 has proposed that the electromagnetic quantum vacuum interacts with the electromagnetic structure of accelerating bodies to cause or, at least, contribute to their inertia, and Puthoff, 2002 proposes that space-warping fields emitted by a body for propulsion can be viewed as perturbing the electrical permittivity (ε) and magnetic permeability (μ) of the vacuum. In this respect, Froning and Roach, 1999, 2002, 2007 have developed fluid dynamic methodologies to make very preliminary simulations of accelerated vehicle flight through the quantum vacuum. The accuracy of such fluid dynamic approximations, of course, depends upon degrees of similarity between air flight in planetary atmospheres and space flight in the vacuum of space. In this respect, Figure 7 from Froning and Roach shows some similarity in densities associated with thermal radiation pressures from air molecule interactions in the atmosphere and zero-point radiation pressures from virtual particle pair creation and annihilation in the vacuum of space. As shown in the lower picture, this allows similarity in the aerodynamic and radiation pressure gradients that form over an accelerating vehicle at 99 percent of speed-of-sound in air and 99 percent of speed-of-light in space.

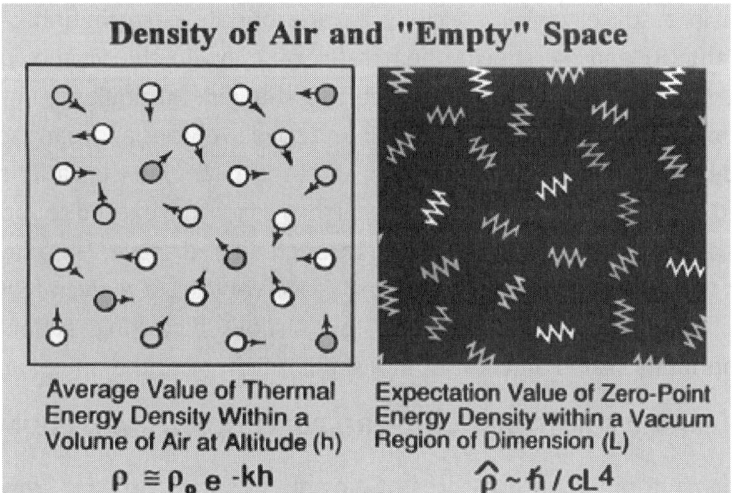
Density of Air and "Empty" Space
Average Value of Thermal Energy Density Within a Volume of Air at Altitude (h)
ρ ≅ ρ₀ e -kh
Expectation Value of Zero-Point Energy Density within a Vacuum Region of Dimension (L)
ρ̂ ~ ħ / cL4

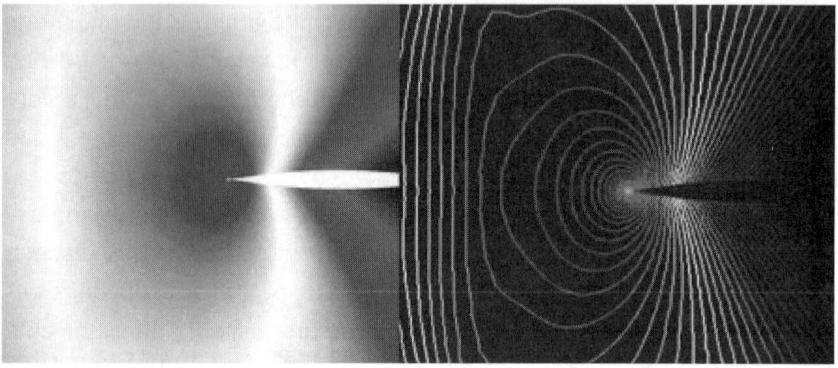

Figure 7. Similarity in Pressures and Gradients at 0.99 Sound Speed in Air and 0.99 Light Speed in Vacuum of Space

Froning and Roach 1999, 2002 showed similarity in acoustic disturbance propagation in air and electro-magnetic disturbance propagation in space. This enabled modeling of space-warping by perturbing vacuum ε and μ in a space vehicle's vicinity by appropriate perturbation of gas constant (R) and air specific heat ratio ratio (γ) in the vicinity of an appropriately scaled air vehicle. They also showed: (a) similarity in flight resistance increase as accelerating air and space vehicles approach sound speed in air and light-speed in space; and (b) similarity in the variation of air and space vehicle flight resistance as they reach and exceed sound and light speed during their acceleration through air and space. This enabled space-warping actions by accelerating space vehicles at a given flight to light-speed ratio to be modeled by airflow distorting actions of appropriately scaled air vehicles at a similar flight to sound-speed ratio.

ACCELERATED FLIGHT THROUGH ZERO-POINT QUANTUM VACUUM

The thermal radiation pressures of air are "positive" exerting "inward-pushing" pressures over the entire surface of any accelerating vehicle – such as the upper vehicle shown in Figure 8. Airflow is both compressed and expanded as the speed of sound is approached, reached, and exceeded and an adverse pressure gradient forms about the vehicle. Here, higher than ambient pressures acting upon the front of the vehicle cause a resisting "push" and lower than ambient pressures acting on the rear of the vehicle cause a resisting "pull". And it may be noted that air pressures are resisted by a very thin repulsive field region that is caused by electrical repulsion between electrons of the outer- most atoms of the vehicle's skin and the most nearby electrons of the atoms of passing air. In air-less vacuum, such a thin repulsive field vanishes, and Haisch, 1994 shows that an electromagnetic interaction between the accelerating vehicle and its quantum vacuum medium causes an inward-pushing electromagnetic radiation pressure gradient that is somewhat similar to an aerodynamic pressure gradient. And this flight resistance can be viewed as vehicle inertia.

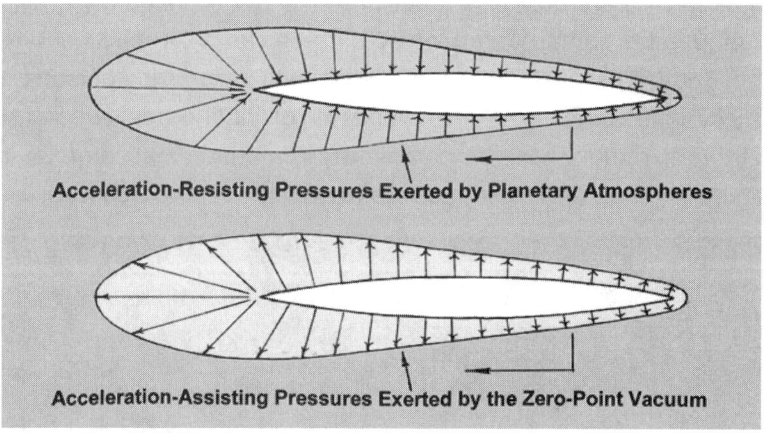

Figure 8. "Resisting" and "Assisting" Vehicle Pressures Exerted by Planetary Atmospheres and the Quantum Vacuum

A repulsive field similar to that generated by electronic interaction between air vehicle skin and passing air can be provided the lower vehicle of Figure 8 – which is accelerating with respect to the quantum vacuum - by embodying space-warping, particle-repelling field generators in its outer skin. But the quantum vacuum has a "negative" pressure which can be viewed as acting in the opposite direction that inward-pushing positive pressure fields act. So, with repulsive field pressures already acting "outward" from the vehicle skin, there is nothing to prevent "outward- pulling" (rather than inward-pushing) pressures to act over the entire ship. And this would cause higher-than-ambient, outward-pulling, acceleration-assisting force to act upon the front of the vehicle; and lower-than-ambient, outward pulling pressures to act upon its rear – that also results in a acceleration-assisting force. Such acceleration-assisting force would maximize in the transluminal speed range where pressure gradients become the most intense. Such force would be diminished by entropy-increasing energy dissipation in shock waves – such as emitted Cherenkov radiation. However, a negative pressure vacuum should still assist repulsive field propulsion systems. It must be admitted that fluid dynamic analogues of vehicle flight through warp-able space and perturb-able vacuum by simulations using Computational Fluid Dynamics (CFD) are only first-order approximations inadequate for any kind of precise computation for actual vehicle or propulsion system design. But it is believed that they are useful in introducing persons such as engineers to features and problems of field-propelled flight by visualization rather than the complex tensor mathematics of General Relativity. Figure 9 and Figure 10 show samples of such visualizations.

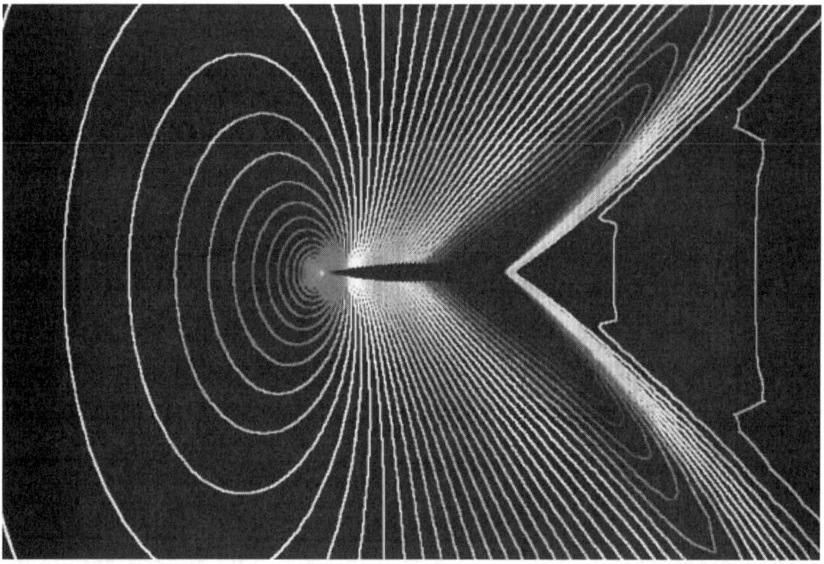

Figure 9. Zero-Point Radiation Pressure Gradient Caused by Accelerating Ship at 99 percent of the Speed-of-Light

Figure 10. Zero-Point Radiation Pressure Gradient Caused by Accelerating Ship at Twice the Speed-of-Light

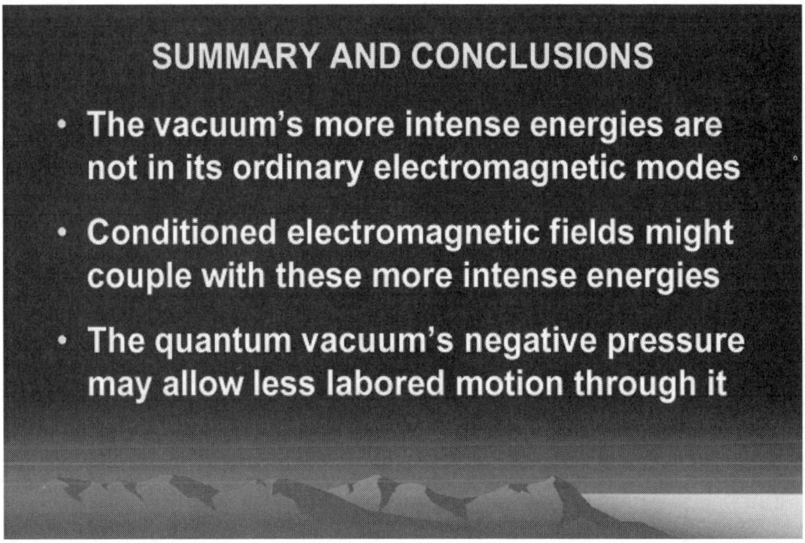
SUMMARY AND CONCLUSIONS
• The vacuum's more intense energies are not in its ordinary electromagnetic modes
• Conditioned electromagnetic fields might couple with these more intense energies
• The quantum vacuum's negative pressure may allow less labored motion through it

REFERENCES

Barrett, T.W., "Topological Foundations of Electromagnetism", World Scientific, 2008

Froning, H.D., "Propulsion Requirements for a Quantum Interstellar Ramjet", Journal of the British Interplanetary Society, Vol 33, No 7 (1980)

Froning, H.D., Roach, R.L., "Fast Space Travel by Vacuun Zero-Point Field Perturbations", Proceedings of Space Technologies and Applications Forum (STAIF1999, Edited by M.S. EL Genk, American Institute of Physics, Melville, New York, 1999

Froning, H.D., Roach, R.L., "Preliminary Simulations of Vehicle Interactions with Zero-Point Vacuum by Fluid Dynamic Approximations", AIAA 2000-3478, 36th AIAA/ASME/SAE/ASEE Joint Propulsion Conference, Huntsville Alabama (2000)

Froning, H.D., Roach, R.L., "Fluid Dynamic Simulations of Warp Drive Flight through Negative Pressure Zero-Point Vacuum", Proceedings of Space Technologies and Applications Forum (STAIF2007), Edited by M.S. EL Genk, American Institute of Physics, Melville, New York, 2007

Haisch, B., Rueda, A., Puthoff, H., "Inertia as a zero-point field Lorenz Force", Phys. Rev. A, Vol 49, p678

Puthoff, H.E., Little, S.R., Ibsen, M., "Engineering the Zero-Point Field and Polarizable Vacuum for Interstellar Flight", Journal of the British Interplanetary Society, Vol 55, p137 (2002)

Wheeler, J.A., "Superspace and the Nature of Quantum Geometrodynamics", Topics in Non-Linear Physics, p615-644. Proceedings of the Physics Section, International School of Non-Linear Mathematics and Physics, Springer-Verlag (1968).

Reducing Coulomb Repulsion and Field Energy for Fusion by SU(2) and SU(3) Electromagnetic Field

H. David Froning

PO Box 1211 Malibu, CA 90262, USA

e-mail: froning@infomagic.net

Reducing needed electrical input energy for nuclear fusion by use of specially conditioned electromagnetic (EM) fields is explored. Reduced electric input energy would be accomplished by reducing coulomb repulsion between fusion fuel ions by irradiating them with EM fields that are especially conditioned to higher SU(2) or SU(3) symmetry. Such conditioning involves transforming ordinarily polarized EM wave energy into EM wave energy whose polarization and phase is modulated at very high rates. The possible efficacy of such specially conditioned EM energy has been briefly explored for Inertial Electrostatic Confinement (IEC) fusion developed by George Miley at the University of Illinois and briefly considered for the "Polywell" IEC Fusion system developed by Robert Bussard of EMC Corporation, and versions of the Dense Plasma Focus (DPF) system – such as the "Focus Fusion" development by Eric Lerner of Lawrenceville New Jersey. SU(3) EM fields are deemed the most promising for reducing coulomb repulsion and input energy for fusion. But SU(2) EM field theory is advanced enough for computation work and SU(2) field generation hardware is available for testing.

1.Introduction

1.1. The Central Problem of Nuclear Fusion Physics

Nuclear fusion occurs in our sun because of charge-changing actions occurring in SU(2) weak nuclear fields, and charge attracting action occurring between particles in SU(3) strong nuclear fields. And accomplishing nuclear fusion on Earth and in spacecraft was once believed by many to be the most promising way to meet Earth's future terrestrial and spaceflight needs. But on Earth and in space there is a formidable barrier that must be overcome. This barrier is the enormous coulomb repulsion between like-charge nuclei which must be overcome by enormous levels of compressive electrical power and force. So far, this formidable coulomb barrier hasn't been effectively overcome by any of fusion systems that have been proposed the past 50 years.

The general historical trend in fusion energy development has been continued striving for ever more powerful magnetic or electric or laser confinement of fusion plasmas. Unfortunately, each power increase often reveals new unexpected plasma behaviors or instabilities that make the plasmas even more difficult to manipulate or control. So, still more power and new surprises are continually revealed. As a result, the fusion research reactors that get all current government funding are now so large and expensive that many now doubt that nuclear fusion could ever compete with fossil fuels or renewable energies like solar-wind-tidal.

1.2. Prospects for Reducing Fusion Confinement Energy

Short range, ion-attracting SU(3) strong nuclear fields will not cause fusion of ions until separations between the ions are very small. When this occurs, very strong repulsion is occurring between ions as very strong compression is exerted on them from electric (E) or magnetic (B) or electromagnetic (EM) fields. One exemption from this occurs in hydrogen fusion inside the core of the Sun. Here E, B, or EM confining pressures are too low for the SU(3) strong force to initiate fusion of the hydrogen ions. But quark-mutating actions in SU(2) weak nuclear fields convert strongly repelling hydrogen ions into nearly neutral hydrogen- neutron pairs – which can then be easily fused by ion-attracting SU(3) nuclear fields. Deuterium ions are created by this fusion reaction, and these ions are then used in the subsequent chain of fusion reactions that emit charged particles and solar EM energy from the Sun, as hydrogen is transformed into helium 3 inside it. In this respect, Barrett [1] has shown the possibility of transforming ordinary U(1) electromagnetic fields into electromagnetic fields that have the same Lie Group symmetry as the SU(2) and SU(3) nuclear fields that participate in nuclear fusion inside the Sun. This has led the Author to explore the possibility of fusion ions experiencing less charge or less coulomb repulsion if embedded within or irradiated by SU(2) or SU(3) EM fields inside fusion reactors. For if such change in ion charge or repulsion would happen, less input power would be needed for fusion.

2.Specially Conditioned EM Radiation Fields

Ordinary electric (E) and magnetic (B) fields are generally described by vectors and Abelian algebra and they develop and propagate in space and time in accordance with the four Maxwell Equations that were formulated by Clerk Maxwell about 140 years ago. In the 1950"s Feynman and others made Maxwell's classical EM theory compatible with quantum theory and special relativity. This advance was Quantum Electrodynamics (QED) which precisely predicts interactions between radiation and matter. However, Maxwell's classical theory and equations have been used, essentially unchanged, in the design of all electric-magnetic-electromagnetic devices for more than over 140 years.

Maxwell EM theory describes electromagnetism in terms of: electric field strength (E), magnetic flux density (B), and current density (*J*). Electric and magnetic fields are generally described by vectors and Non-abelian algebra. They develop and propagate in accordance with the four Maxwell Equations shown in Table 1. These fields embody U(1) Lie Group symmetry and are sometimes defined mathematically in terms of non-physical potentials: the "magnetic vector potential " (*A*) and a "scalar potential" (ϕ).

Gauss' Law	$\nabla \bullet \boldsymbol{E} = \boldsymbol{J}_0$
Ampere's Law	$\frac{\partial \boldsymbol{E}}{\partial t} - \nabla \times \boldsymbol{B} - \boldsymbol{J} = 0$
Coulomb's Law	$\nabla \bullet \boldsymbol{B} = 0$
Faraday's Law	$\nabla \times \boldsymbol{E} + \frac{\partial \boldsymbol{B}}{\partial t} = 0$

$$E = -\frac{\partial A}{\partial t} - \nabla\phi, \quad B = \nabla \times A$$

Table 1 Maxwell Equations for Ordinary U(1) EM Fields.

Barrett [1] has used group and gauge theory and topology to develop SU(2) EM field theory, and Table 2 shows the Expanded Maxwell Equations associated with SU(2) EM fields. Ordinary U(1: J, E and B vector fields are transformed into SU(2) : J, E and B tensor fields that are described by Non-Abelian algebra. Also, the magnetic vector potential (which is non-physical in U(1) electro-magnetism) is transformed into a physical A tensor field. And SU(2) EM, Maxwell Equations have additional terms that involve A tensor fields interacting with E and B tensor fields in various ways. SU(2) EM fields differ from SU(2) nuclear fields in that they act over much greater than SU(2) nuclear fields act over. Also, SU(2) nuclear field interactions are mediated by 3 bosons – not by the single boson (photon) that mediates SU(2) EM field interactions. But the three (A, E, B) field couplings shown in the SU(2) Maxwell Equations may be somewhat analogous to the actions of the three (W +, W - Z) bosons in SU(2) nuclear fields.

$\nabla \bullet \boldsymbol{E} = \boldsymbol{J}_0 - iq(\boldsymbol{A} \bullet \boldsymbol{E} - \boldsymbol{E} \bullet \boldsymbol{A})$
$\frac{\partial \boldsymbol{E}}{\partial t} - \nabla \times \boldsymbol{B} - \boldsymbol{J} + iq[A_0, \boldsymbol{E}] - iq(\boldsymbol{A} \times \boldsymbol{B} - \boldsymbol{B} \times \boldsymbol{A}) = 0$
$\nabla \bullet \boldsymbol{B} + iq(\boldsymbol{A} \bullet \boldsymbol{B} - \boldsymbol{B} \bullet \boldsymbol{A}) = 0$
$\nabla \times \boldsymbol{E} + \frac{\partial \boldsymbol{B}}{\partial t} + iq[A_0, \boldsymbol{B}] = iq(\boldsymbol{A} \times \boldsymbol{E} - \boldsymbol{E} \times \boldsymbol{A}) = 0$

Table 2 Maxwell Tensor Equations for SU(2) EM Fields

Barrett has made a small start towards SU (3) EM field theory. SU(3) EM theory is, of course, more complex than SU(2) EM theory - with higher order tensors and requiring much more work. For example, more and higher order A, E, B couplings will be associated with SU(3) EM than with SU(2) EM - just as more complex action is associated with the 6 boson mediating action of the SU(3) strong force than the 3 boson action of the SU(2) weak force. So, because of the much lower state of development of SU(3) EM theory, this paper will focus on SU(2) EM radiation possibilities for reducing input field energy for nuclear fusion.

Table 2 shows that SU(2) electromagnetism involves couplings between A and E and A and B fields that do not occur in ordinary electromagnetism. Thus, forces exerted on like- charge ions and like-charge electrons embedded in an ordinary U(1) EM fields can be different if embedded in an SU(2) field. An example of this is Table 3 which is taken from [1]. Shown is the Lorentz force exerted on moving charged particles in presence of SU(2) E and B tensor fields , compared to Lorentz force exerted on the particles by ordinary U(1) E and B vector fields. Here, U(1) Lorentz Force is described by E and B fields and A vector potentials; while SU(2) Lorentz force is described by E and B and A tensor fields. Differences in numbers of Lorentz Force terms and the term's different vector and tensor natures imply that an SU(2) EM field can exert Lorentz force of different intensity and direction on moving plasma ions and electrons, compared to that exerted on the plasma ions and electrons by a U(1) EM field

U(1) Lorentz Force	$\mathscr{F} = e\mathbf{E} + ev \times \mathbf{B} = e\left(-\frac{\partial \mathbf{A}}{\partial t} - \nabla\phi\right) + ev \times \left((\nabla \times \mathbf{A})\right)$
SU(2) Lorentz Force	$\mathscr{F} = e\mathbf{E} + ev \times \mathbf{B} = e\left(-(\nabla \times \mathbf{A}) - \frac{\partial \mathbf{A}}{\partial t} - \nabla\phi\right) + ev \times \left((\nabla \times \mathbf{A}) - \frac{\partial \mathbf{A}}{\partial t} - \nabla\phi\right)$

Table 3 Lorentz Forces Acting on Moving Charged Particles

Barrett has identified two ways of generating SU(2) EM field energy. One, that is described in Barrett [2] is emission of SU(2) EM radiation by driving alternating current through toroidal coils at a resonant frequency determined by the specific toroid geometry. Figure 1 shows an A tensor field pattern of a transmitting toroid. It is composed of two U(1) A vector potential fields, ϕ1 and ϕ2 overlapping in polarity across the hole of the obstruction of the toroid. Also shown are the different resonant Frequencies (when phase is such that ϕ1 - ϕ2 is maximal at each of the possible harmonic frequencies) that can occur for a toroid.

At every resonant frequency, the alternating difference in the overlapping U(1) vector potential fields is maximized for a given toroid geometry and alternating current frequency. This results in a concomitant maximization of an SU(2) tensor potential field.

Toroid testing by Hathaway and Froning at Hathaway Laboratories in Toronto Canada is described in [3]. There was not enough testing to conclusively prove that SU(2) EM tensor fields were created at the resonant frequencies that were measured by available instrumentation. But resonant frequencies (which were measured between 400 KHz-110 MHz) were in good agreement with resonant frequencies predicted from Barrett's theoretical work. And, at all resonant frequencies, relatively intense and very focused magnetic fields formed over the upper surfaces and below the lower surfaces of numerous toroids that were tested.

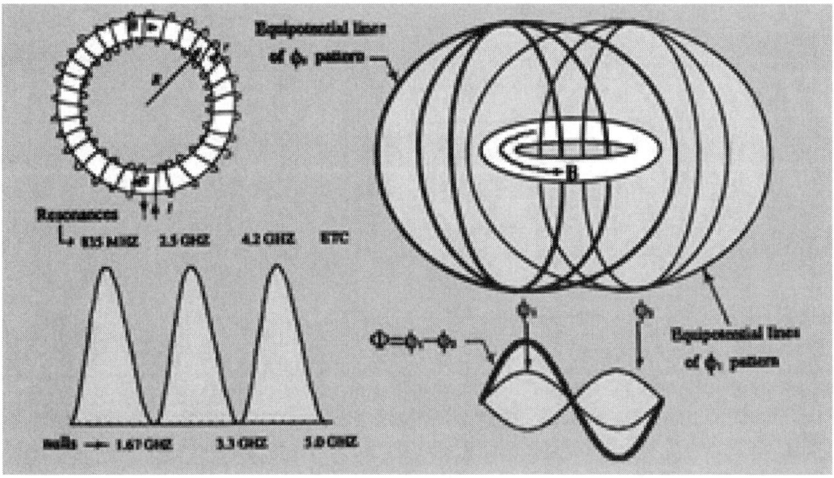

Figure. 1 U(1) A vector potential patterns surround transmitting toroids and result in maximum SU(2) A tensor field strength at the nodes associated with three different resonant frequencies.

Barrett has identified another way of generating SU(2) EM field energy. This special modulation is taken from pages 46 and 61 of [4] and is shown in Fig. 2. It uses a waveguide paradigm which has oscillating U(1) EM wave energy being transformed into SU(2) EM wave energy by phase and polarization modulation. Fig. 2 shows an adiabatic system where oscillating wave energy enters from the left and divides into 3 parts; is phase and polarization-modulated; and exits from the right.

Part of the input wave energy is unchanged; another part provides phase modulation (∂A/∂t) and combines with another part that has passed through a "polarization rotator". This results in two orthogonally polarized waveforms (with one waveform being the unchanged fraction of input wave energy. The phase modulated and polarization modulated waveforms are superimposed at an output and combined into a single beam of SU(2) EM radiation of continually varying polarization. The lower right of Fig. 2 is polarization-modulated EM radiation swiftly sweeping through many polarizations (between linear, elliptical and circular ones) during one cycle of polarization modulation.

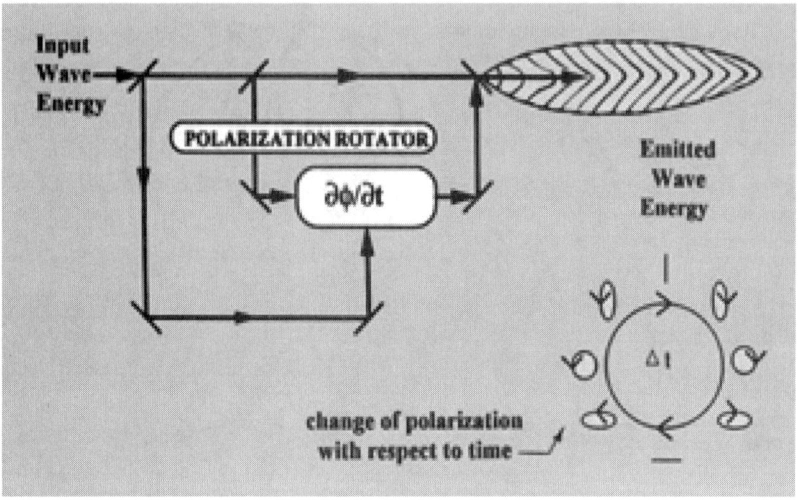

Figure 2 Generation of Polarization-Modulated EM Radiation

Fig. 3 from [1] shows the swift, variegated change in E field amplitude and direction during one cycle of polarization- modulation. Similar B field change occurs at 90 degrees to the E field change. Such rapid variation in field intensity cannot be approached with fixed linear or circular or elliptical polarization. Higher angular dynamics of such a polarization-modulated EM beam (as compared to lower angular dynamics of ordinarily polarized EM beams) exert different forces and moments on particles like electrons or fusion ions. However, ability of such high angular dynamics to modify charge distributions or to modify repulsion-attraction between charged particles is not yet known.

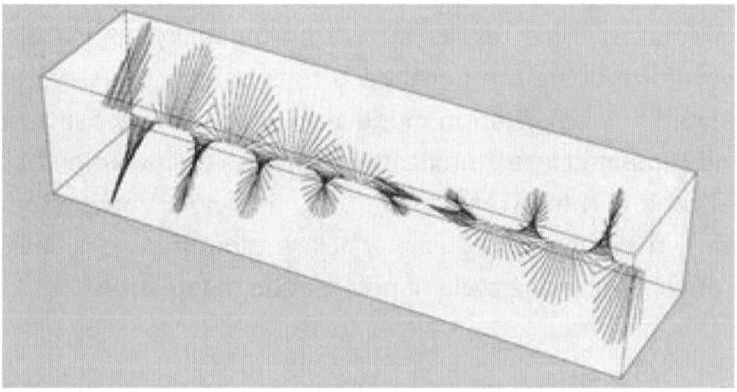

Figure 3. Rapid change in an E field's amplitude and direction during one polarization-modulation cycle by an SU(2) EM Beam.

Ion-attracting SU(3) nuclear fields consummate every nuclear fusion reaction. So, ion-attracting SU(3) EM fields would be expected to show the most promise for reducing coulomb repulsion of ions and ,thus, input electrical power for fusion. Unfortunately, only SU(2) EM theory is sufficiently advanced for making SU(2) EM field computations and hardware testing. And this has enabled preliminary tests with encouraging results. However, a start on SU(3) EM field formulation has been begun. For example, Fig. 4 from Barrett [4] shows an additional phase modulation of polarization-modulated EM wave energy that may it approximate the form of SU(3) EM wave energy.

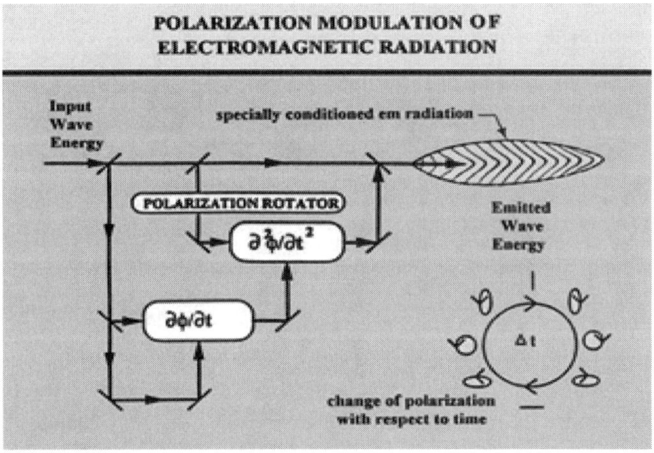

Figure 4. Added phase modulation that creates SU(2) EM wave energy that may closely approach wave energy of SU(3) EM form

Developing SU(2) EM field theory to its present state has required much labor by a single individual (Barrett) and there are still SU(2) EM issues to be resolved. So, development of even more complex SU(3) EM field theory will surely require even greater effort. SU(3) EM Experimental challenges will also be encountered. For example, phase modulations and polarization rotations needed to emit the SU(2) polarization-modulated radiation shown in Fig. 1, are developable with undulator and oscillators that are state-of-the-art in performance and design. But development of the much higher-order phase modulations indicated in Fig. 4 may will require undulator and oscillator bandwidths that are achievable only with high performance free electron lasers at certain Universities and National Laboratories.

3. Use of SU(2) and SU(3) EM Fields for Fusion

Reducing input power for fusion by use of SU(2) EM fields has been explored in most depth for "Inertial Electrostatic Confinement" (IEC) systems. The specific one considered was that one pioneered by George Miley at the University of Illinois. Fig 5. shows an IEC system in operation at the University of Illinois. This system is seen to include multiple ion beams – composed of swiftly-moving streams of ions that are converging to the center of a reactor. These ion beams are emitted from "ion guns" mounted on the periphery of the reactor, and their inward motion is accelerated by a negatively-charged electrode – a spherical wire-woven "grid" allows positively-charged ions to pass through. And inside this grid, ions converge and criss-cross as hundreds to thousands of fusions/sec. cause the central glow

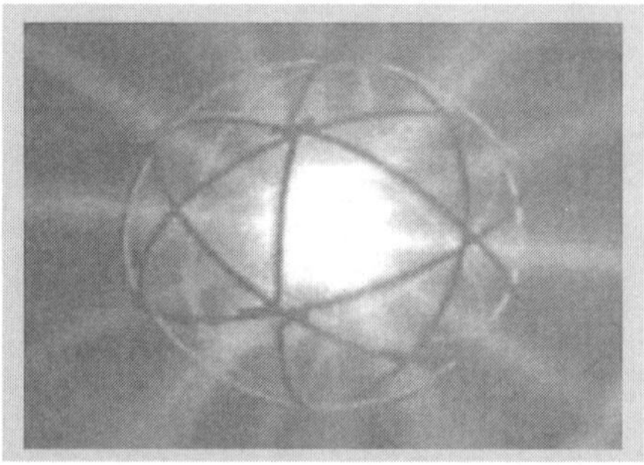

Figure 5. IEC fusion reactor in operation at University of Illinois

Figure 6. Ion gun embodied in IEC Reactor at University of Illinois

Fig. 7 from [5] shows an IEC ion gun. It contains flowing gaseous deuterium fuel; an electric power system; a "helicon" radio-frequency (RF) antenna whose electromagnetic energy heats gaseous fusion fuel to form fusion ions. The major

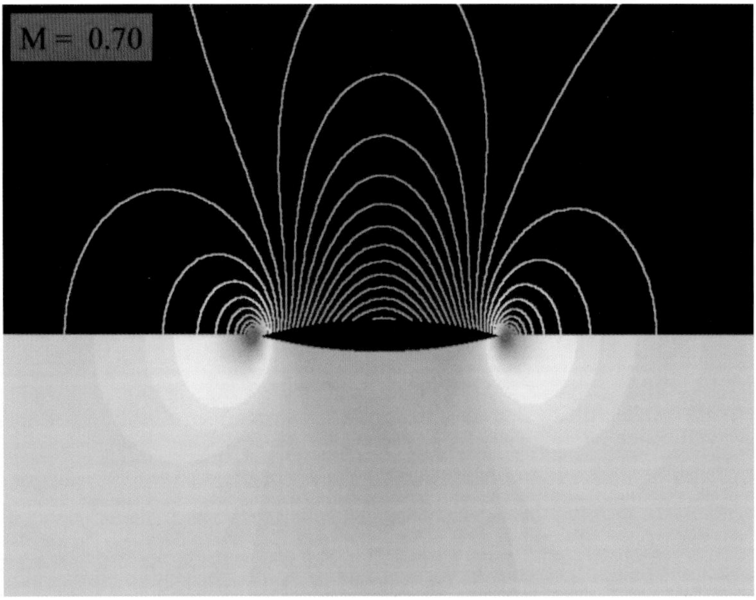

Craft in black moving at 0.7c to left; slight vacuum disturbance from EM field generation

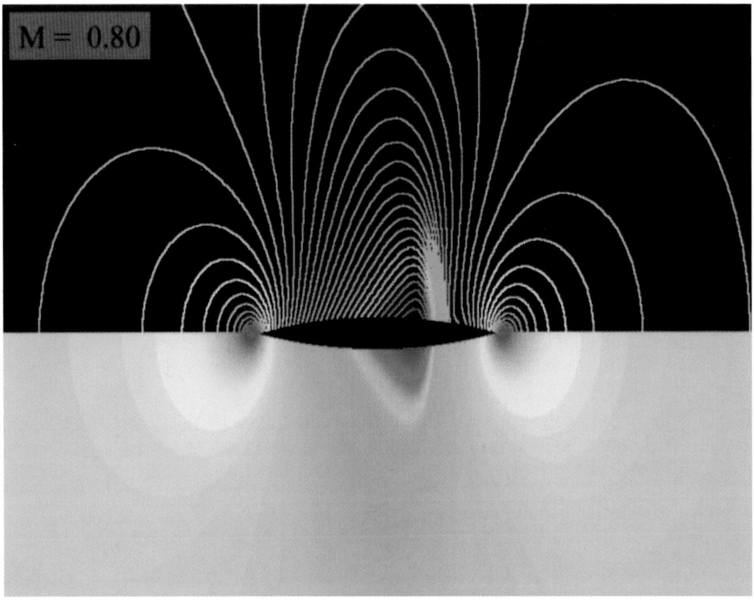

EM fields projected fore and aft; beginning of shock wave; craft at 80% of lightspeed

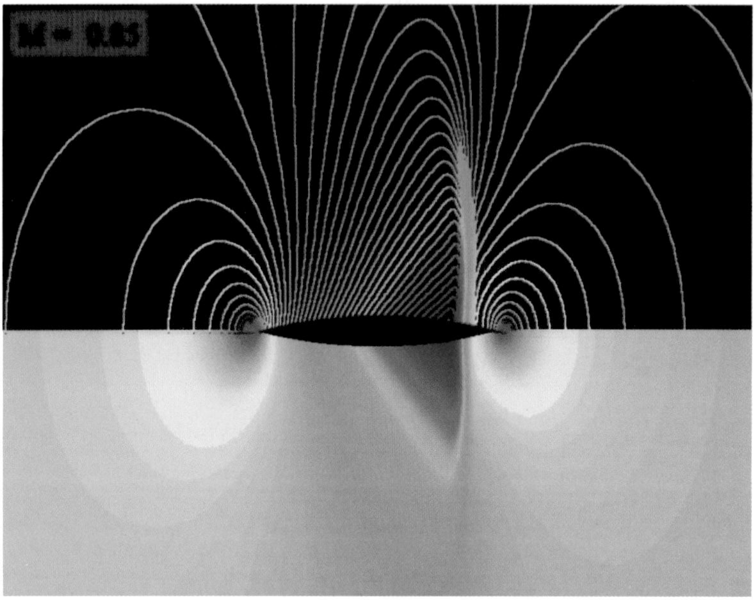

Craft at 85% of lightspeed (0.85c); Reader can flip pages to create a sense of motion

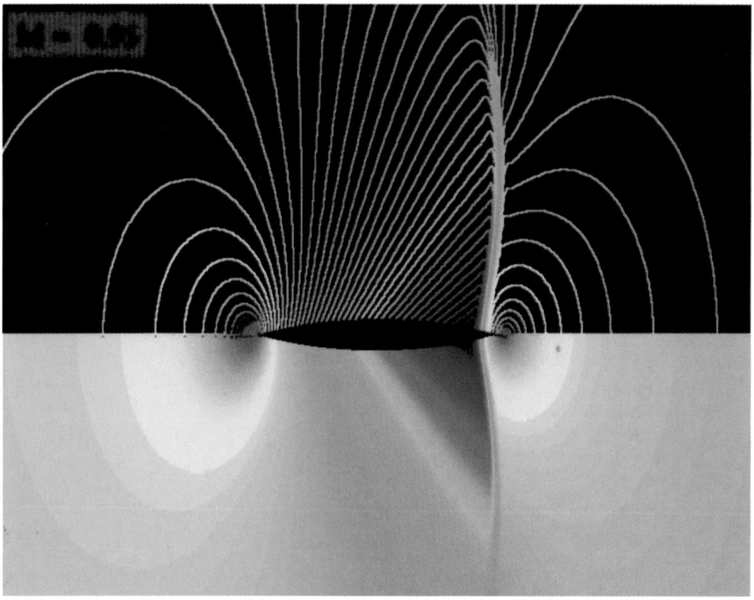

Craft at 95% of lightspeed (0.95c); EM fields projected fore and aft

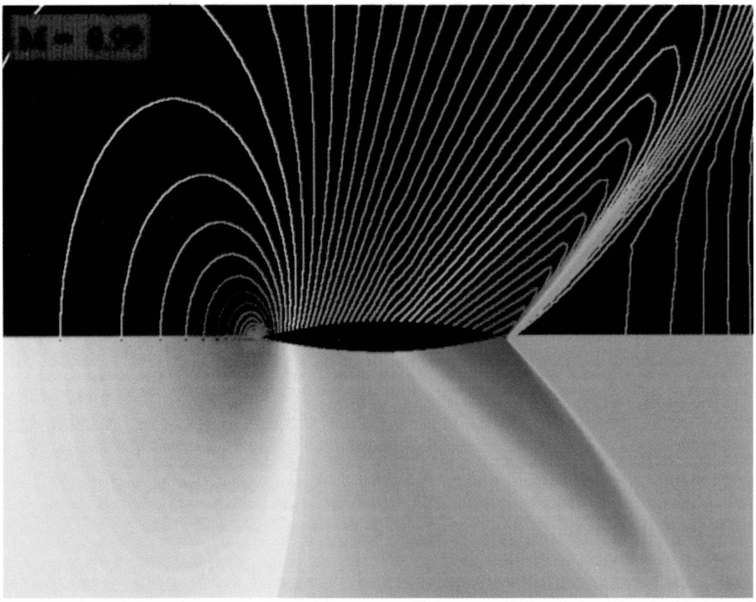

EM projected disturbance only in front (left) of craft; Craft at 99% of lightspeed (0.99c)

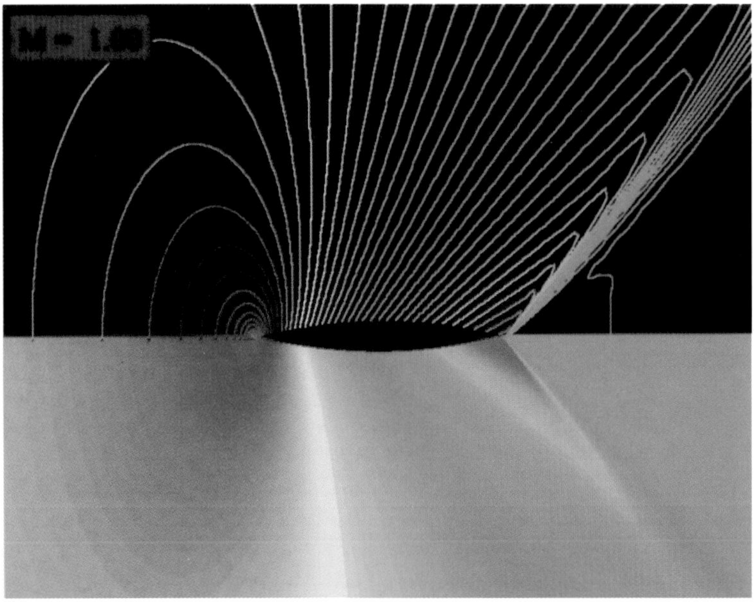

Craft at lightspeed 1.0c which is 100% of the speed of light, with specially conditioned EM disturbance projecting ahead of the craft (on left leading edge)

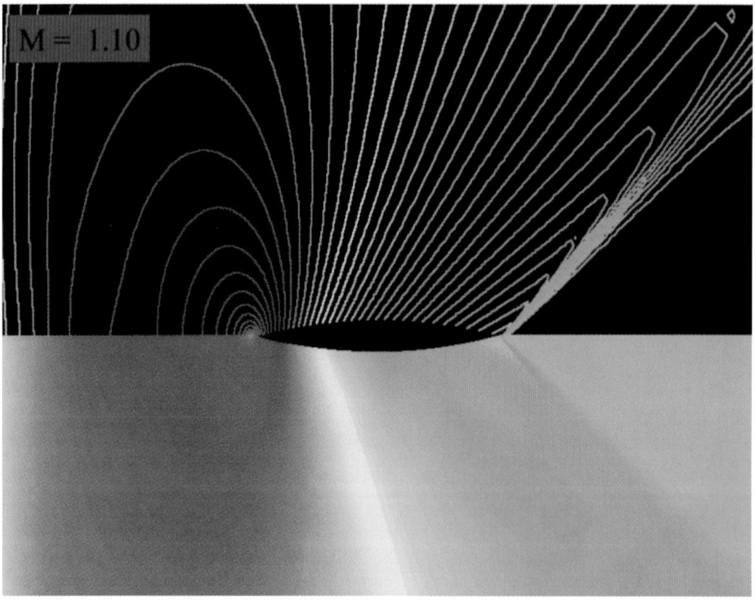

Craft at lightspeed 1.1c which is 110% of the speed of light

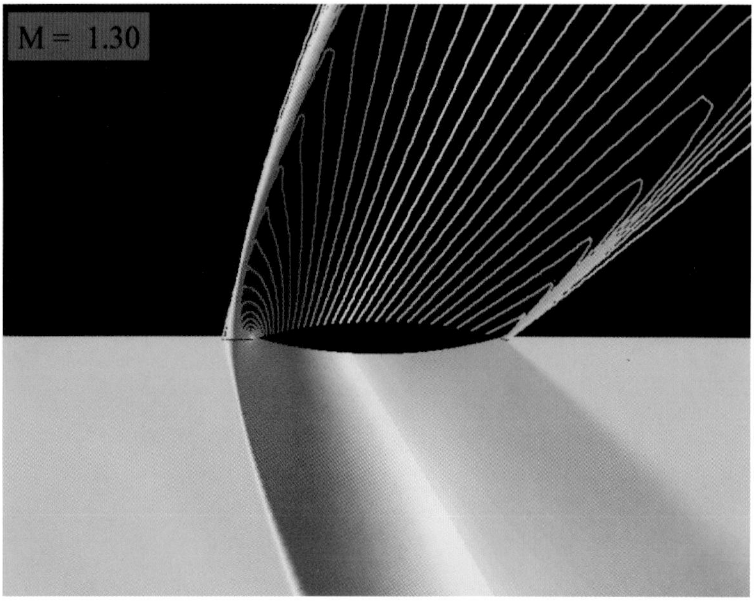

Craft exceeds lightspeed at 1.3c; Note quantum vacuum pressure gradients fore and aft

Craft exceeds lightspeed at 150% of c; pressure gradients are more inclined toward rear

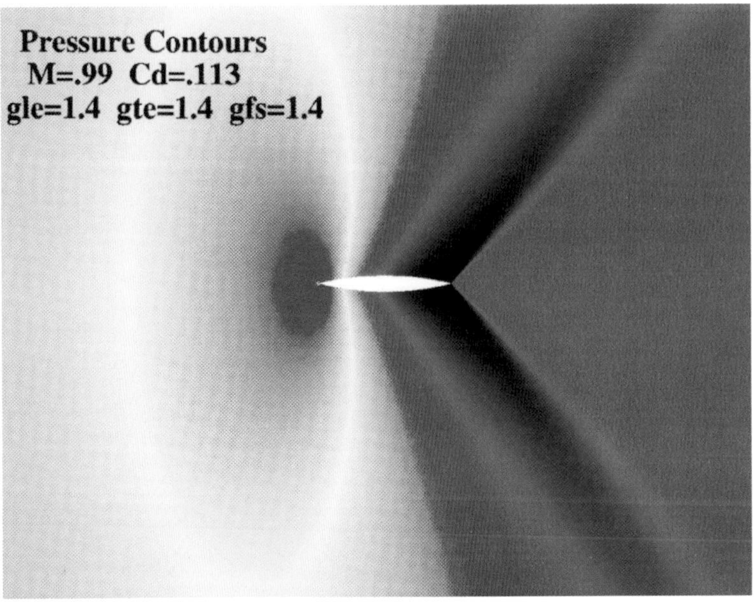

Craft 0.99% of c; Different computational fluid dynamics (CFD); undisturbed medium

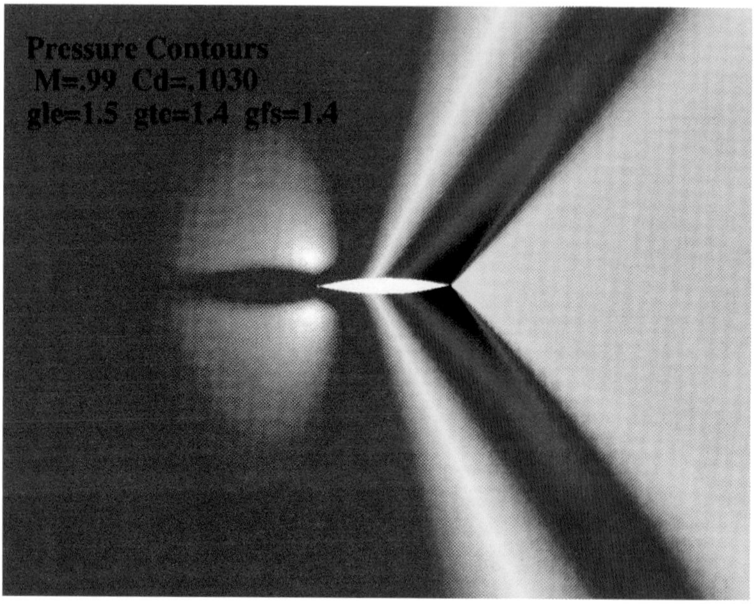

Craft 0.99% of c, which can be Mach 0.99 as well; See AIAA 95-2368 (p. 318 of this book)

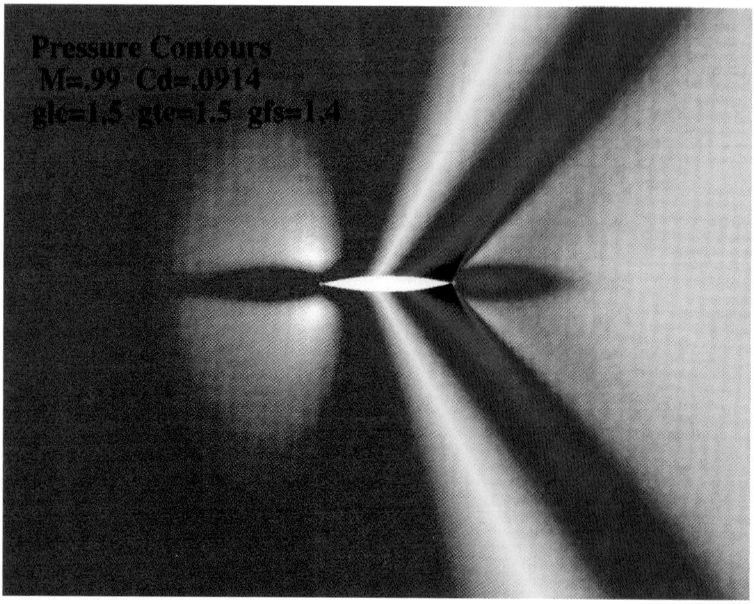

Craft 0.99c; Cd reduced to 0.0914, gle and gte increase to 1.5; fore and aft EM effect; Drag reduction and possible impulsion can be from perturbing fluid or vacuum fields

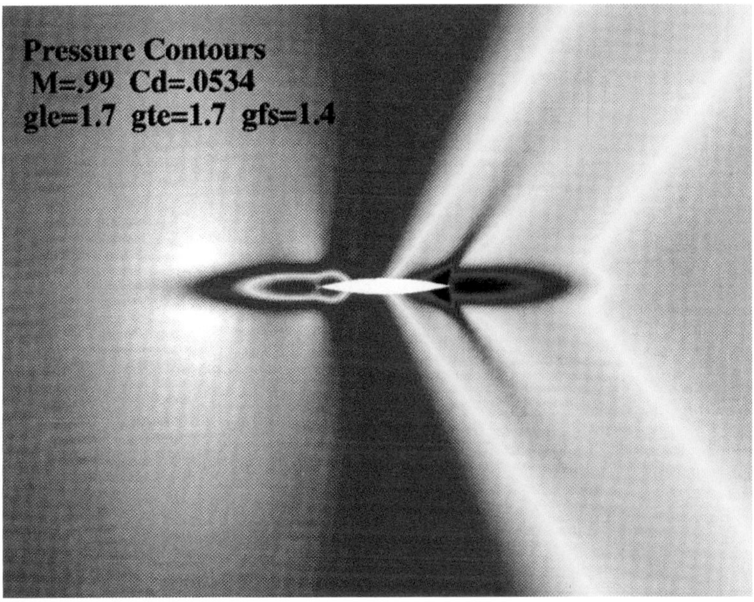

Craft 0.99c; Cd reduced to 0.0534, gle and gte increase to 1.7; fore and aft EM effect

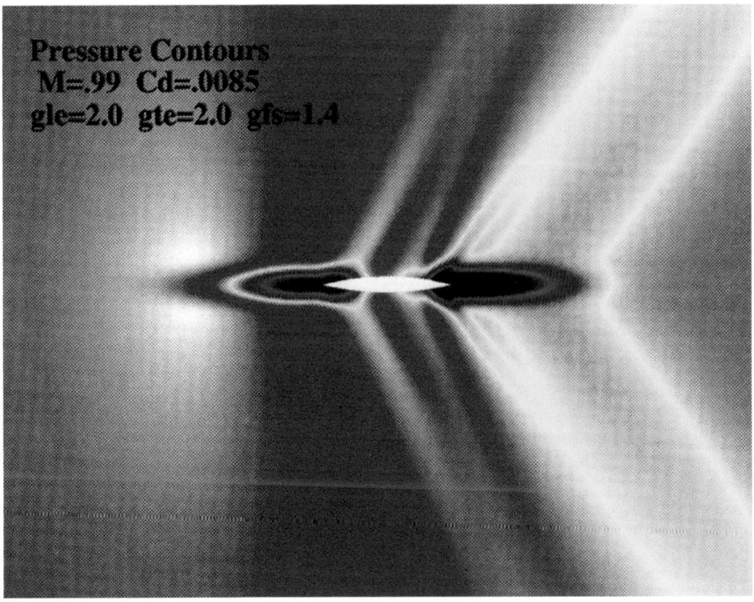

Craft 0.99c; Cd reduced to 0.0085, gle and gte increase to 2.0; fore and aft EM increase

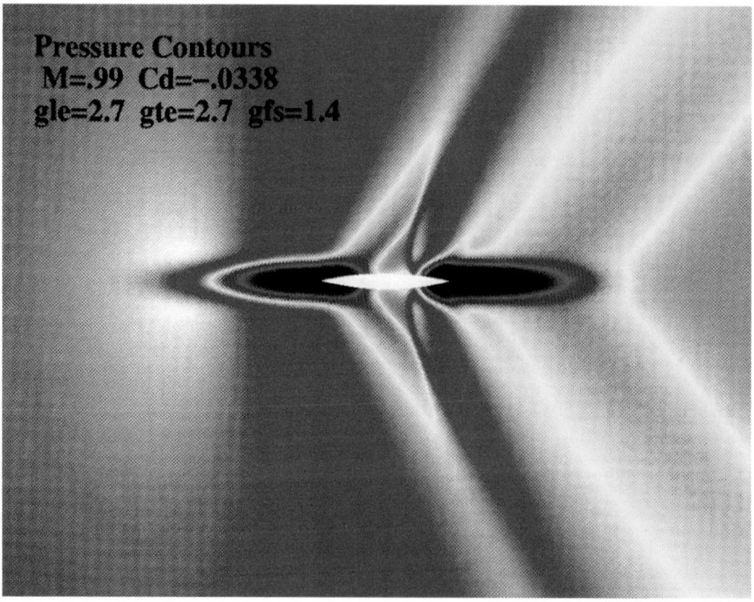

Craft 0.99c; Cd reduced to -0.0338, gle and gte increase to 2.7; fore and aft EM increase

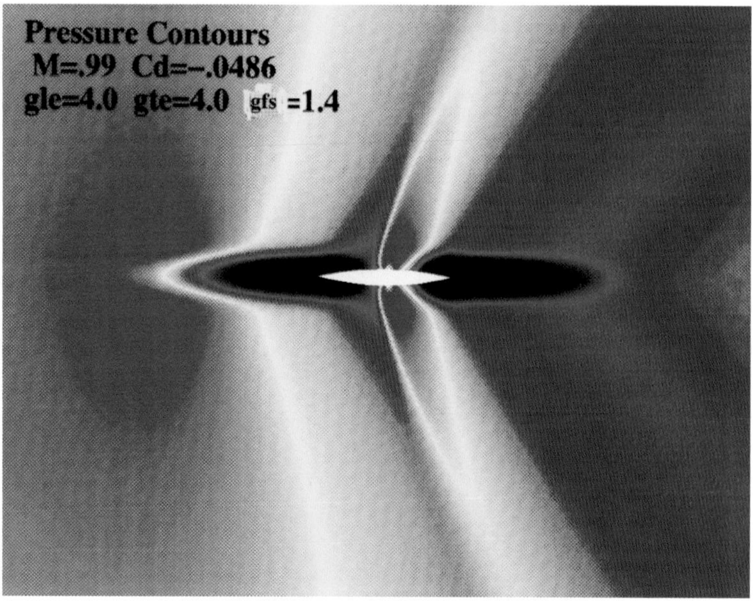

Craft 0.99c; Cd, gle, gte simulate EM effect fore, aft, diminishing $\mu_o\varepsilon_o$ in black

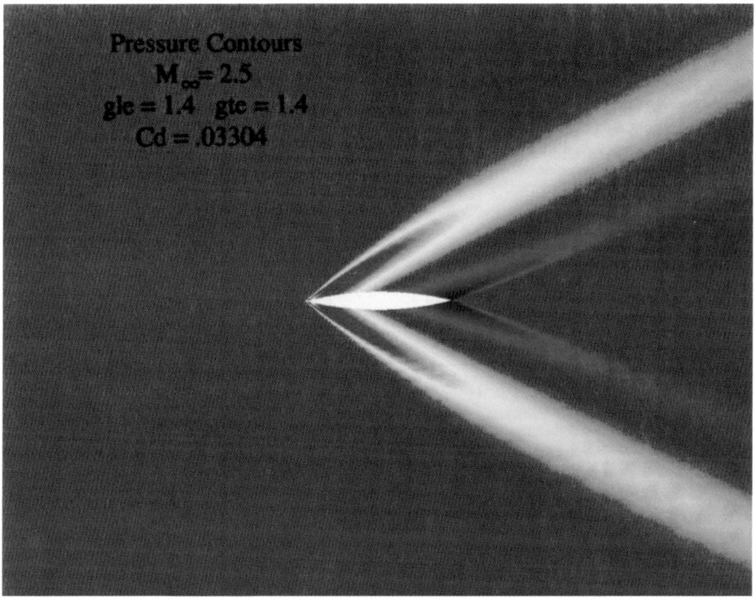

Craft 2.5c; Cd reduced to 0.03304, gle and gte at 1.4; No EM projected; large resistance of vacuum in forefront (at leading edge on left of craft)

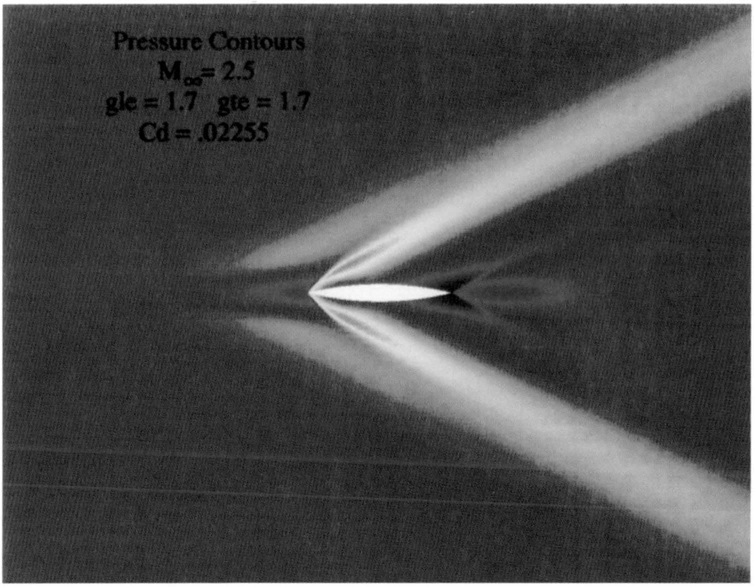

Craft 2.5c; Trends in perturbation of fluid fields are similar to trends for vacuum fields

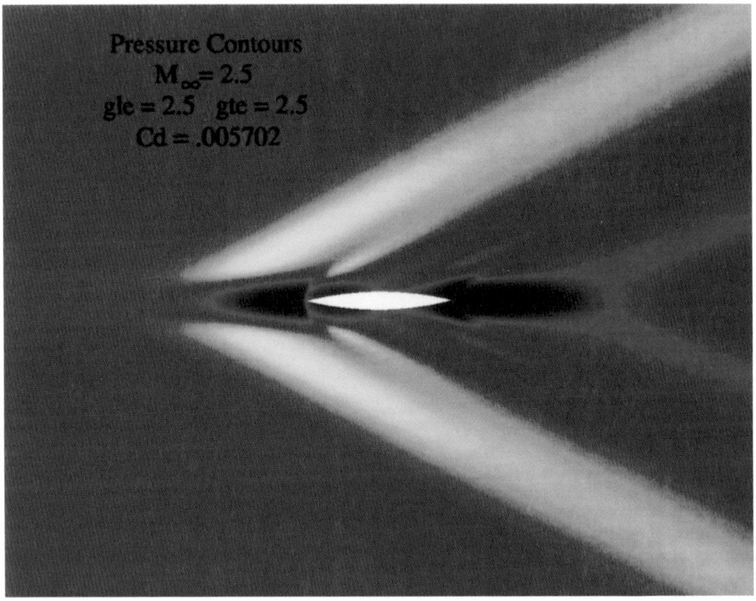

Craft 2.5c; Cd reduced to 0.0057, gle and gte at 2.5; fore and aft EM projection working

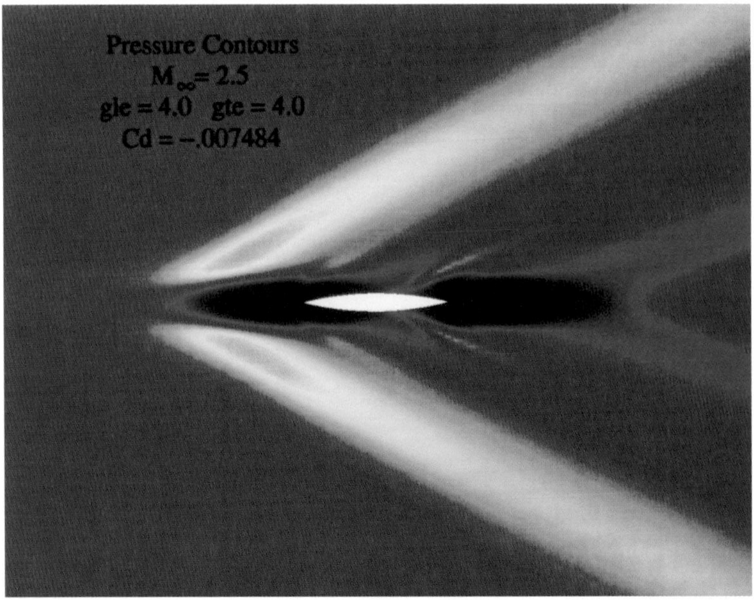

Craft 2.5c; Cd reduced to 0.00748, gle and gte at 4.0; fore and aft EM projection bigger

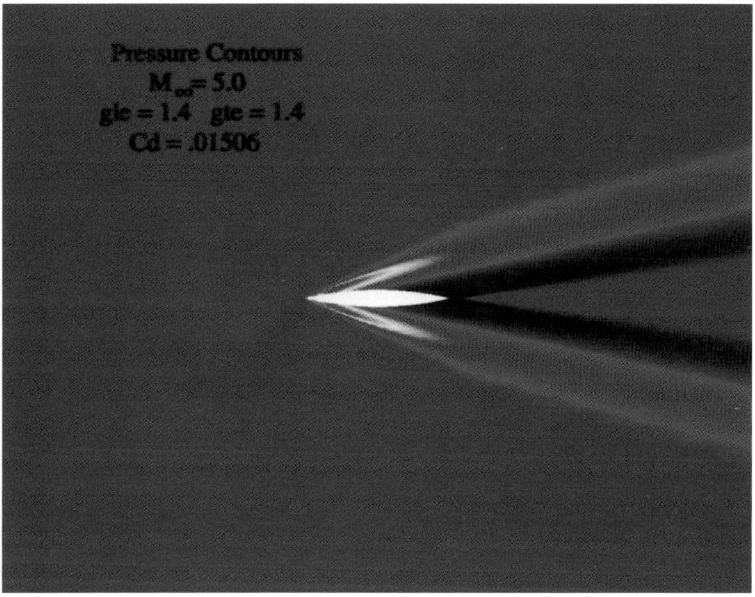

Craft 5.0c; Cd reduced to 0.015, gle and gte at 1.4; No EM projected; large resistance of vacuum in fore (at left, leading edge on left of craft)

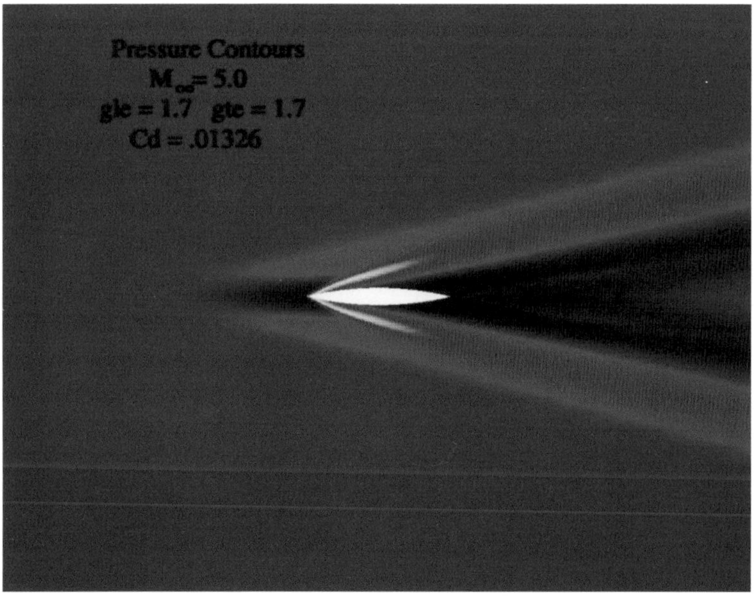

Craft at 5.0c; apparent beneficial effects can be seen from perturbation of fore medium

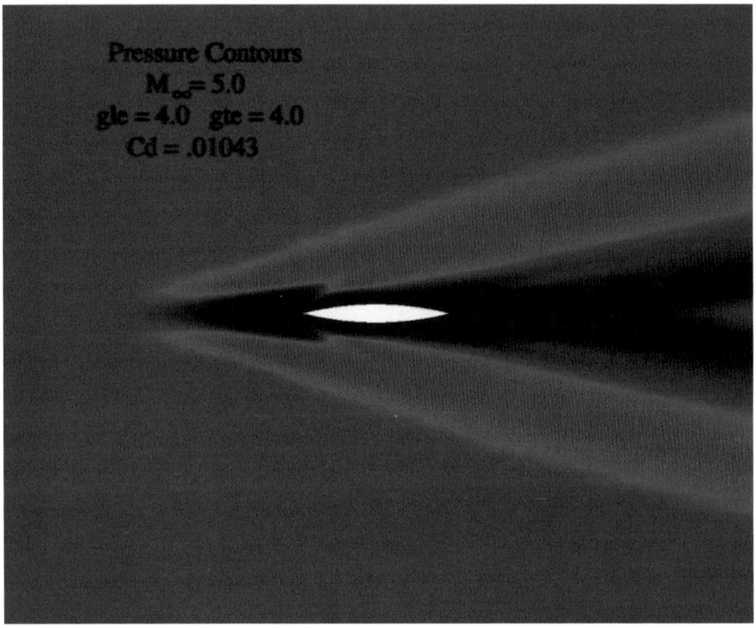

Craft at 5.0c; EM disturbance beneficial effects may include inertia reduction

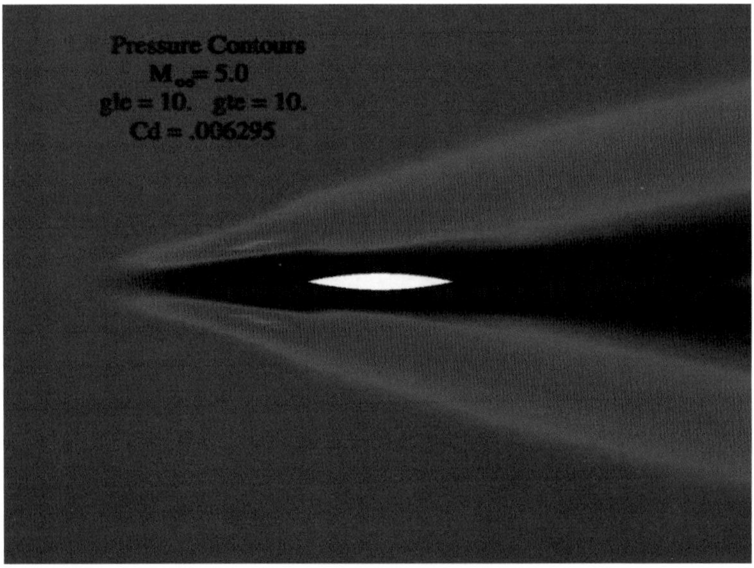

Craft at 5.0c; strong EM interaction with the vacuum creates impulsion from favorable distribution of zero-point radiation pressures surrounding the craft

ion gun modifications would be emission of polarization-modulated SU(2) RF radiation instead of fixed-polarization U(1) RF energy that is currently emitted. Modifications would be replacing the existing RF antenna elements with orthogonally-polarized antenna elements; and driving the antenna elements with state- of-the-art phase modulators in the high (10^{-12} to 10^{-15}) Hz range.

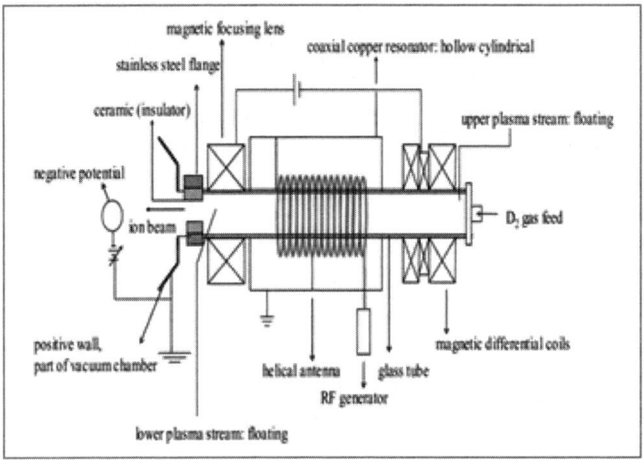

Figure 7. Schematic of one of the Ion guns of an IEC system

It is expected that emitted SU(2) or SU(3) wave energy would ionize fusion gases in ion guns with about the same efficiency as currently emitted RF energy does. But field energy intensity falls-off with distance, and it is not yet known how much this reduction would effect SU(2) or SU(3) RF field ability to modify ion charge or repulsion near the reactor center. So, more wave energy than that needed for fusion gas ionization may be needed.

The most favorable location for SU(2) or SU(3) field energy deposition may be inside the spherical, negatively-charged IEC wire grid – where ions are being forced close together for fusion. Irradiating ions near the center of the IEC reactor would require as many as four narrow EM beams that are equal or smaller in width than the IEC ion beams. The converging SU(2) or SU(3) EM beams would be mounted on the reactor periphery like the IEC ion beams are. They would also probably operate at high frequency to enable EM beams that are equal or narrower in width than the IEC ion beams – to eliminate large enlargement of the central grid so that both ion and EM beams can pass through.

Many antenna concepts for irradiating fusion ions with SU(2) or SU(3) radiation in the central region would have to be examined. One antenna concept would be an "undulator" like the one used in the free electron laser shown in Fig, 8. Fig. 8 shows dipole magnets of an undulator straddling the axis of an electron or ion beam. This beam axis could become the axis of a very narrow beam of SU(2) or SU(3) EM radiation that irradiates a reactor region. Undulators can radiate EM radiation from either electron or ion beams. So, one possibility would be investigating the modification of an IEC ion gun so that its emitted ions would enter an undulator. Such a modified gun would include an aft ion beam would include an aft ion beam section and a forward undulator section. Adding an undulator section to an existing ion gun would, of course, not be the best possible combination of ion and EM beams. But it might be least expensive for early testing.

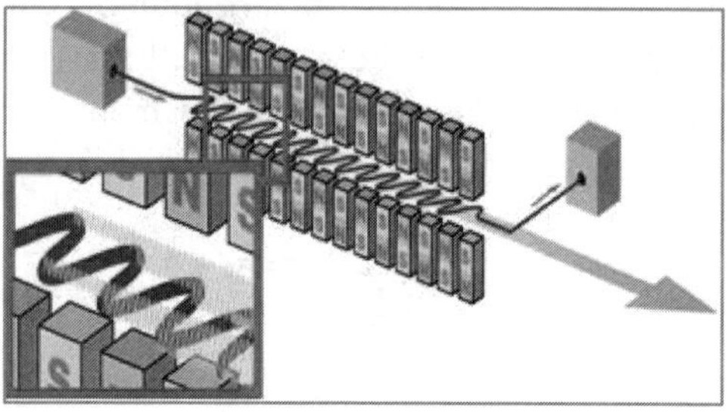

FIGURE 8. VERY PRECISE AND HIGHLY MODULATABLE ELECTROMAGNETIC BEAM FORMED BY ELECTRON BEAM PASSING THROUGH DIPOLE MAGNET.

Reducing input power for several other fusion systems have been very briefly explored. One is an "EXL" variant of an IEC system by Robert Bussard [6]. This EXL variant is shown on the right in Fig. 9, It replaces the negatively-charged grid with a spherical distribution of electrons that repel (push) inward streaming ions from an ion gun – compressing them into the reactor core where about 10^9 fusions/s have been achieved. The electron distribution is confined by inward-pushing magnetic field pressures from pulsed current flowing through the unique geometry of Bussard's specially configured toroid coils in Fig. 10.

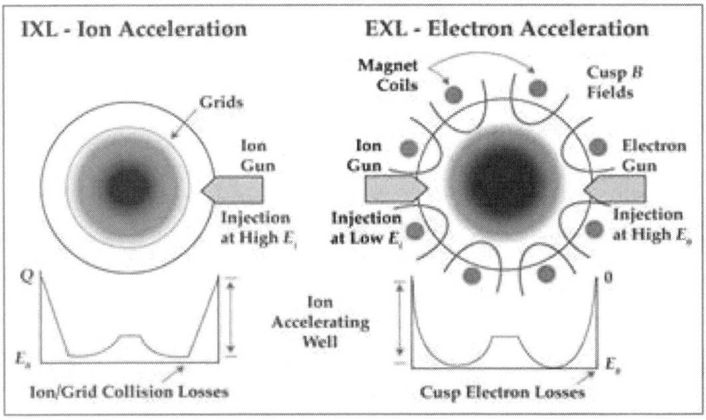

Figure. 9. Comparison of IXL with Robert Bussard's EXL

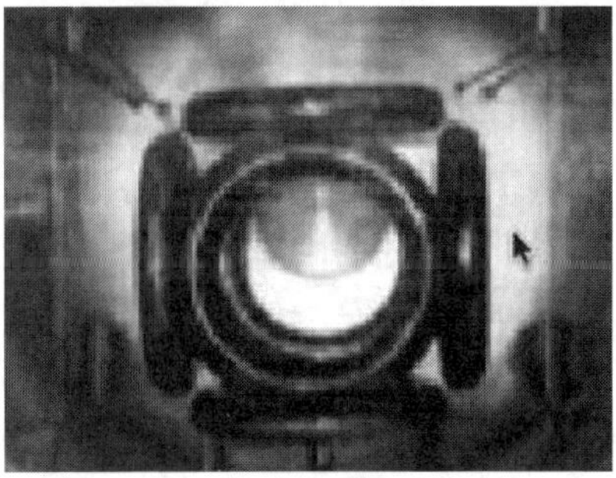

Figure 10. Toroidal geometry for EXL magnetic field generation

Details of Bussard's toroidal coil design and its modulation are not known. But it is conceivable that Barrett's idea of emitting SU(2) magnetic field energy at resonant frequencies (as shown in Fig. 1) might improve electron-confining efficiency of Bussard's confining magnetic field. Like Miley's gridded IEC system (called IXL in Fig. 9) Bussard's EXL system uses an ion gun and accomplishes ion fusion in the reactor center. Thus, irradiating this region with Narrow SU(2) EM beams might be a possibility for Bussard's EXL system – as well as for Miley's IXL one.

Reducing input power for one other fusion system has also been considered. This system is referred to as "Focus Fusion" by its developer, Eric Lerner of Lawrenceville Plasma Physics Inc. It is a variant of a device called "Dense Plasma Focus" which was studied by the US Air Force for fusion propulsion in the 1980's. Fig. 11 from [7] shows one of the DPF fusion concepts developed by the Air Force.

As in an ion gun, gaseous fusion fuel is ionized by a strong discharge and the formed plasma sheath is accelerated in a chamber. At the chamber's open end the sheath reverse direction and enormously intense magnetic fields form in a narrow pinch region, as shown in Fig. 12, where fusion occurs.

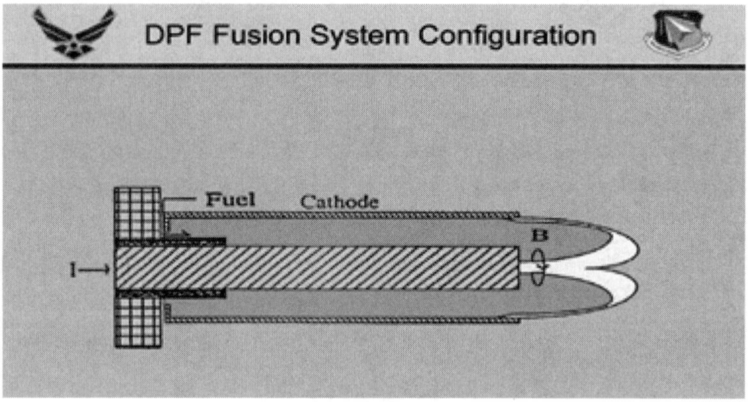

Figure 11. Dense Plasma Focus fusion propulsion for flight

The Lawrenceville Plasma Physics Inc. website (https://lppfusion.com) describes significant advancement in DPF state-of-art with their Focus Fusion (FF) designs – with plasmoid densities of $8x10^{19}$ ions per cm^3 quoted. As with Bussard's EXL system, detailed FF information is needed before any SU(2) ; SU(3) EM benefits are claimed. But, it is known that accelerating, turning and compressing of DPF or FF fusion plasmas are strongly influenced by Lorentz Forces - which would be different for SU(2) or SU(3) A, E and B fields.

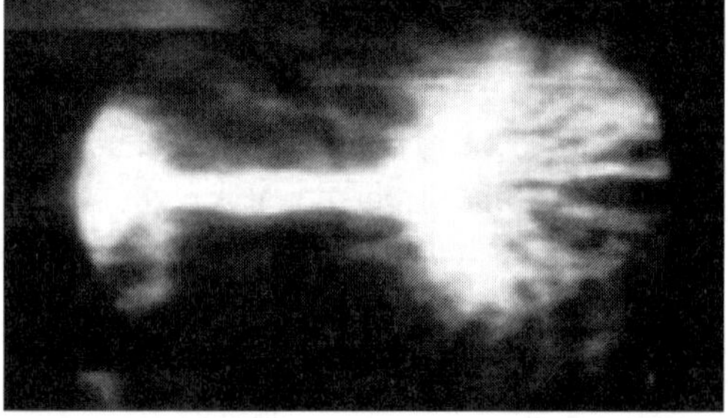

Figure 12. Dense Plasma Focus (DPF) Fusion Tests by Nardie et al. at Stevens Institute in the 1980's

4. Ideas on Theoretical and Experimental Work

4.1. Theoretical SU(2) and SU(3) EM Field Development

As previously mentioned, Dr. Terence Barrett (though his own consulting company "BSEI" has developed SU(2) EM radiation field theory sufficiently for numerical studies to be performed and SU(2) EM field generators to be designed. This has enabled preliminary experiments to be performed with encouraging results. But some unresolved SU(2) EM field issues remain. Furthermore, more complex SU(3)EM field theory - which may be even more promising for reducing input energy for fusion – will require even more effort than a lone person like Barrett could ever accomplish without academic institution support. It is, thus, believed important that faculty and students of academic institutions should become involved in advancing SU(2) EM field theory and that academia also become involved in exploring the possibility of developing SU(3) EM field theory as well.

It is estimated that the initial scope of this theoretical SU(2) and SU(3) work would require a level of effort of about 3 years at a reasonably modest cost with existing university faculty, graduate students and computational facilities – together with Barrett consulting help. This would: (a) enable code development for numerical modeling of the SU(2) electromagnetic processes associated with the fluid dynamics and plasma dynamics involved in typical power and propulsion systems; and (b) lay out the beginnings of SU(3) EM field theory – with Barrett's help.

4.2. SU(2) EM and SU(3) EM Field Experimental Work

Two testing options are possible for relatively quick and inexpensive experiments to confirm or refute possibility for SU(2) EM fields modifying ion coulomb repulsion One experimental example would be use of an available IEC reactor and ion gun system at one of several universities that possess IEC systems and modification of the helicon RF antenna of one of the IEC guns to emit SU(2) EM wave energy. Using measured fusion intensity with fixed-polarization RF discharge as reference, any change (or lack of change) in fusion intensity with a polarization-modulated RF discharge could be directly determined.

Another test option could use relatively inexpensive plasmas formed by RF discharges in gases like Argon or Xenon. Such tests can be run in plasma chambers that, like the University of Illinois IEC ion guns, use helicon RF antenna systems for gas ionization. Fig. 1 shows such a plasma chamber. It is currently used for ion thruster research at The Australian National University (ANU).

Figure. 13. Plasma chamber with a Helicon RF Antenna system

Just as small differences in fusions for U(1) and SU(2) RF discharges must be detected in IEC reactors, so small differences in ion and electron behaviors would have to be detected in plasma chambers for U(1) and SU(2) RF discharges. This appears possible in the ANU chamber. For it incorporates an innovative "movable energy analyzer" that allows plasma properties to be accurately and rapidly mapped throughout the entire chamber.

As with theory work, a 3 year level of effort is recommended for experimental work. It must be admitted that this theoretical and experimental work is exploratory and hence of somewhat "high risk" in that specific accomplishments or success can't be assured. But, at the very least, advances in EM field theory will occur, new plasma physics be learned, and the experiments would significantly extend plasma physics state-of-the-art.

5. Conclusions

A possibility for reducing input power for fusion is revealed in work by T.W. Barrett - who has derived EM fields with the same SU(2) and SU(3) Lie Group symmetry as the SU(2) weak and SU(3) strong nuclear fields that bring about fusion in the Sun. And it is suggested that the unique ***A, E,*** and ***B*** tensor field couplings embodied within SU(2) and SU(3) electromagnetism could conceivably modify ion charge and repulsive force much like SU(2) weak and SU(3) strong fields do in achieving fusion despite modest ion confinement energies in the Sun's core.

Reducing fusion input power for IEC fusion systems has been explored in 2 ways. One is creating SU(2) or SU(3) radiation within the RF discharges that create ions in IEC ion guns. The other is irradiating fusion ions in the central fusion region with very narrow SU(2) or SU(3) EM beams. Finally, SU(3) EM is the most promising – but not yet sufficiently developed for experiments. Theoretical SU(2) and SU(3) EM research would require about a three year academic effort, which should be relatively modest in cost. And a similar three year period would be required for SU(2) experimental work. This research would be high-payoff if successful, But, at the least, it would advance electromagnetic field theory and extend the stare-of-the-art of plasma physics.

Acknowledgments

I would like to acknowledge the extensive, pioneering work of Dr. Terence Barrett, who is the originator and developer of the advanced electromagnetic field theory described in this paper. I am also grateful to Dr. Barrett for many stimulating discussions on experimental issues associated with his work. I also want to acknowledge all the nuclear fusion knowledge that I have had the privilege of gaining from Professor George Miley on our many collaborations on Inertial Electrostatic Confinement and Dense Plasma Focus fusion systems – collaborations based on Professor Miley"s very innovative and ground-breaking fusion energy work.

References

[1] Barrett, T.W. , „"Topological Foundations of Electromagnetism", World Scientific, 2008

[2] Barrett, T.W., " The impact of topology and group theory on future progress in electromagnetic research", Progress in Electromagnetic Research Symposium (PIERS 1998) Nantes, France, 1998

[3] Froning, H.D. and Hathaway, G.W., "Specially Conditioned EM Radiation Research with Transmitting Toroid Antennas", 2001-

3658, 37th AIAA / ASME/SAE/ASEE Joint Propulsion Conference and Exhibit , Salt Lake City Utah, 2001

[4] Barrett, T.W. " Topological foundations of electromagnetism" Annales ds la Foundation Louis de Broglie – special issue on George Lochak pour non 70eme anniversaire , 26 ,55-79, 2001

[5] Miley, G.H. , Shaban, Y. . Yang,Y. , "RF Gun Injector in Support of Fusion Ship II Research and Development", Proceedings of Space Technology and Applications Forum – STAIF 2005 , Edited by E.L. Genk, American Institute of Physics, Melville, New York, 2005

[6] Duncan M.. "Should Google Go Nuclear", askmar.com, video of talk at <http://video.google.com/>

[7] Froning, H.D. , Czysz, P. , "Advanced Technology and Break- through Physics for 2025 and 2050 Military Space Vehicles", Proceedings of Space Technology and Applications Forum-STAIF 2006, Edited by E.L. Genk, American Institute of Physics, Melville, New York, 2006.

Less Labored Acceleration and Faster-than-Light Travel in Higher Dimensional Realms

H. David Froning[1], Morgan Boardman[2]
[1]University of Adelaide, Adelaide SA, Australia,
[2]Morningstar Applied Physics, Vienna VA, USA
E-mail: [1]dfron4@gmail.com, [2]boardmanj@gmail.com

ABSTRACT: The simplest possible higher dimensional realm, with one more dimension than a 2-D space-time, is used to explore possibilities for less-labored ship acceleration than is possible in ordinary space-time. The higher realm has an additional dimension (iz = ikτ) that is orthogonal to a 2-D Minkowski x-ct space-time plane. Like time t is, τ (which is = q/I coulombs of charge per ampere of current) is in time units and k is the speed that tau unfolds along the ikτ coordinate. Thus, kτ, like x and ct, is in units of distance. The higher dimensional realm and its higher system states is assumed the result of an interaction between a ship-generated, specially-conditioned electro-magnetic (EM) field with a gravity or Higgs field. The formed higher-D realm gives "room" for the system's energetics vector E/c and momentum vector p to rotate at constant magnitude inside it – to allow unlabored system acceleration to nearly infinite speed. It is shown that the ship-generated field interaction must: modify system mass and time; propagate charge at light-speed or even higher; and have ship EM field dielectric εμ approach undisturbed dielectric ε0μ0 of surrounding vacuum at the ship's outer boundary.

KEYWORDS: Higher dimensional realms; Lie Symmetry; Minkowski spacetime; unlabored motion

INTRODUCTION

Since the first aircraft flew above Earth in 1903, there was steady progress in increasing flight speed and distance. And this progress was greatly accelerated by the development of chemical rocket propulsion. So, by the time the Pioneer 11 space craft crossed the orbit of Neptune almost 3 billion miles from Earth in 1983, flight speed increased about a thousand times and flight distance about 150 billion times during the first 80 years of flight. But there has been no such progress since, as chemical rockets reached a plateau in performance and more energetic nuclear rocket development has very little support in today'' challenging economic times.

High rocket thrust and propellant consumption is a direct consequence of the conservation laws of physics where system momentum (p = mV) and system energy ($E = mc^2$) conservation requires there be no net rate of change dp/dt in a system's existing p state when undergoing an acceleration (dV/dt). Thus, any dp/dt must be accompanied by an opposite -dp/dt that requires rocket engine thrust. Similarly, system energy conservation requires no net change dE/dt in a system's existing E state when undergoing the acceleration. So, needed dE/dt must be accompanied by an equal and opposite dE/dt in order that there be no net change in the system's E .

And generating the needed dE/dt requires combustion of flowing rocket fuel. But, if needed acceleration dV/dt does not require significant change in the system's existing p or E state, there will be less force dp/dt to be exerted and less power dE/dt to be expended. And hence, less needed rocket thrust and less rocket propellant.

This could happen if ship field interactions increase the ship's degrees of freedom so its momentum and energy could develop in a new direction iz that is orthogonal to the existing spatial (x) and temporal (ct) directions that its p and E are developed in. And if the new components of the ship's momentum (pr + i pi) and energy (Er + iEi) could be modulated by the ship's EM field so magnitudes of ship p and E remain nearly the same during acceleration as their magnitudes (p0 and E0) before acceleration, unlabored acceleration will result.

Finally, it should be emphasized that the deeper, higher dimensional realm that will be described is not a pre-existing geometrical arena. Rather, it's a higher domain formed by interactions of higher symmetry fields.

However, such a higher geometrical realm is needed to display the new and more complex system actions and states that would be associated with extraordinary things such as unlabored motion and faster-than- light flight.

HIGHER-D REALM AND REGION OF UNLABORED MOTION INSIDE IT

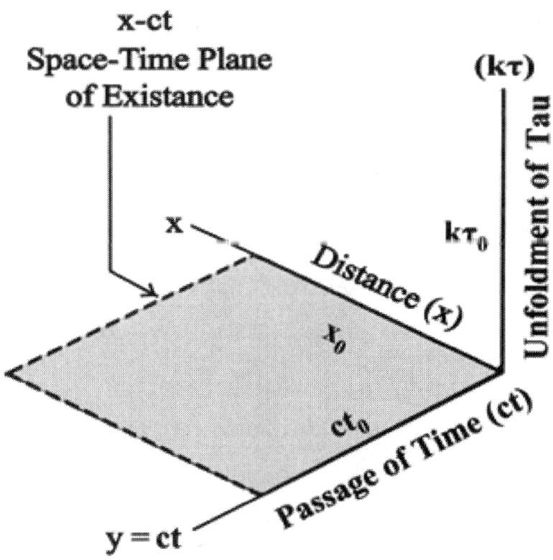

4 D space-time reduced to 2-D space-time by neglecting effect of anything not on the path of a ship's flight. Ship time travel Ct_0 set equal to distance x_0 traveled by ship during t_0 by light propagating at speed c

Fig. 1 An x-ct space-time plane of existence, where only labored, slower-than-light travel is allowed.

Figure 1 was created by considering only rectilinear motion and assuming no influence from any effect that's transverse to ship rectilinear motion along the single distance coordinate (x).

The simplest possible higher dimensional realm, which possesses only one more coordinate (iz) than a 2-D x-ct Minkowski space-time plane in Fig. 2 shows one of the 4 quadrants of the simplest possible higher dimensional realm that was created above the x-ct space-time realm by the height (iz) and depth (-iz) of a an iz coordinate that extends above and below it. Not shown is: a 2nd quadrant, to allow travel in an opposite (-x) distance direction; and a 3rd and 4th quadrant to allow travel in the (x) and (-x) directions and the (-iz) direction. And if backward-in-time travel and negative entropy were allowed in a higher realm, there would be four more quadrants in the (-ct) direction.

Much depends on what the iz coordinate represents and the metric used to define it. Space travel in the x direction is the product of elapsed time (t) times the ship's velocity (V) and is in distance units. Time travel in the y direction is elapsed time (t) times light propagation speed (c), and is distance units. System tau travel in the iz direction represents an electric or electromagnetic action that unfolds in the iz direction. And tau (τ) is charge/current (q/I) in coulombs per ampere and is in units of time. Thus, system "tau" travel is (q/I) times a propagation speed (k) in the iz direction, and is in distance units like x and ct are. Thus, just as the x-ct plane is a Minkowski space, so the higher dimensional x-ct-iz realm is considered a Minkowski space. And for initial studies, propagation speed (k) of charge in the iz (ikτ) direction was assumed equal to light propagation speed c.

Variables that are real quantities in x-ct Minkowski space-time become complex quantities in higher dimensional x-ct-ikτ Minkowski space-time-tau with the real and imaginary components shown in Table 1. In this respect, the "real" component of a quantity is what is perceived of it in x-ct spacetime and the "imaginary" component of the quantity is what is perceived of it along the iz coordinate of the higher dimensional realm.

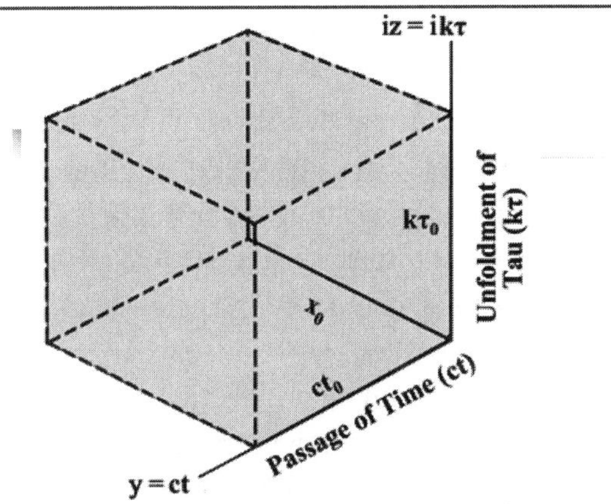

Deeper higher dimensional realm formed by coordinate ikτ that rises above and goes below an x-ct space time of being

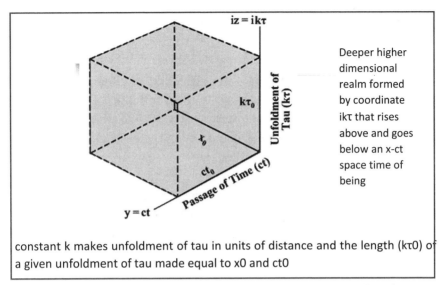

constant k makes unfoldment of tau in units of distance and the length (kτ0) of a given unfoldment of tau made equal to x0 and ct0

Fig. 2 Deeper, higher dimensional realm of existence, a x-ct-ikt Minkowski space.

• *Distance in space :*	x	$x = x_r$
• *Passage of time :*	t	$t = t_r + i\,t_i$
• *Velocity :*	V	$V = V_r + i\,V_i$
• *Light Speed :*	c	$c = c_r + i\,c_i$
• *Relativistic Dilation :*	β	$\beta = \beta_r + i\,\beta_i$
• *Mass (Energy/c^2) :*	m	$m = m_r + i\,m_i$
• *Momentum (mV) :*	p	$p = p_r + i\,p_i$
• *Momentum (mV) :*	$(m_rV_r - m_iV_i) + i\,(m_iV_r + m_rV_i)$	
• *Energy: (mc^2) :*	E	$E = E_r + i\,E_i$
• *Electrical Permittivity :*	ε	$\varepsilon = \varepsilon_r + i\,\varepsilon_i$
• *Magnetic Permeability :*	μ	$\mu = \mu_r + i\,\mu_i$
• *Vacuum Dielectric (εμ) :*	$(\varepsilon_r\mu_r - \varepsilon_i\mu_i) + i\,(\varepsilon_i\mu_r + \varepsilon_r\mu_i)$	

Subscript : r = real part; i = imaginary part

Table 1. Variables defined by complex quantities in higher-D realm

Special Relativity (SR) applies in the higher-D x-ct-ikτ realm as well as in the x-ct one. Here, dx/dt must not exceed c in the x-ct plane while dx/idτ must not exceed c on the x-ik plane and, as shown in the forbidden flight regions in Figure 3. So, V which is dx/(dtr+idti) mustn't exceed c on any plane between the x-ct and x- ikτ ones. However It is seen that planes of existence in such close proximity to the higher-D realm's x-iz plane that time interval dtr can be so small compared to distance travel dx that Vr greater than c can be reached without crossing any barrier. In fact, in the x-iz realm itself, distance dx + idz can be traveled while no time dtr would elapse in space-time at all. So, although there is no possibility for accelerating systems to achieve faster-than-light travel in lower-D x-ct space-time, there does appear some possibility for this in the higher dimensional x-ct-iz realm.

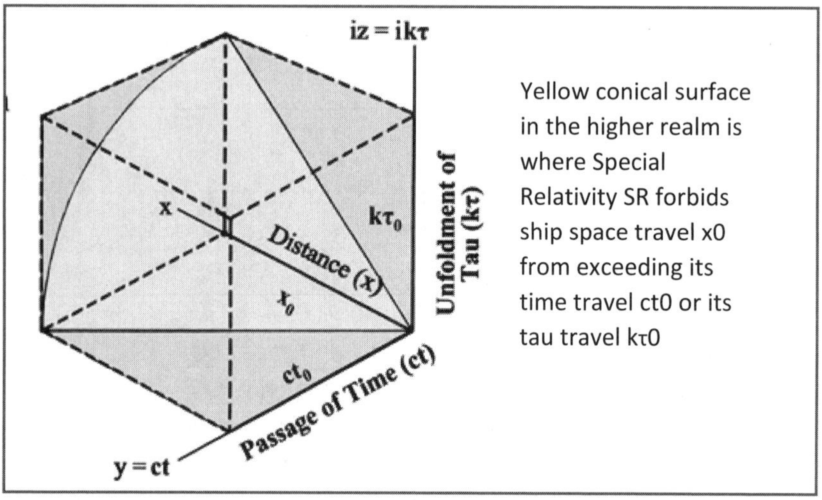

Fig. 3 Flight region that is forbidden by Special Relativity inside the deeper, higher-D x-ct-ikτ realm.

Fig. 4 shows world lines (paths traced out in space, time and tau in the higher x-ct-ikτ realm) for the simple, special case of constant speed flight at zero, light and infinite velocity relative to the earth. These world-lines were taken from a family of constant-speed world lines generated by: a given "tau travel" distance (dz) along the x coordinate that is equal to a given "rho travel" distance (dr) in the x-ct plane: where $dr = [(cdt)^2+(dx)^2]^{1/2}$.

Ship World Lines in Higher-D Realm

Curved surface containing straight world lines for constant-speed flight at zero and light and infinite velocity relative to the Earth

No world line on curved surface encounters a Special Relativity forbidden region until ∞ speed is reached.

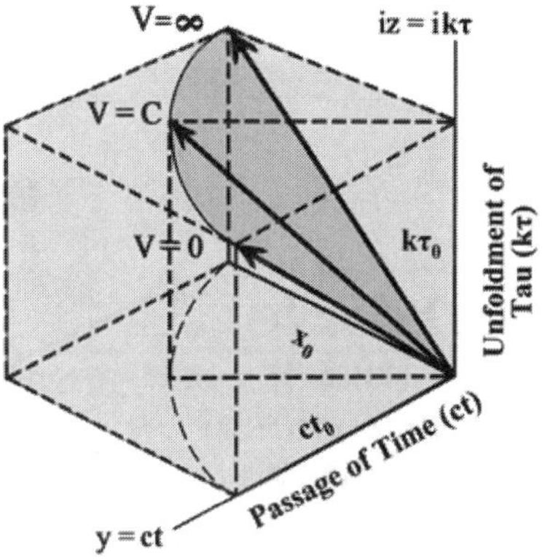

Fig. 4 Worldlines of systems undergoing constant speeds in the higher dimensional x-ct-iz realm

It is seen that all possible worldlines are traced-out on a conic surface that rises above the flat surface of x-ct space-time; and no world-line on this conic encounters any SR forbidden conic region until infinite system speed is reached. So, faster than light flight is allowed upon this curved conic surface in the deeper x-ct-iz realm, even though it is forbidden on the flat x-ct space-time plane. Curved world-lines for unlabored system accelerations can also be traced out on this same curved conic. They, of course, curve because of a change in velocity.

REQUIREMENTS AND POSSIBILITIES FOR UNLABORED SYSTEM ACCELERATION

A system's momentum ($p = mV$) has the same proportionality to its energetics ($E/c = mc$) as its space travel (x) has to its time travel (ct) – with each proportionality being equal to V/c. Since p acts in the x direction and its E/c is associated with its time-varying vibrational states, a system's p and E/c states in an x-ct realm look like its x and ct travel in that realm except for being in momentum instead of distance units. Similarly, a system's p, E/c and ikρ states look much like its x, ct and ikτ travels in an x-ct-ikτ realm, as is shown in Fig. 5.

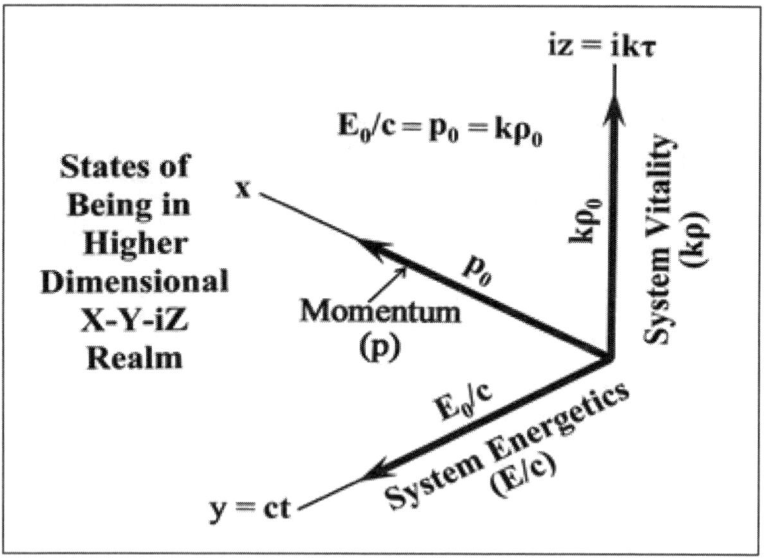

Fig. 5 System states: p, E/c and ikρ in a higher dimensional realm

Unlabored acceleration from zero to infinite speed can be viewed as conservation of a system's energy by rotation of its energetics vector E/c at its initial magnitude E_0/c and conservation of its momentum by rotation of its momentum vector p at its initial magnitude $k\rho_0$. These rotations in the higher realm are shown in Fig. 6.

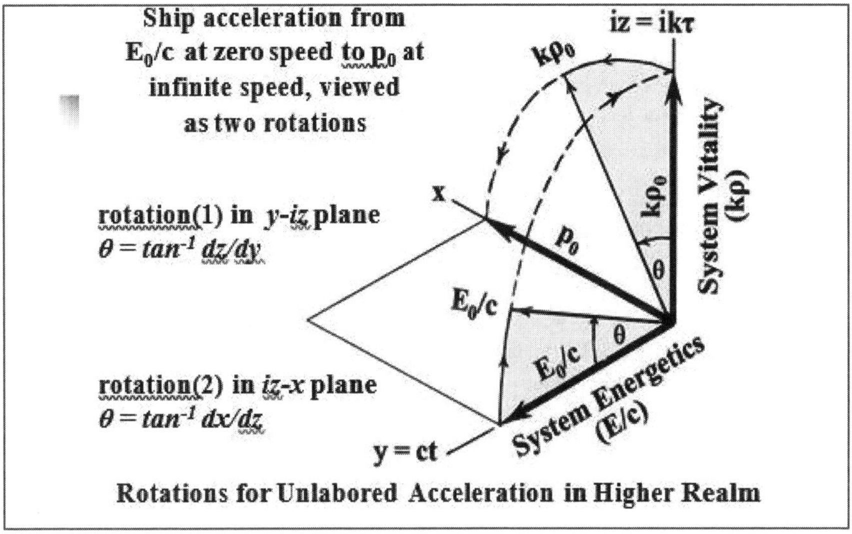

Fig. 6 Ship acceleration from near zero to infinite speed viewed as 2 rotations in a higher realm.

A 90 deg rotation of the system energetics vector E_0/c occurs In ct-Iz plane, with no moving in space or exerting of force, transforming E_0/c into $ik\rho_0$ in the iz direction. A 90 deg. rotation in the iz-x plane then transforms system momentum state $ik\rho_0$ into momentum state p_0 - which moves at infinite speed in space relative to Earth.

Slow speed at a system's destination requires that its unlabored acceleration from very slow to very high speed be followed by unlabored deceleration to very slow speed - as shown in Fig. 7. Though viewed from a different angle, its seen that continued rotation of system momentum vector occurs in the x-iz plane - from p_0 on the x-coordinate to - $ik\rho_0$ on the - iz coordinate. This is followed by a final 90 deg. vector rotation in the ct- iz plane from - $ik\rho_0$ state on – iz coordinate back to the system's initial low-speed E_0/c state in x-ct space-time.

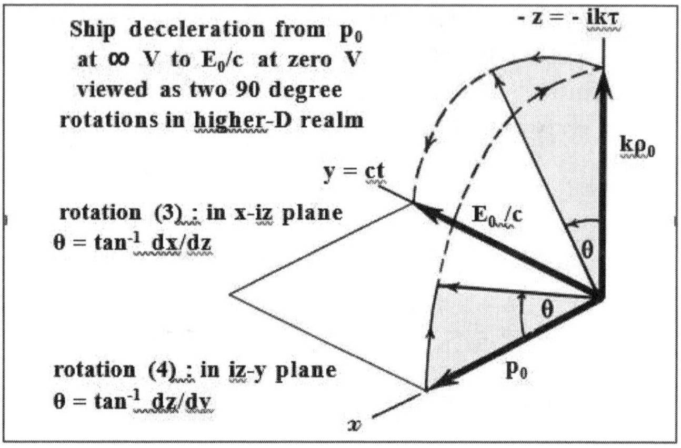

Fig, 7 Continued rotations in higher dimensional realm during deceleration from near-infinite speed

In Star Trek-like science fiction flight, people and things stay about the same on starship bridges – with time seeming to pass at the same constant rate. But people, things and clock ticks don't stay the same in ships undergoing unlabored accelerations in the higher-D x-ct-iz realm. Fig. 8 shows the different components of a system's time and substance changing significantly in the moving ship frame during its unlabored acceleration.

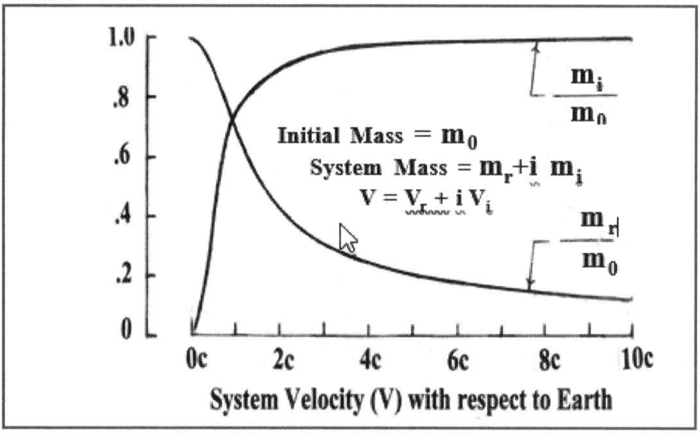

Fig. 8 Mass components of an accelerating system's substance decreasing and increasing with increase in speed.

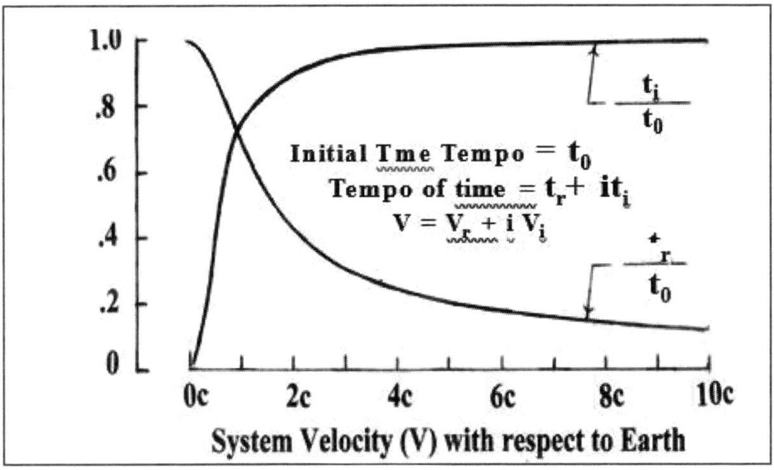

Fig. 9 Time components of an accelerating system's substance decreasing and increasing with increase in speed

CREATING HIGHER-D REALMS AND HIGHER-D SYSTEM STATES BY SPECIAL CONDITIONING OF EM FIELDS TO HIGHER LIE SYMMETRY STATE

Fig. 8 and 9 indicate that mass and time must be significantly changed in systems undergoing unlabored acceleration to superluminal speed. One envisioned concept is an "electrogravitic" coupling between a system-generated electro-magnetic (EM) field and gravity. And a new one may be an "electro-Higgs's coupling between an EM field and a Higgs field that gives rest mass to quarks, leptons and W and Z bosons of a system's matter.

But, ordinary EM fields do not seem very promising for such field couplings. Here, ordinary electric (E) and magnetic (B) fields are usually described by vectors and Abelian algebra and develop in space and time in accordance with the four Maxwell Equations formulated by Clerk Maxwell about 140 years ago. In the 1950's Feynman and others made Maxwell's EM theory compatible with quantum theory and special relativity and quantum electrodynamics resulted. Quantum electrodynamics precisely predicts interactions between matter and radiation. But, Maxwell's classical electromagnetic theory and equations are still used, essentially without changed, in the design of every electromagnetic device since the 1880's. Maxwell theory describes electromagnetism in terms of: electric field strength (E), magnetic flux density (B), and total current density (J). Electric and magnetic fields develop and propagate in accordance with the 4 Maxwell Equations shown in Table 2. Its E and B vector fields embody U(1) Lie Group symmetry, and are sometimes defined mathematically in terms of non-physical potentials that are called a magnetic vector potential (A) and electric scalar potential (ϕ).

Table 2. Maxwell's Equations for U(1) Symmetry Electromagnetic Vector Fields

Gauss' Law	$\nabla \bullet \boldsymbol{E} = \boldsymbol{J}_0$
Ampere's Law	$\frac{\partial \boldsymbol{E}}{\partial t} - \nabla \times \boldsymbol{B} - \boldsymbol{J} = 0$
Coulomb's Law	$\nabla \bullet \boldsymbol{B} = 0$
Faraday's Law	$\nabla \times \boldsymbol{E} + \frac{\partial \boldsymbol{B}}{\partial t} = 0$

$$E = -\frac{\partial A}{\partial t} - \nabla \phi, \quad B = \nabla \times A$$

Barrett [1] has used group and gauge theory and topology to develop SU(2) EM field theory, and Table 3 shows the Expanded Maxwell Equations associated with SU(2) EM fields. Ordinary U(1): J, E and B vector fields are transformed into SU(2): J, E and B tensor fields that are described by Non-abelian algebra, while the magnetic vector potential (which is non-physical in U(1) electromagnetism) is transformed into a physical A tensor field.

And SU(2) EM, Maxwell Equations have additional terms that involve A tensor fields interacting with E and B tensor fields in various ways. Like gravity and U(1) EM fields, SU(2) EM fields act over much longer distances than SU(2) nuclear fields do. Also, SU(2) EM fields are mediated by a single boson (a photon) while SU(2) nuclear fields are mediated by 3 bosons. However, mediating actions associated with the 3 different couplings of A, E, and B fields with each other (as is shown in the 4 SU(2) Maxwell Equations of Table 3) may be somewhat analogous to mediating actions associated with the W +, W -, Z bosons of SU(2) nuclear fields.

Table 3. Extended Maxwell's Equations for SU(2) Symmetry Electromagnetic Tensor Fields

$\nabla \bullet \boldsymbol{E} = \boldsymbol{J}_0 - iq(\boldsymbol{A} \bullet \boldsymbol{E} - \boldsymbol{E} \bullet \boldsymbol{A})$
$\frac{\partial \boldsymbol{E}}{\partial t} - \nabla \times \boldsymbol{B} - \boldsymbol{J} + iq[\boldsymbol{A}_0, \boldsymbol{E}] - iq(\boldsymbol{A} \times \boldsymbol{B} - \boldsymbol{B} \times \boldsymbol{A}) = 0$
$\nabla \bullet \boldsymbol{B} + iq(\boldsymbol{A} \bullet \boldsymbol{B} - \boldsymbol{B} \bullet \boldsymbol{A}) = 0$
$\nabla \times \boldsymbol{E} + \frac{\partial \boldsymbol{B}}{\partial t} + iq[\boldsymbol{A}_0, \boldsymbol{B}] = iq(\boldsymbol{A} \times \boldsymbol{E} - \boldsymbol{E} \times \boldsymbol{A}) = 0$

Barrett has made a start towards SU(3) EM field theory - which, of course, is more complex than SU(2) EM theory. SU(3) EM fields embody higher order tensors and SU(3) Maxwell Equations would have more and higher order A, E, B couplings than SU(2) EM. But since SU(3) EM theory requires much more development, this paper focuses on SU(2) EM field descriptions and mentions an experiment with SU(2) EM radiation fields.

Table 3 shows SU(2) EM couplings between A and E and A and B fields that do not occur in ordinary U(1) EM. Thus, forces exerted on moving charged particles in an U(2) EM field can be different than forces on the particles when moving in an ordinary U(1) EM field. This is shown in Table 4 which is taken from [1]. Shown is Lorentz force exerted on a moving charged particle in an SU(2) E and B tensor field – as compared to Lorentz force exerted on the moving particle in an ordinary U(1) E and B vector field. Here, U(1) Lorentz Force is described by E and B fields and A vector potentials; while SU(2) Lorentz force is described by E and B and A tensor fields. Differences in numbers of Lorentz Force terms and their different vector and tensor natures will cause an SU(2) EM field to exert a Lorentz force of different intensity and direction on moving charged particles, as compared to Lorentz Force exerted on the particles by an ordinary U(1) EM field of equal strength.

Table 4. Lorentz Forces Acting on Moving Charged Particles

U(1) Lorentz Force	$\mathscr{F} = e\mathbf{E} + ev \times \mathbf{B} = e\left(-\frac{\partial \mathbf{A}}{\partial t} - \nabla\phi\right) + ev \times \left((\nabla \times \mathbf{A})\right)$
SU(2) Lorentz Force	$\mathscr{F} = e\mathbf{E} + ev \times \mathbf{B} = e\left(-(\nabla \times \mathbf{A}) - \frac{\partial \mathbf{A}}{\partial t} - \nabla\phi\right) + ev \times \left((\nabla \times \mathbf{A}) - \frac{\partial \mathbf{A}}{\partial t} - \nabla\phi\right)$

Barrett [3] has identified one way of radiating SU(2) EM field energy. It is by flowing alternating current at radio frequencies through a toroid coil at one of the possible resonant frequencies that are possible for a given toroid and coil configuration. Figure 9 shows the two A field patterns that form about a transmitting toroid. They are viewed as two U(1) A vector potential patterns (ϕ1 and ϕ2) which overlap in polarity. A resonant frequency occurs when the difference (ϕ1-ϕ2) in overlapping vector potential amplitude maximizes for the toroid at one of its alternating current frequencies. A single SU(2) tensor field forms about the toroid at this frequency. And, as indicated, many resonant frequencies are possible for the toroid and its coils in Figure 10

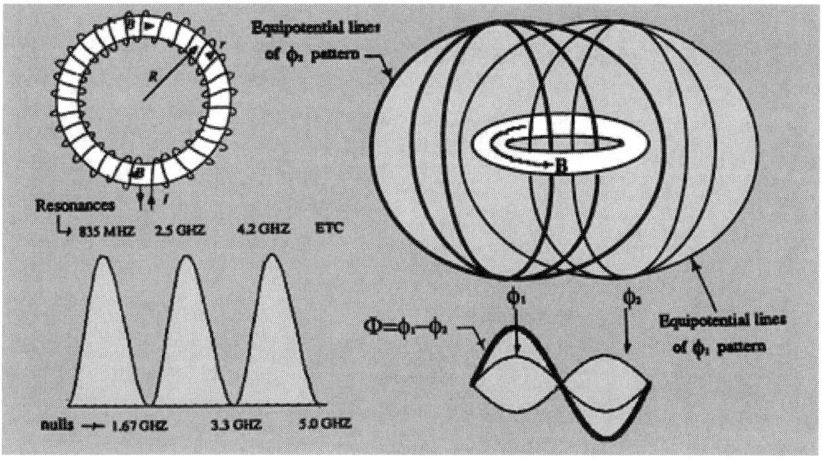

Fig. 10. A vector potential patterns emitted by alternating current flow in toroid coils. Phase difference (ϕ1-ϕ2) of the patterns maximize when a "resonant" frequency is reached. Shown are 3 different resonant frequencies for a given toroid coil geometry. A slightly different SU(2) EM field will be emitted at each different resonant frequency.

Toroid testing at radio frequencies [1] was not exhaustive enough to conclusively prove that SU(2) EM fields were emitted at resonant frequencies But the measured resonant frequencies were in good agreement with those predicted from Barrett's SU(2) toroid theory. And, more intense and highly focused magnetic fields were always measured above and below toroid surfaces when the toroids were radiating at a resonant frequency.

Barrett (1) has identified another way of generating SU(2) EM field energy. The description of this way is taken from pages 46 and 61 of [2]. It is shown in Fig. 11, which uses a waveguide paradigm to show oscillating U(1) EM wave energy being transformed into SU(2) EM wave energy by phase and polarization modulation. Input wave energy enters from left, is polarization-modulated, and is emitted as SU(2) wave energy to the right.

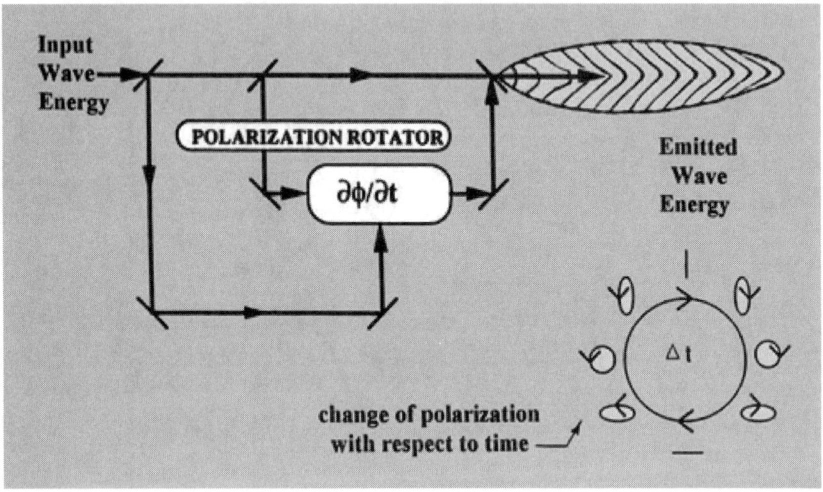

Fig. 11. Generation of Polarization-Modulated EM Radiation

SU(2) EM fields may not have sufficiently high Lie group symmetry to couple strongly with gravity or Higgs fields to cause the significant modulations of system time and mass shown in Fig.8 and 9 during unlabored acceleration. So, higher symmetry SU(3) EM fields might be very promising. Fig. 12 from Barrett [2] shows an additional phase modulation of polarization-modulated waves that might result in SU(3) EM field generation.

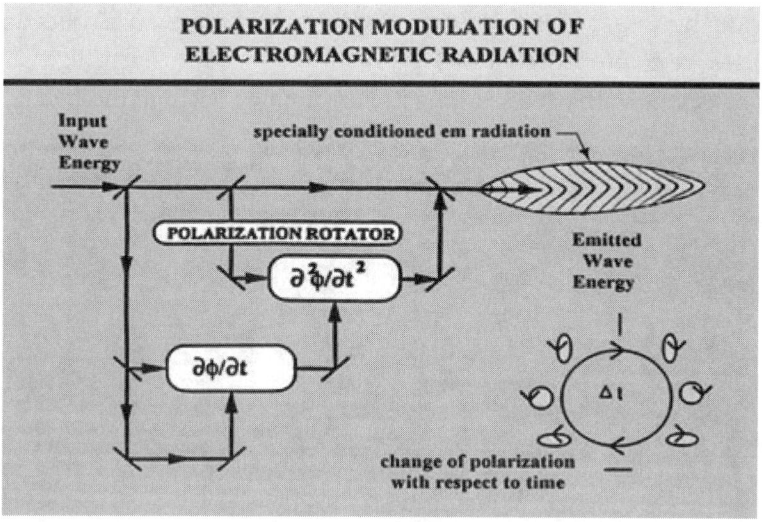

Fig. 12 Added phase modulation of EM wave energy that may result in emission of SU(3) EM wave energy

Developing SU(2) EM field theory to its present state required extensive effort by Barrett, So, development of even more complex SU(3) EM field theory would require even greater effort. SU(3) EM hardware challenges may also be encountered. For example, Fig. 12 indicates that a much higher degree of phase modulation would be associated with SU(3) field emission. Barrett [3] indicates that such level of phase modulation would require undulator or oscillator bandwidths in the 20-200 THz range. Such performance has not yet been approached by any high-bandwidth device, even those in undulators of the highest performing Free Electron Lasers of today.

Effective SU(2) or SU((3) EM field generation is complicated by need to avoid excessive disturbance of surrounding vacuum. Light-speed in undisturbed vacuum dielectric ε0 μ0 is c . But light speed (Vc) can be much higher than c in the ε μ formed about a system by its EM field. can be faster than c if ε μ is less than ε0 μ0. So, ship V in the ε μ that its EM field system creates could be less than light-speed Vc in this disturbed dielectric – but much greater than c relative to Earth. Thus, if ε μ associated with the ship-generated EM field ε μ is much less than ε0 μ0 outside the ship, and if its ε μ value rapidly approaches ε0 μ0 outside the ship (as in colored region of Fig.13) the ship could reach high super-luminal speed without disturbing its surrounding vacuum too much.

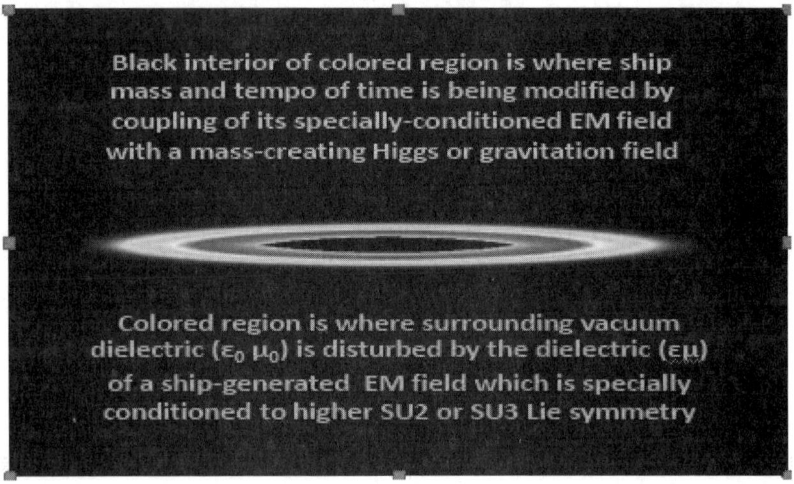

Fig. 13 Ship EM field generation and unlabored motion causing little disturbance of surrounding vacuum

Fig.14 is an example of EM field generation causing weak disturbance propagation into the surrounding undisturbed vacuum. A CFD calculation assumes very low ε μ in the EM field generation region (black). So, ship speed is only 0.7 light-speed Vc in the disturbed region, but its 10 times light-speed c in undisturbed vacuo .

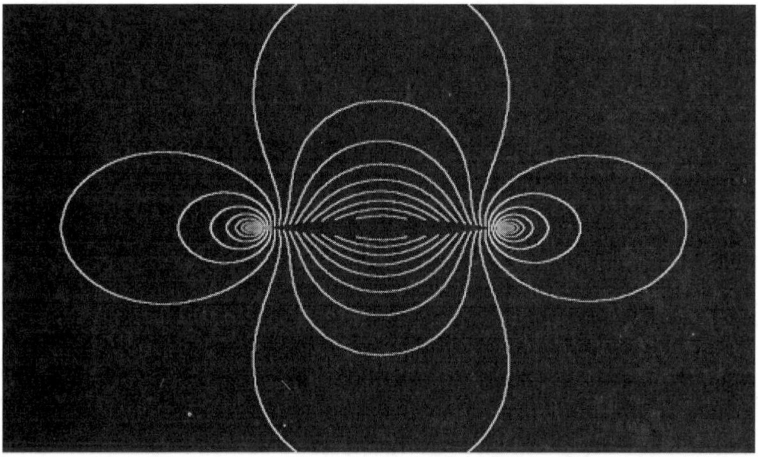

Fig. 14 Vacuum slightly disturbed by EM field generation in black region

CONCLUSIONS

Rapid transits may be possible in a higher dimensional realm than 4-D space-time. This higher realm may be a consequence of a coupling between a gravity or Higgs field and a ship-generated EM field. If this action adds a component i kEi to a ship's energetics Er and i pi to its momentum pr . And if the ship's resulting energetics Er/c+ iEi/c and momentum pr + ipi can be modulated by field propulsion systems to maintain their magnitudes near E_0/c, and p_0 at acceleration start, force exertion and energy expenditure will be slight during acceleration.

After its rapid transit to a distant world in the higher-dimensional realm, a starship could slow and return its crew to the lower-dimensional realm of space and time to begin the distant world's exploration. But, possibly, the expedition's most expansive exploring and its highest experiencing may be what is seen and felt during: modulation of the starship's mass-energies and momentums during its unlabored motion in the higher-D realm.

It is recommended that further work to explore less labored motion in higher-dimensional realms include a tensor (matrix) formulation of a 5-D higher dimensional realm with at least One more dimension than 4-D space-time and at least one more degree of translation and rotational motion. It is also recommended that work begin on a tensor (matrix) formulation of a ship-generated, time and mass-modifying EM field of SU(2) or SU(3) Lie group symmetry to explore the possibility of its modulations to cause less-labored ship acceleration.

ACKNOWLEDGEMENT

We want to acknowledge the insightful suggestion by John Brandenburg that our added higher dimension "tau" (τ) can be best thought of as representing the electrical essence "charge"(q), which is just as fundamental a property of physics as distance (x) and time (t). In modeling, we defined tau (τ) as ql which is in units of time, like (t) is. Multiplying τ times k (charge propagation speed) then placed kτ in units of distance, like ct and x is.

REFERENCES

[1] Barrett, T.W., Topological Foundations of Electromagnetism, World Scientific, 2008

[2] Barrett, On the distinction between fields and their metric, Annales de la Foundation Louis de Broglie,14,1,1989

[3] Barrett, T.W., Private communication

APPENDIX A: ENERGY AND MOMENTUM COMSERVATION IN HIGHER REALM

This appendix adds some discussion to that of the paper on issues associated with nearly unlabored system acceleration by its momentum and energy conservation in a higher-dimensional realm. This is mainly by use of Table 6, which summarizes governing equations for achieving such conservation by an accelerating system. It is seen that unlabored system acceleration (dV/dt) relative to earth is equal to its mass rate of change times its starting mass-velocity (m0V0) divided by mass squared, which tends to greatly increase with increased speed. Table 1 also reminds us that all quantities are complex, so computations can be somewhat more complicated.

Table 1. Requirements for unlabored system acceleration

Table 1. Requirements for unlabored system accekeration

- **Momentum Conserving Flight ; No force, dp/dt exerted on ship during its acceleration, dV/dt relative to Earth; where:** $dp/dt = d(mV)/dt = 0$ **also :** $dV/dt + m(dV/dt) = 0$; **and** $mV = m_0V_0$

  so: $dV/dt = -(m_0V_0/m^2)\ dm/dt.$

- **Energy Conserving Flight ; No expenditure of ship energetics d(E/c)/dt during ship's acceleration (dV/dt) relative to the Earth**

  so: $d(E/c)/dt) = c\ (dm/dt) = 0$

  where : $p = p_r + ip_i$; $V = V_r + V_i$
  $m = m_r + im_i$; $E = E_r + iE_i$; $t = t_r + it_i$

APPENDIX B: IS "BACKWARDS IN TIME TRAVEL" POSSIBLE IN A HIGHER-D REALM?

Recent work indicates that round-trip flybys of distant stars could conceivably be accomplished in short Earth times and in almost no ship time by one revolution of an ascending-descending spiral trajectory around the cone of unlabored motion shown below. The trip would begin and end at the cone's vertex (Earth). There is no time travel into the past. But backwards-in-time travel seems to occur in 2 higher realm quadrants, which would cancel ship forward-in-time travel in the 2 other quadrants. So, it planned to explore this issue in more depth.

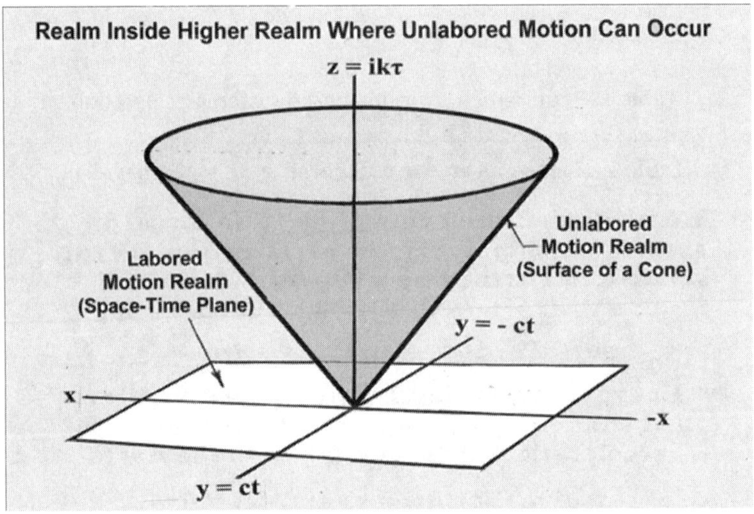

A SPACE EXPLORATION INITIATIVE FOR FUTURE NEEDS

H. D. Froning, Jr.

McDonnell Douglas Space Systems Company
Huntington Beach, California 92647

Abstract

Eighty nine years ago, the Wright Brothers accomplished the first successful powered flight above the earth, reaching a speed of about 30 miles per hour and covering a distance of about 150 feet.[§] Since then, we have made more than a thousand-fold increase in flight speed during our first 89 years of flight, while increasing distance covering capability by more than 200 billion times. Furthermore, if we could maintain this past rate of progress, we will have surpassed light speed and reached distant stars before the next 89 years of flight are past. This paper describes the kinds of breakthroughs that would enable the extraordinary flight progress of our past to be maintained during the next 89 years and proposes an international spaceflight initiative that would strive to make them come to pass. It also argues that the kinds of breakthroughs needed to revolutionize spaceflight and take us to the stars will also be needed to help meet the formidable future problems facing earth.

Introduction

Although there is general public approval for the exploration of space, some feel that it should be embarked upon only after critical environmental, economic and social needs are met. By contrast, Table 1 contends that earth's future environmental, economic and social needs can only be met by the kinds of scientific and technological breakthroughs that are needed to revolutionize the exploration of space.

[§] **Ed. Note**: This was first published in 1994 and the Wright Brothers performed their first successful flight on December 17, 1903.

Beamed Energy - systems that use beamed laser or RF energy from ground source to heat propellant to generate thrust (e.g. lightcraft)
Electric Sail - system that uses a number of long/thin high voltage wires to interact with solar wind to generate thrust.
Fusion - systems that use fusion reactions indirectly (fusion power system to drive EP), or directly (fusion reaction provides kinetic energy to reactants used as propellant)
High Energy Density Materials - materials with extremely high energy densities to greatly increase propellant density and potential energy.
Antimatter – system that converts large percentage of fuel mass into propulsive energy through annihilation of particle-antiparticle pairs.
Advanced Fission – enhanced propulsion ideas that utilize fission reactions to provide heat to propellants (and in some cases utilize magnetic nozzles)
Breakthrough Propulsion – area of fundamental scientific research that seeks to explore and develop deeper understanding of nature of space-time, gravitation, inertial frames, quantum vacuum, and other fundamental physical phenomenon with objective of developing advanced propulsion applications.

Table 1. Breakthroughs That Would Revolutionize Flight Would Meet Future Human Needs as Well**

Essential Needs for Future Life

Although ever-increasing amounts of energy will be needed for ever-increasing numbers of people on earth, continued burning of fossil fuels and fission of nuclear fuels will eventually cause unacceptable deterioration in atmospheric quality and life. Furthermore, significant amounts of material from other bodies within our solar system may also be needed to prevent eventual depletion of the soils, forests, and mineral resources of earth.

Energy from fusion of relatively "clean" nuclear fuels such as deuterium and helium 3 (mined on the moon) would result in cause much less radioactivity than nuclear fission and much less atmospheric pollution than energy from combustion of fossil fuels. Solar energy beamed from orbiting satellites or the moon has also been proposed as a source of non-polluting power for earth. However, propulsion breakthroughs may be required for economic transport of helium 3 from the lunar surface to earth, or for economic transport of heavy solar energy components to geostationary orbits or to the moon.

Incredibly, almost unlimited energies are contained within the seeming emptiness of space itself. For quantum physics reveals the existence of more than enough zero-point electromagnetic fluctuation energy within every cubic foot of space to meet all conceivable future energy needs of earth. If such

** **Ed. Note**: Table 1 has been updated with the latest NASA breakthroughs that are needed, in keeping with the spirit of the author's interest to inspire his audience with advanced technology.

invisible energies could somehow be harvested, they would meet future terrestrial energy needs, and possibly provide propulsive power for low cost space transportation as well. And such transportation would enable economical acquisition of things such as extraterrestrial materials to preserve critical resources of earth.

Minimal Needs for Future Life on Earth

Once bare essentials for life are met, the next level of need is purposeful activity to give people a sense of worth. Here, new occupations are needed to enable earth's ever-increasing numbers of individuals to live a useful and productive life, and the history of air transportation is an example of how scientific discovery and technological advancement can help meet such human needs.

Aviation progressed steadily after the Wright Brothers' first powered flight. But, it did not really "takeoff' until a new science "quantum physics" enabled solid-state devices, for avionics and micro- computers, and until a new technology "jet propulsion" enabled economical flight over long distances at high speed. This resulted in an air travel revolution which began approximately 60 years after the Wright Brothers' first flight and spawned a vast and still growing infrastructure of hotels, resorts, and travel based activities that have provided expansive new experiences and productive careers for millions of people on earth Similarly, spaceflight has progressed steadily since Sputnik achieved orbit in 1957, with things such as satellite communications providing telecommunications to almost everywhere on earth. Nevertheless, space travel has not yet been revolutionized like air travel has because it still has a very high cost. Here, current rocket costs to place a person in low earth orbit are roughly a million dollars per trip, and there are estimates that hundreds of millions of dollars per person would be required to take people on much longer interplanetary rocket trips [1].

In this respect, single stage to orbit vehicle developments such as the National Aerospace Plane (NASP) may eventually enable as much as a 50 to 100-fold reduction in earth-to-orbit travel costs and this might enable profitable space commerce (such as space tourism) within close proximity to earth [2]. But, foreseeable advances in space rocketry and space sailing cannot provide the cost reductions needed for things such as commercial interplanetary travel or profitable acquisition of extraterrestrial resources for earth.

Therefore, entirely new modes of propulsion will be required for the cost reductions that would revolutionize space travel and provide new economic opportunities and careers (like air travel did) for millions of people on earth.

Higher Needs for Future Life

Serious problems facing the United States include increased illiteracy, crime, unethical behavior and, to growing numbers of people, a general dissatisfaction with merely material things. Such lowering of the human spirit can only be reversed by endeavors that lift it above the ordinary and mundane, while inspiring it to more than mere maintenance of material life. In this respect, spaceflight has lifted the human spirit, fired the imagination more than any other technical endeavor of today; with youth interest in science and engineering directly following U.S. dedication to the exploration of space, with growth of technical graduates closely matched growth of space funding during the 1960 Apollo program decade. Then, decrease in graduates closely followed diminished space activity in later years.

It follows that a bold, long-term spaceflight program should, more than any other technical endeavor, motivate youth to greater achievement – by participating in an adventure of discovery and exploration, and the challenge of worthwhile difficult things.

What if We Could Maintain the Aerospace Progress of Our Past

The Wright Brothers accomplished the first successful controlled and powered flight above the sands of Kitty Hawk, North Carolina in 1903. Since then, as shown in Figure 1, many significant advancements have enabled enormous progress during the first 89 years of flight.

The Wright Brothers' first powered flight above the earth was at a speed of about 30 miles per hour over a distance of about 150 feet. Today, 89 years later, mankind's longest flight is still under way. For the Pioneer 10 spacecraft, traveling faster than while increasing distance covering capability by more than 200 billion times. Figures 2 and 3 show that if we could maintain this past rate of progress during the past 89 years of flight, we will have surpassed light speed and reached distant stars before the next 89 years of flight are past.

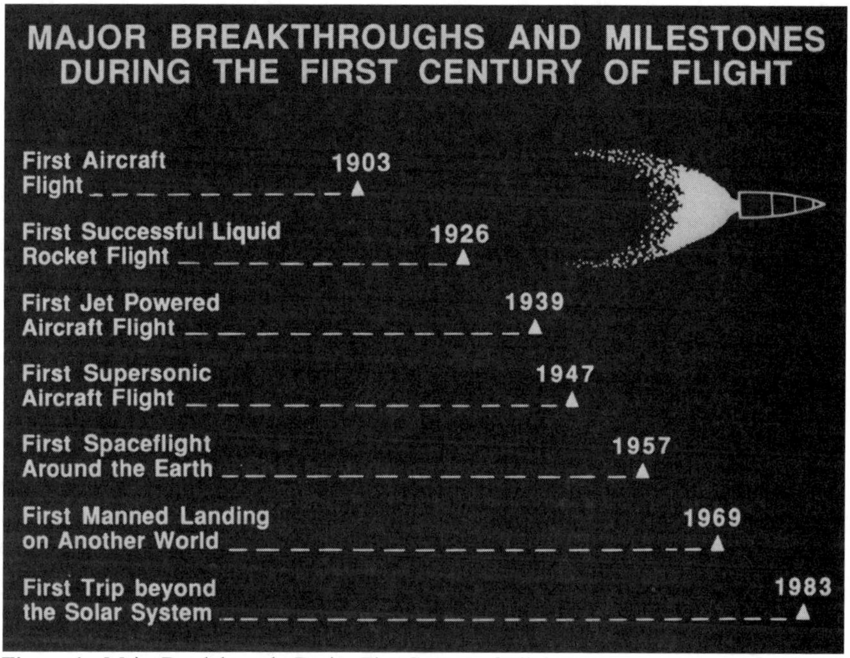

Figure 1. Major Breakthroughs During First Century of Flight

Such achievements certainly appear preposterous at this time. For just as faster-than-sound travel and manned expeditions to the moon were not foreseen in 1903, so things such as faster-than-light travel and intergalactic expeditions cannot be envisioned (except in science fiction) today.

The Possibility of Impossible Things

Robert Browning once said, "By far the greatest obstacle to the progress of science and to the undertaking of new tasks and provinces therein, is found in this-that men despair and think things impossible," and there has never been a lack of prestigious people to declare the impossibility of future things. In 1885, Lord Kelvin, President of the Royal Society, declared that "Heavier than air flying machines are impossible." Charles Duell, Director of the U.S. Patent Office in 1899 announced that, "Everything that can be invented has already been invented." While in 1923, Robert Millikan, winner of the Nobel Prize in Physics proclaimed that, "There is no likelihood man can ever tap the power of the atom."

Just as one must be cautious before accepting authoritative statements on impossibility of future things, so we must avoid presumptuous prophecies of what might come to pass. Nevertheless, Table 2, reviews past advancements and the flight progress they allowed, together with examples of the kinds of future advances that might enable comparable progress for future flight.

As advancement from Newtonian to quantum physics was needed for a better understanding of matter over microscopic scales of time and space, so Table 2 indicates that further advancement may be needed for deeper insights as to the essence space, itself, over microscopic scales; and as past advancements in scientific understanding enabled technologies (such as solid-state electronics) that exploit electromagnetic phenomenon within specially crafted mass, so future advances might enable technologies that would exploit the electromagnetic phenomenon within non-material media (such as vacuum fields).

Similarly, as past material developments have enormously improved thermostructural material properties, so future developments should enable further thermostructural advancements and perhaps breakthroughs in electrical and magnetic properties as well.

Furthermore, just as new sources of power (solar and nuclear energy) were developed during the first 89 years of powered flight, so new sources (such as the "zero-point" electromagnetic energies of the vacuum) might be perfected during the second; and as advancement from propeller propulsion to jet propulsion was needed to revolutionize air travel and enable what many once thought impossible – supersonic speed, so advancement to a new mode of impulsion (such as field propulsion) may be needed to revolutionize space travel and enable what many now believe impossible – superluminal (faster-than-light) flight.

Possible Breakthroughs for Future Flight

Although we cannot forecast the precise breakthroughs that will be accomplished during the next 89 years of flight, we can consider advancements that are typical of the kinds of breakthroughs that might occur. Thus, this section describes four examples of breakthroughs that could enable our future rate of aerospace progress to be comparable to that of our past. These are: fusion propulsion, vacuum energy, field propulsion and superluminal flight. However, other breakthrough possibilities may be as likely to occur. For example, "magnetic monopole" propulsion [3] and "negative matter" propulsion [4] are other breakthrough possibilities that have been considered for spaceflight art.

Fusion Propulsion

Thermonuclear fusion releases millions of times more energy (in terms of calories per gram of mass) than chemical combustion of hydrogen fuel and has long been considered for meeting earth's future energy needs and for propulsive flight. Here, most fusion research has been directed towards large systems to meet earth's enormous future energy needs and, unfortunately, fusion with such systems remains very difficult to achieve. Thus, nuclear fusion has generally been perceived as being extremely difficult for propulsive use as well. However, recent studies in the United States and the United Kingdom [5, 6] indicate that small fusion reactors, operated in a pulsed mode, may be developable for space propulsion much sooner than more massive machines operated in a steady-state mode, can be developed for terrestrial energy use.

Studies indicate that fusion propulsion systems could enable a 10- to 20-fold improvement in engine thrust per fuel flow rate compared to chemical systems and reduce Mars trip times by about a factor of 5 to only about 50 days [7]. Furthermore, neutron flux and radioactivity from fusion of fuels such as deuterium and helium 3 may be mild enough that ground testing of fusion propulsion would not cause the environmental concerns than nuclear fission propulsion would.

Vacuum Energy

Although the vacuum of space seems inert and "empty, "quantum theory reveals that it is teeming with vigor and vitality over scales of time and distance that are too short and small for the material senses to perceive. Over these microscopic scales the vacuum's vigor and vitality is manifest in "zero-point" electromagnetic energy pulsations throughout the entire breadth, length, and depth of cosmic space. Estimates of the zero-point energy in the vacuum [7, 8 and 9] indicate that a single cubic foot of so-called "empty space" contains more than enough zero-point energy to meet earth's total energy needs for more than a million years. Furthermore, the enormous power achievable with such stupendous energy could rapidly accelerate starships to nearly the speed of light [10].

Just as distinguished scientists once rejected the idea of tapping the nuclear energies of the atom, so most scientists today doubt the feasibility of ever tapping the vacuum energies of space. Nevertheless, some scientists [11, 12] describe some of the possibilities being considered for extracting zero-point energy from the vacuum of space.

Field Propulsion

Propellant mass currently constitutes as much as 90% of each stage of a multi-stage rocket ship, with structures and systems accounting for most of the rest. This, typically, results in payload being less than 1% of earth launching weight for travel to distant places such as Mars. Thus, field propulsion systems, which would develop force by the action and reaction of fields (instead of by combustion and expulsion of mass), could potentially enable enormous payload increase for a given launching weight.

But although there are known ways of developing thrust by actions and reactions between electric and magnetic fields, the developed thrust would be insufficient for future flight. Thus, much more powerful field actions and reactions would be needed for rapid transits to the further reaches of space.

It is conceivable that innovative arrangements of magnetic and electric elements, together with breakthroughs in things such as superconductivity would enable sufficient impulsion for first generation field propulsion ships. But further perfection of field propulsion would probably require more efficient and powerful interactions than those achievable with current electromagnetic knowledge and hardware art.

Critical questions for field propulsion flight would be: Are there ways of interacting much more effectively with the electromagnetic currents that flow through the vacuum of space? Does the invisible vacuum and its associated space-time metric contain much more energetic substructures than ordinary electric and magnetic currents and electromagnetic fields? And, if so, can such substructures be effectively coupled to and interacted with by energies that can be emitted by material means? References [13, 14] examine the general problems of coupling propulsively with the vacuum, and its associated spacetime metric, while Reference 15 outlines the general form of field theories that would permit such coupling to occur.

Figure 2 visualizes first generation field propulsion systems as enabling travel throughout the solar system at a small fraction of light speed, while second generation systems could accelerate to almost light speed for interstellar travel within the lifetimes of starship crews. And if third generation systems could enable light speed to be surpassed, round trips to distant stars could be accomplished within the lifetimes of starship crews and those on earth who made possible their trips.

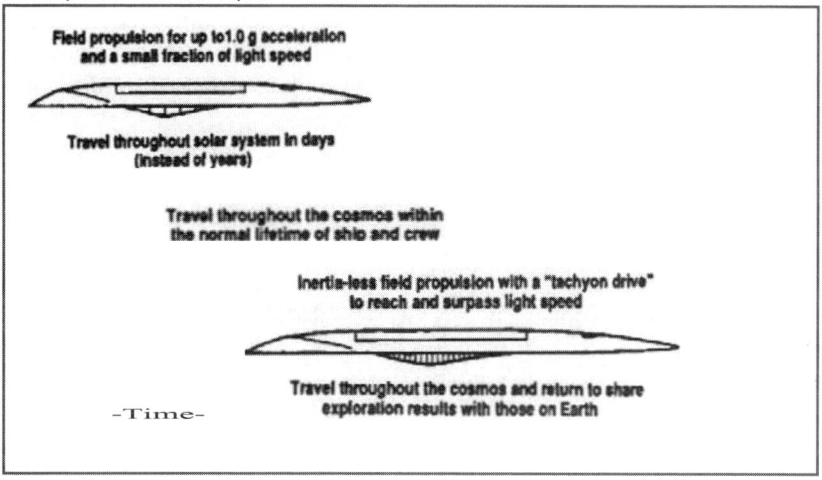

Figure 2. Evolution of Field Propulsion Flight

Superluminal Flight

Superluminal ships would be able to leave one location and arrive at another before light could travel between the two locations in normal space [16]. Thus, they would be somewhat analogous to supersonic aircraft that travel faster than their sound can. Even hypothesizing superluminal ships is controversial since the overwhelming consensus of scientific opinion is the impossibility of faster- than-light flight Nevertheless, only such ships could reach the further reaches of cosmic space and return to share the fruits of their expeditions with the people of earth.

Just as Einstein's Special Relativity predicts resistance to accelerated vehicle motion (its inertia) becomes infinite at the speed of light, so simplified (Prantl Glauret) theories of aerodynamics predict resistance to vehicle motion (its drag) becomes infinite at the speed of sound. But, in actuality, there is a change in flow about a vehicle as it approaches sonic speed (which is not accounted for in the simplified theories) and this change prevents flight resistance from becoming infinite at the speed of sound. Thus, some analogous change in field state (which is not accounted for in Special Relativity) would have to be made to occur in order that resistance to vehicle motion not become infinite at the speed of light.

It, of course, is impossible to transform to faster-than-light state if such a state is absolutely forbidden by valid law. But references [16] reveal faster-than-light solutions to the Einstein's General Relativity solutions that entail travel through tunnels formed by warping space and held open with special fields. Furthermore, references [17] suggest that warping space by means of highly charged, highly excited nuclei might be possible with conceivable advancements of our technical art.

Such realms, where travel over enormous distances can be accomplished in negligible intervals of time, can be visualized as existing "above" the spacetime realm of existence of our material world. Therefore, just as aircraft that exceed takeoff speed climb above the length and breadth of earth, so Figure 6 visualizes spacecraft that exceed light speed during their rapid transits to distant worlds as climbing above the length and breadth of space and time.

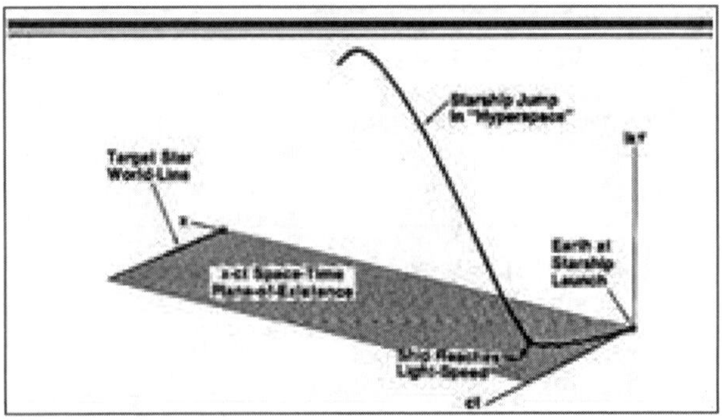

Figure 3. - Faster than light travel in Hyperspace

A Typical Schedule for Future Flight

Since one can't forecast or legislate when technical breakthroughs will be made, it is obviously impossible to lay out precise program schedules and milestones for the future of flight. Just as Figure 1 shows significant milestones that have been associated with the first century, can result in effects preceding causes in some propulsion would enable the first rapid transits to Mars, while reduced travel times and costs can be made. Relativistic time-lines in hyperspace seem to suggest that even time-travel may be possible with proposed trips made into the past or future if faster-than-light travel segments are possible. Such causality paradoxes can be possible by field propulsion and thereby could enable space commerce and fast trips to outer planets. It is also seen that achievement of high insufficient time for anomalous time ordering of acceleration and speed could enable interstellar events to occur. In this respect, probe missions and faster-than-light science describes faster-than-light solutions to the Einstein Relativity equation technology that could enable the first manned round-trip journeys to distant stars before the next century and possibly instantaneous travel within realms of existence in distant parts of our universe.

Setting Goals for Future Flight

Space goals have been an important part of the exploration of space. Those that have been visionary, and well-articulated have stirred the human spirit and enabled enormous advancement of our spaceflight art.

It is, therefore, proposed that an ambitious long-term space goal be set forth that will meet human needs and challenge the human spirit for as long as the next 100 years. Furthermore, it is suggested that the long-term goal be far beyond our present technical reach, requiring about the same rate of extraordinary aerospace advancement that we accomplished in the past. Less ambitious near-term goals would also be established. These would include currently envisioned goals that involve exploration of the Moon and Mars) since they would be necessary stepping stones to the long-term one.

Table 3 presents space exploration goals that include long-term activities and emphasize human needs, together with goals for nearer term activities that involve our next steps in space. These goals also include a long-term spaceflight intention – landing on a planet in another solar system within 100 earth years after the first landing on another celestial body (the moon). Such an ambitious interstellar goal may seem absurd and, indeed, it might ultimately be proved unattainable by any achievable mode of flight. But, this won't be known for sure for a very long time. Therefore, since "a man's grasp should always exceed his reach" [19]; let's "reach" the highest that we can envision in furthering the progress of flight.

Creation of new sources of clean energy, new modes of propulsion, and new means of exploiting the resources of space:

- To ensure adequate energy and resources for satisfactory quality of life on Earth for the next 100 years
- To create new Industries, careers, and opportunities for the future generations of people In the United States
- To return to the moon to stay and land people on Mars by 2019 (50 years after the first manned landing on the moon)
- To explore the entire solar system and land people on a planet in another solar system by 2069 (100 years alter the first manned landing on another celestial body— the moon)

Table 3. Example of Goals That Include an Initiative for Interstellar Flight

Starflight may not take eons of time to achieve, for if we could maintain the average rate of aerospace progress we accomplished during the first 89 years of flight, we will have surpassed light speed and reached distant stars before the next 89 years of flight have passed. Such progress would require breakthroughs in energy, propulsion and our understanding of so-called" matter, time and space" and these breakthroughs are also needed to help solve the serious environmental and economic problems facing the people of earth.

Space exploration plans must emphasize our next steps in space. However, it is recommended that these plans also set forth our intention to embark upon our first interstellar journey before the next 89 years of flight have elapsed. It is also recommended that a small fraction of spaceflight funding should be dedicated towards breakthrough research that would help make such star flight come to pass.

It, of course, is conceivable that flight progress will be eventually be stopped by insurmountable barriers that can never be overcome; that new sources of energy or new modes of impulsion will never be discovered; that light speed will never be surpassed or overcome; and that we will, before ever limited to an infinitesimal part of the vastness of space. But, since there is not yet overwhelming evidence that this is so, it seems too soon to give up our battle to overcome all barriers to flight. Therefore, until we have no hope whatsoever of traversing the stupendous gulfs of cosmic space, let our spaceflight goals and plans reflect our intention to strive for and to reach the further stars.

References

1. Stine,G.H., Space Tourism, the Unbelievable Market, The Journal of Practical Applications in Space, Vol. 1, No.4, pp. 71-76 (1990)

2: Woodcock, G.R., Space Transportation for Settlement of Mars, Proceedings of the Case for Mars International Conference (1990)

3. Khalil, A.,Futuristic Approach for Interstellar Travel - Advanced Magnetic Monopole Propulsion Concept, AIAA 28th Aerospace Sciences Meeting, AIAA 90-0615, (1990)

4.Forward, R.L., Negative Matter Propulsion, AIAA/ASME/SAE/ASEE 29th Joint Propulsion Conference, AIAA88-3168, July (1988)

5. Froning, H.D. and Mead, F.B., Propulsion for Rapid Transits Between Earth and Mars, Case for Mars N International Conference, University of Colorado, Boulder, CO, Air Force Astronautics Laboratory ALPAS 90- 067, (1990)

6. Bond, R.A., et a!, The Starlight Fusion Propulsion Concept, 41st. Inter Astra Federation, IAF 90-233, (1990)

7. Wheeler, T.A., Superspace and the Nature of Quantum Geometro dynamics, Topics in Nonlinear Physics, pp.615-644. Proceedings of the Physics Section, International School of Nonlinear Mathematics and Physics, Springer Verlag, (1968)

8. Aichison, T.J.R., Nothing's Plenty - the Vacuum in Modem Quantum Field Theory, 18. Contemporary Physics, Vol. 26, No. 4, (1985)

9. Puthoff, H.E., Ground State of Hydrogen as Froning, H.D., Requirements for Rapid Transport to the Further Stars, J. Brit Interplanet Soc. Vol. 36, pp. 227-230 (1983) a Zero-Point Fluctuation- Determined State, Phys. Rev. D, Vol. 35, No. 10, pp. 3266- 3269, (1987)

10. Froning, H.D., Use of Vacuum Energies for Interstellar Spaceflight, 36th Inter Astro Federation Congr., IAA-85-492, (1985)

11. Hansson, P.A., On the Use of Vacuum for InterstellarTravel,38th Inter Astro Federation Congr., IAA-87-611, (1987 Hansson, P.A., On the Use of Vacuum for Interstellar Travel, 40th Inter Astro Federal Congr., IAA89-667, (1989)

12. King, M.B., Tapping the Zero-Point Energy, Paraclete Publishing, ISBN 0-9623356-0-6, Provo, UT, 1989

13. Millis, M.G., Exploring the Notion of Space Coupling Propulsion, NASA Symposium "Vision 21" Space Travel for the Next Millennium, NASA Lewis Research Center, Cleveland, Ohio, (1990)

14. Froning, H.D., Field Propulsion for Future Flight, AIAA/ASMEISAE/ASEE 27th Joint Propulsion Conference, AIAA91-1990, June 1991

15. Cravens, D.J., Electric Propulsion Study, Air Force Astronautics Laboratory Report TR-89-040, September (1989)

16. Morris, Thome, et al., Wormholes, Time Machines, and the Weak Energy Condition, Physical review Letters, Vol. 61, No. 88, pp. 14-46

17. Forward, R.L., Space Warps: A Review of One Form of Propulsionless Transport, AIAA Joint Propulsion Conference, AIAA89-2332, July (1989)

18. Contemporary Physics, Vol. 26, No. 4, (1985) a Zero-Point Fluctuation-Determined State, Phys. Rev. D, Vol. 35, No. 10, pp. 3266-1987

19. Poem by Robert Browning entitled Andrea del Santo (The Faultless Painter), 1855

Specially Conditioned EM Radiation Research with Transmitting Toroid Antennas

H.D. Froning Jr. Flight Unlimited Flagstaff, AZ, USA

G.W. Hathaway. Hathaway Consulting Services Toronto, ON, Canada

37thAIAA/ASME/SAE/ASEE Joint Propulsion Conference and Exhibit 8-11 July, 2001 Salt Lake City, UTAH

Abstract

Experimental work to: (a) determine EM field characteristics associated with EM radiation created by alternating current flowing through toroidal coils at resonant frequencies, and (b) determine if the specially conditioned EM fields associated with such radiation could cause a discernible gravity modification, is described. This experimental work was the result of collaboration between Flight Unlimited (FU) and Hathaway Consulting Services (HCS), performed at the laboratories of HCS in Toronto, Canada during a test period in 1998 and during a test period in 2000. Tested toroid configurations included circular toroids with differing diameters and winding densities; and asymmetrical toroids for focusing EM radiation into narrower and more intense beams. The toroid configurations and the AC power and instrumentation systems available at HCS limited the experimental work to the relatively low radio frequencies (400 kHz to 110 MHz) of the electromagnetic spectrum.

Introduction

Just as airflight was not revolutionized until propeller propulsion was superceded by a new mode of impulsion (jet propulsion) so spaceflight may not be revolutionized until jet propulsion is superceded by a new mode of impulsion (field propulsion). Field propulsion would develop thrust by actions and reactions of fields instead of by combustion and expulsion of mass. And field actions and reactions that would greatly reduce propellant (the major portion of rocketship mass) and engine thrust requirements would be those that would reduce the resistance of gravity and inertia to ship acceleration.

One conceivable way of reducing the resistance of gravity and inertia is by accomplishment of a favorable coupling between those fields which underlie electromagnetism and gravity. But no significant coupling of ordinary em fields with those that give rise to gravity may be achievable because their essence is completely dissimilar. Yang (I) notes that "nonabelian fields which probably give rise to gravity are of more intricate topology and higher internal symmetry than the "abelian U(I)fields that underlie ordinary electromagnetism. In this respect, Barrett (2,3) has identified two ways of transforming ordinary EM fields into specially conditioned EM fields of nonabelian form and higher than U(I) symmetry. One identified way of creating such fields is modulating the polarization of EM wave energy emitted from microwave or laser transmitters. Such polarization modulation creates EM fields of nonabelian form and SU(2) symmetry within beams of radiated power that can be focused into very narrow beams of very high energy density. Thus an experiment to detect possible gravity modifications within narrow polarization modulated laser beams has been submitted to the NASA Breakthrough Propulsion Physics (BPP) program. This experiment is described in (4).

Another way of transforming ordinary EM fields into specially conditioned EM fields of nonabelian form and SU(2) symmetry is with toroidal coils through which alternating current is flowing at resonant frequencies. Barrett (3) shows that such specially conditioned EM radiation includes not only electric and magnetic field energy – but A Vector potential field energy as well. Barrett predicts that A Vector field intensity maximizes at discreet resonant frequencies. Thus, if an A Vector potential field underlies the essence of gravitation, gravity modification might be possible in the vicinity of toroids transmitting at such frequencies.

Fabrication and testing costs were significant for polarization modulated laser beams. However, they were found to be relatively modest for toroidal coils-configured for operation in the lower (radio-frequency) range of the em spectrum. Thus a cooperation between Hathaway Consulting Services (HCS) and Flight Unlimited (FU) was established to: test Barrett's hypotheses as to specially conditioned EM radiation emitted from toroidal coils; and to determine if gravity modification could occur within such radiation. Probability of gravity modification by radio frequency radiation from inexpensive toroids was deemed to be very low. But it was hoped that the tests would reveal interesting electromagnetic phenomenon and extend our knowledge of electromagnetics.

TRANSMITTING TOROID ANTENNAS

EM wave propagation by transmitting toroid antennas has been examined by various investigators for more than a decade. Examples are U.S. Patent No. 4,751,515 awarded to Corum for an "electrically small, efficient electromagnetic structure that may be used as an antenna or waveguide probe" and U.S. Patent No. 5,442,369 awarded to Van Voorhies for an antenna that "has windings that are contra wound in segments on a toroid form and that have opposed currents on selected segments". In this respect, Barrett (4) has shown that specially conditioned EM fields of SU(2) symmetry and nonabelian form can be created by transmitting toroid antennas -- as in Figure 1

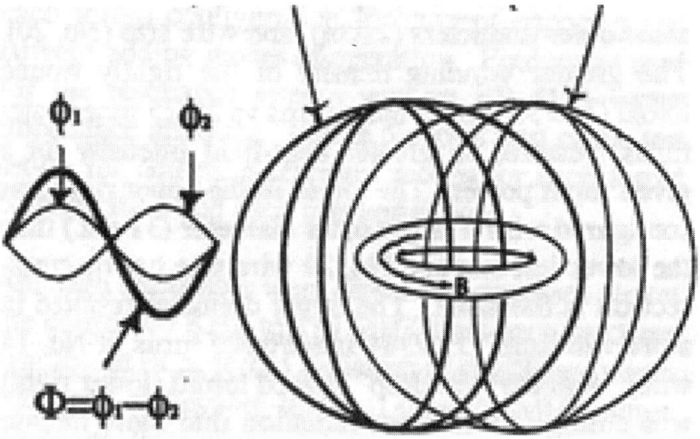

Fig. 1 A Vector Potential Patterns

The magnetic and electric fields which encompass a transmitting toroid are accompanied by A Vector potential fields, and the alternating current flow produces overlapping A Vector potential patterns which encircle the toroid ring, as shown in Figure 1. These A Vector patterns combine into "phase factor" waves which represent disturbances in A Vector potential. The maximum disturbances in A Vector potential occurs as phase factor wave intensity peaks at the resonant frequencies where A Vector potential patterns are exactly out-of-phase, and a predicted pattern of these disturbances is in Figure 2.

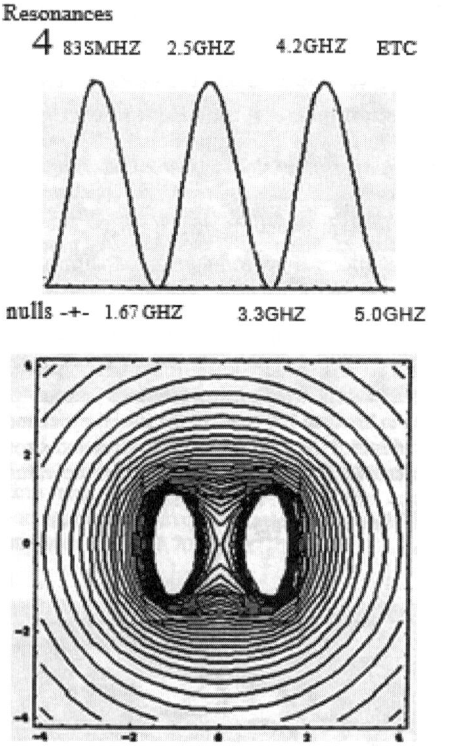

Fig. 2 – "Phase Factor" Wave Patterns

Resonant frequencies are determined by the shape and dimensions of the toroid, and by the propagating direction and speed of the alternating electric current thru its windings. And, if an A Vector potential field underlies the essence of gravitation, the probability of gravity modification in the toroid vicinity would be highest at resonant frequencies.

INITIAL TOROID EXPERIMENTS

Initial experimental work involved:(a) fabrication of transmitting toroid antennas that, according to (4), should emanate specially conditioned EM radiation; and testing of the toroids at low power levels at the laboratories of HCS in Toronto. The general goals of this initial work, which was performed on March 6 and 7 during 1998, were: perfection of techniques for fabricating toroidal coils, detection of resonant phenomenon indicative of A Vector potential resonances with such coils; and identification of problems associated with operating toroidal coils over wide frequency ranges and at significant power levels.

Most of the goals of the initial work were achieved. Toroid antennas with conventional and caduceus windings were successfully fabricated, and although no instrumentation (such as Josephson Junction arrays) were available for directly detecting A Vector fields, measured resonances (reversals in phase and amplification of signal strength) were in good agreement, as indicated in Figure 3 [*missing from this document – Ed.],* with Barrett's predictions predicated on occurrence of A Vector fields. Heat generated by current flow within the relatively thin windings of the toroids and their relatively fragile styrofoam interiors limited input power to less than 100 watts in the initial experiments. This identified the need for thicker wires and stronger structures for higher toroid power and temperature.

FOLLOW-ON TOROID EXPERIMENTS

Because results of the initial toroid experiments were somewhat encouraging, it was decided to have a follow-on experimental program, which included toroids configured for much higher power levels, at HCS between June 9 and June 15, 2000. It included: (a) signal phase/amplitude tests to precisely determine the resonant frequency characteristics of each different transmitting toroid configuration; (b) magnetic field measurements to map EM field intensity in the vicinity of each toroid; (c) propagation characteristics of toroid radiation; and (d) limited gravitometer testing to search for a gravitational disturbance at a one location near one of the transmitting toroids.

Toroid Configurations Tested

To our knowledge, transmitting toroid antennas built and tested by most other investigators have been designed for communication purposes -- with wires loosely wound (widely separated) around the toroid's ring in order to maximize far field intensity and range. By contrast, our tested toroids were "tightly wound" to maximize near-field intensity for possible gravity modification – not far field range for communication. Our tested toroids were "contra-wound" in a caduceus pattern to allow two types of modulation. One, in which current flowed in opposite directions in crossing wires, resulted in an "opposing" or "bucking" mode which caused opposing magnetic fields that cancel themselves along the toroid ring centerline. The other, in which current flowed in the same direction -- resulted in an "adding" mode. Figure 4 shows the 4 different toroid configurations that were tested during the follow-on experimental program.

The loosely-wound toroid (upper left) was built for comparing its near-field intensity with that of the tightly wound toroid (lower left). Both toroids had similar cross sections (approximately 4.0 cm) and the loosely-wound toroid (upper left) was built for comparing its near-field intensity with that of the tightly wound toroid (lower left). Both toroids had similar cross sections (approximately 4.0 cm) and the and the wire size number 20 same outer diameter (21 cm)

The greater winding density of the tightly wound toroid (350/333 inner/outer turns vs 26/25 inner/outer turns) resulted in greater near-field intensity for a given input power. The toroid in the upper right was configured with a larger outer diameter (31 cm.) than the lower left one and No. 20 wire size but its cross- section is the same. The larger diameter resulted in more windings (398/384 inner/outer turns of No. 14 wire). And the "tear drop" shaped toroid (lower right) was configured to focus radiation into more intense and elongated beams. Its length, breadth and thickness was 26.5, 18.0, 2.5 cm. It had a hole diameter of 7.3 cm and 95/88 inner/outer turns of No. 14 wire.

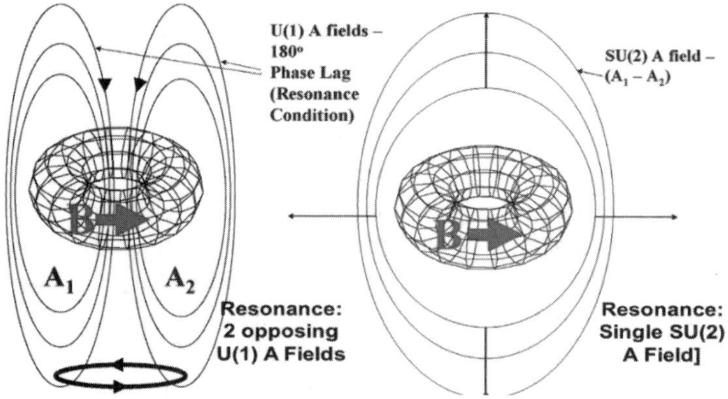

Fig. 3 Effect of Counter Rotating EM Fields in U(1) which Creates an SU(2) A Field

And ,because of their stronger structure (hard maple wood) and larger wire diameter, the tear drop and larger diameter toroids could withstand the heating associated with 1.0 kW of radiated empower As in the first test series, resonant conditions (revealed by reversal in signal phase and rise in signal amplitude) were searched for at all ac frequencies between 400 kHz and 110 MHz. This was done for each toroid configuration for current opposing and current adding modes of operation. Equipment used for the resonance sweeps was an HP 4l93 vector impedance analyzer. Figure 5 shows part of the test set-up for detecting resonant modes for each toroid configuration and each operating mode.

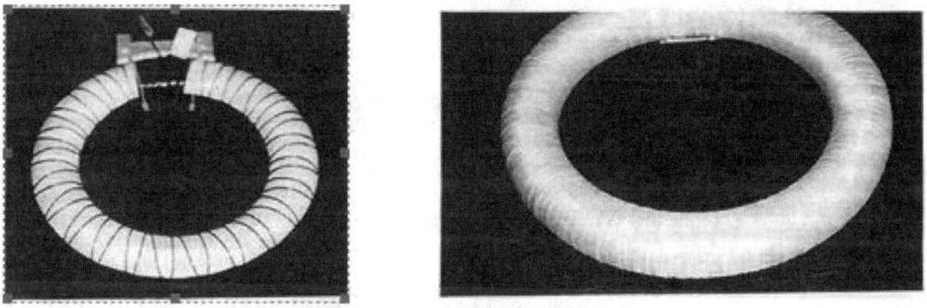

Fig. 4 - Toroid Configurations Tested

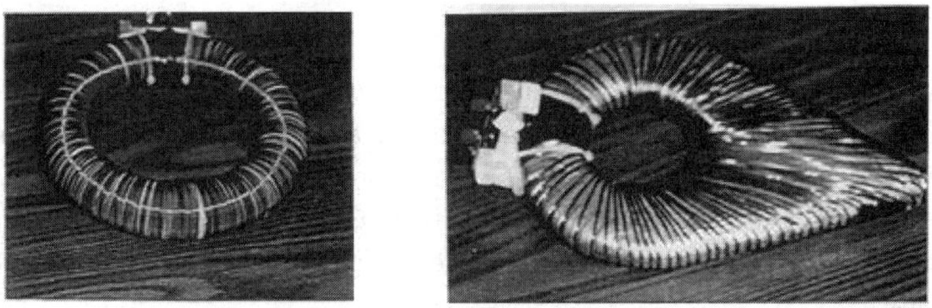

Although resonances were detected throughout almost the entire 400 kHz to 110 MHz frequency spectrum available at HCS, toroid radiation of significant power was only achievable in the 1.0 to 20 MHz range. Resonant frequencies selected for measuring field characteristics of each toroid were therefore within this range. Selected resonant frequencies for the large diameter toroid were 2.36 and 17.30 MHz for current-adding and current-opposing modes of operation, while those for the medium diameter toroid were 2.36 and 18.30 MHz. Selected resonant frequencies for the tear drop toroid was 5.66 and 3.94 MHz for current-adding and current-opposing, while the current-opposing, resonant frequency selected for the loosely wound toroid was 19.70 MHz.

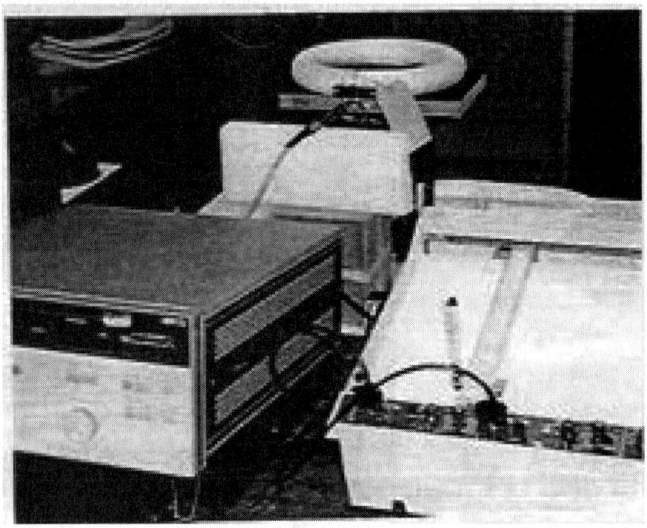

Fig 5. Resonance Sweep setup.

Toroid Field Intensity Measurements

Magnetic field intensity was measured out to 50 cm from each toroid center, and along the upper and lower surface of each toroid as well. For an applied power of 10 W, the magnetic field component of each toroid's radiation was measured by a small magnetic pick-up coil shown in Figure 6, which converted the actual magnetic field intensity into an equivalent electric field strength (in microvolts per meter).

Fig 6. Magnetic field Probe.

Fig 7 Toroid range set-up

Variation of the large diameter toroid's field strength with range (out to 10 meters) was measured with various types of antennas outside the HCS facility with the test set up as indicated in Figure 7. Data consistent with expected near-field signal strength variation with range was measured when the toroid was radiating in a current-adding resonant mode at 1.20 MHz. But measurements in a current-opposing resonant mode were anomalous -in that no significant signal strength variation with range was detected.

Search For Gravity Modification

Final toroid testing activity was searching for gravitational field modifications within the specially conditioned EM field regions surrounding toroids radiating at resonant frequencies. Gravitational disturbances were searched for with a "Prospector Model 420" gravitometer, manufactured by W. Sodin Ltd, which is capable of detecting changes as small as one-millionth of one percent of ambient gravity. This gravitometer's stainless steel shell and aluminum base does not provide complete magnetic field shielding. But its Dewar-enclosed, all-quartz mechanical balance system is not influenced by ordinary EM emissions. Unfortunately, preceding test activities took longer than expected, leaving time to search for gravity modification for only one of the toroids (the large diameter one) at only one location with respect to the gravitometer. The limited time remaining also required a very rapid toroid/gravitometer set up. This was achieved by the positioning shown in Figure 8.

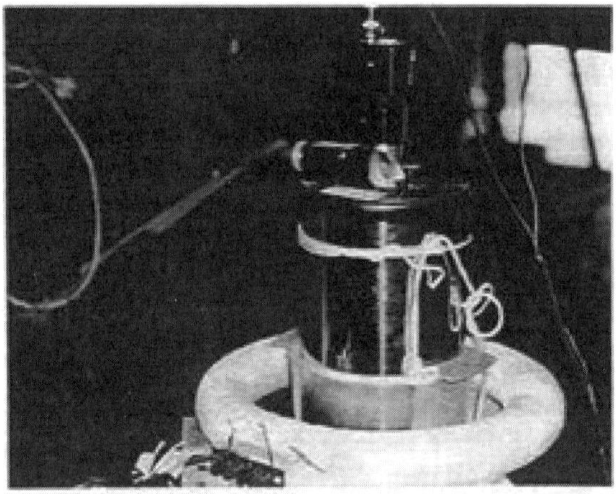

Fig. 8 Toroid Gravitometer setup

TOROID TESTING RESULTS

Resonant frequencies between 400 KHz and 110 MHz were obtained for each toroid. And, for the purposes of mapping magnetic field intensity in each toroid's vicinity, one resonant frequency was selected for each operating mode for each toroid. Field intensity out to 10m from the large diameter toroid was also measured together with the influences of magnetically shielded structures on its field intensity. Finally, the effect of large diameter toroid field intensity on gravity modification was explored. The results of these efforts are summarized in the following sections.

Toroid Resonance Determination

Resonant frequencies for current-adding and current-opposing operating modes were obtained for each toroid. Figures 9 and 10 show examples of the resonances obtained for the large diameter and tear-drop toroids throughout the 400 kHz-110 MHz radio frequency spectrum available at HCS.

Toroid Field Patterns

Magnetic field intensity variation (vertical, Y-Axis in Figure 11 – Ed. Note) with radial distance for the three circular toroids was similar with intensity maximizing near the inner surface of each toroid' s ring. And, as would be expected, intensity diminished rapidly with increasing distance above and outside each toroid. Figure 11a and 11b show no definite trend with respect to the influence of toroid diameter. Higher magnetic field intensity is achieved by the smaller diameter toroid in a current-opposing mode of operation while higher magnetic field intensity is achieved by the larger diameter toroid in a current- aiding mode. Figure 11c shows a definite trend, with increased windings over a given toroid geometry resulting in increased magnetic field intensity.

Significant focusing of the electromagnetic energy radiated from the asymmetrical "tear drop" toroid was accomplished. Figure 11d shows that magnetic field intensity is enormously greater at given distances forward of the center of the toroid's hole than for the same distances aft of the hole center. Figure 9d also shows a top and front view of the tear drop toroid field pattern for a given magnetic field intensity. It is seen that more electromagnetic energy is focused into the forward direction than into the aft or side directions and that the toroids flattened shape (its reduced thickness) causes less radiation to be dissipated in directions transverse to the toroid plane.

One interesting discovery was formation, in the circumferential direction, of standing EM waves along the upper and lower surfaces of transmitting toroids. No standing wave measurements were made on the loosely wound toroid. However, numbers of magnetic field peaks and nodes measured circumferentially on the top and bottom surfaces of the other circular toroids were 8 for the medium diameter toroid and 10 for the large diameter one. And at least 6 magnetic field peaks and nodes were measured on the top and bottom surface of the tear drop toroid.

Attenuation of Toroid Field Intensity

Field strength attenuation with range from the center of the large diameter toroid is shown in Figure 12. It was about as expected when radiating at a resonant frequency of 1.2 MHz in a current adding mode of operation, with a steep signal drop greater than $(r)^3$ out to about 1.0 meter and with an expected near-field $(r)^2$ variation between 3 and 10 meters from the toroid. But measurements made with the toroid radiating at a resonant frequency of 17.3 MHz in a current opposing mode indicated no significant variation in signal strength at distances 3 to 10 meters from the toroid. *After continual measurements and re-measurements with various types of antennas, we have no definitive explanation for lack of signal strength reduction with increasing range – other than the possibility of operating slightly off resonance and a significant drop in signal strength.*[††]

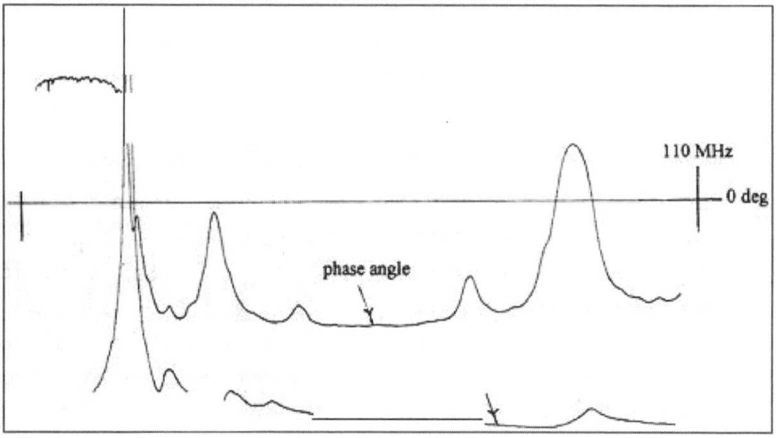

Fig. 9 Resonance for Large diameter toroid

[††] **Ed. Note:** Here Froning is disclosing a very valuable, well-known predicted property of scalar waves, which is the type of non-Hertzian scalar potential varying wave that emanates from a "contra-wound" coil in a caduceus pattern. See the IRI Report #303 "Scalar Potentials, Fields, and Waves" for a comprehensive reference guide to this phenomenon. https://www.integrityresearchinstitute.org/catalog/physicsReports.html

Additional anomalous behavior may also have been observed for the large-diameter toroid, in that almost identical signal strength was measured at a given location and distance from the toroid when radiating in free space and when radiating from within a magnetically shielded (mu metal) enclosure. These results are considered inconclusive because stray signals were detected from power supply leads which were outside the shielded enclosure. Since measured signals were almost identical for both shielded and un- shielded conditions, and since it is unlikely that almost all of the measured free space signal was from the power supply leads, it is conceivable that some of the toroid's EM wave energy was propagated through the magnetically shielded mu metal walls.‡‡

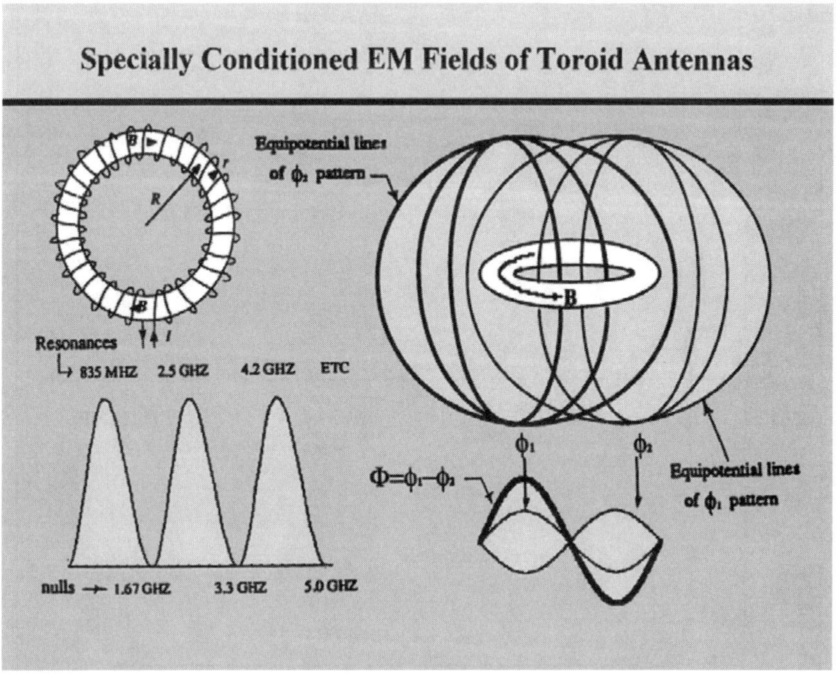

‡‡ **Ed. Note:** The propagation of scalar waves right through magnetic shielding like mu metal (highest permeability nickel alloy metal available today), without attenuation, is also expected for non-Hertzian scalar waves that are non-electromagnetic in nature.

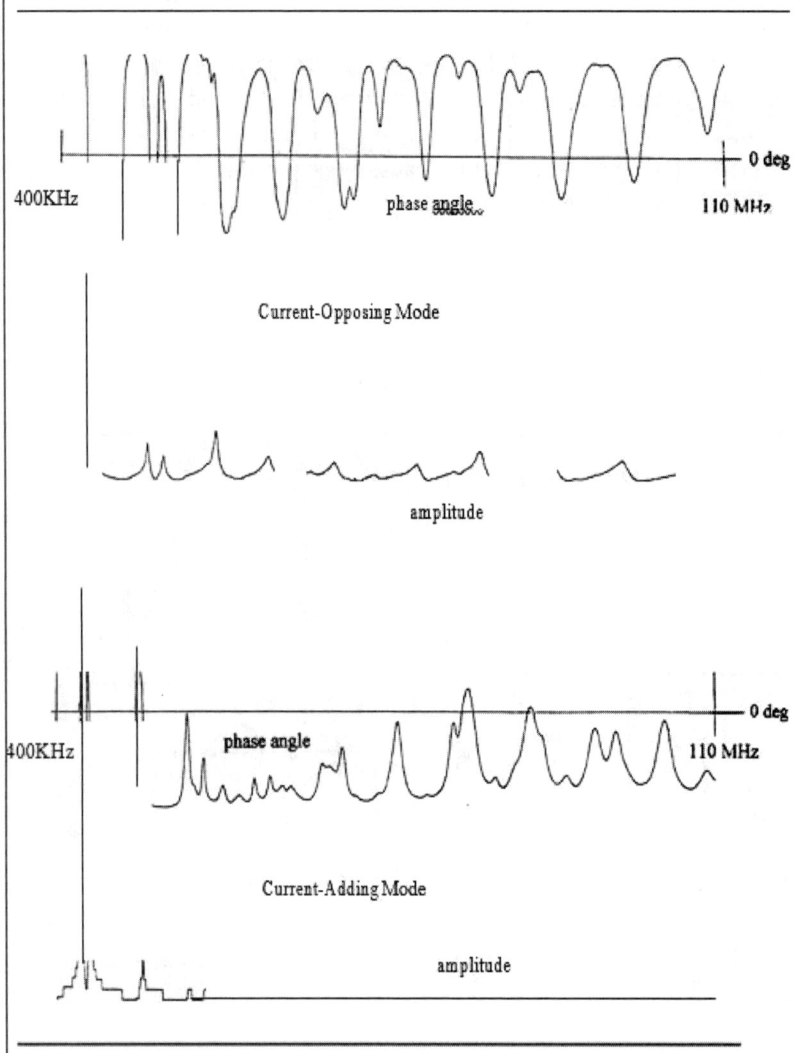

Fig. 10-Resonances for "Tear-Drop" Toroid

Fig. II -Magnetic Field Patterns in Toroid Vicinity

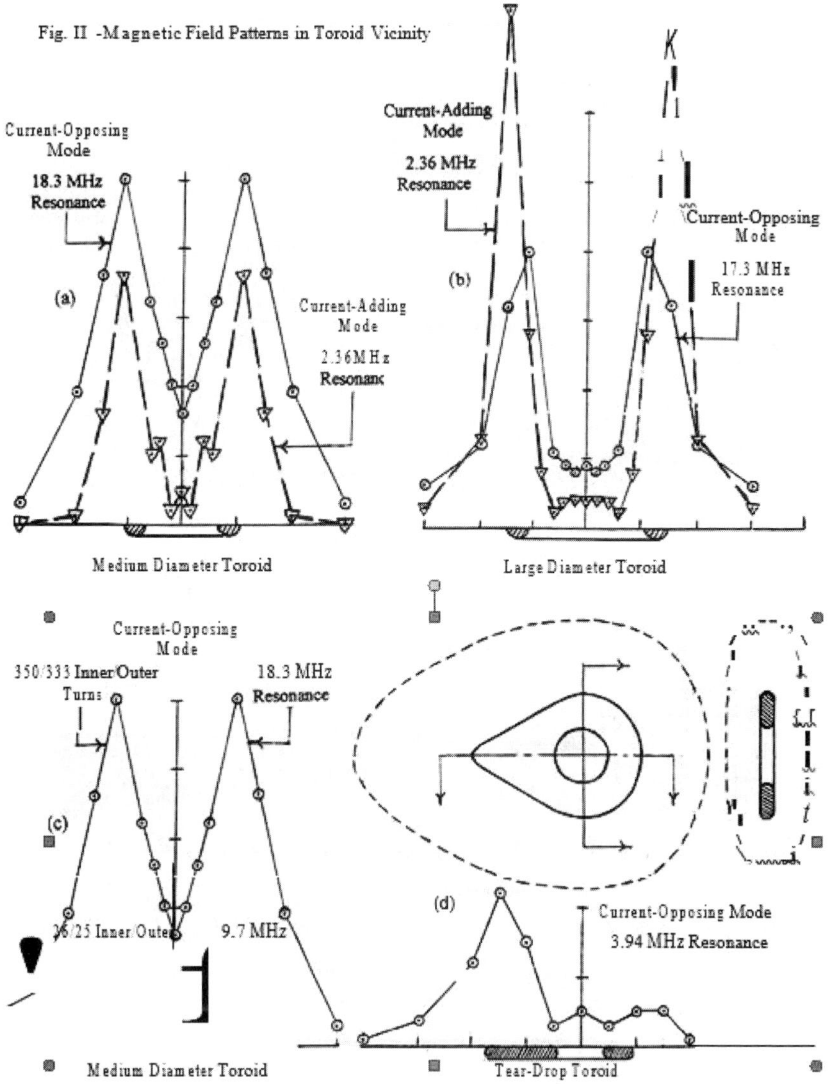

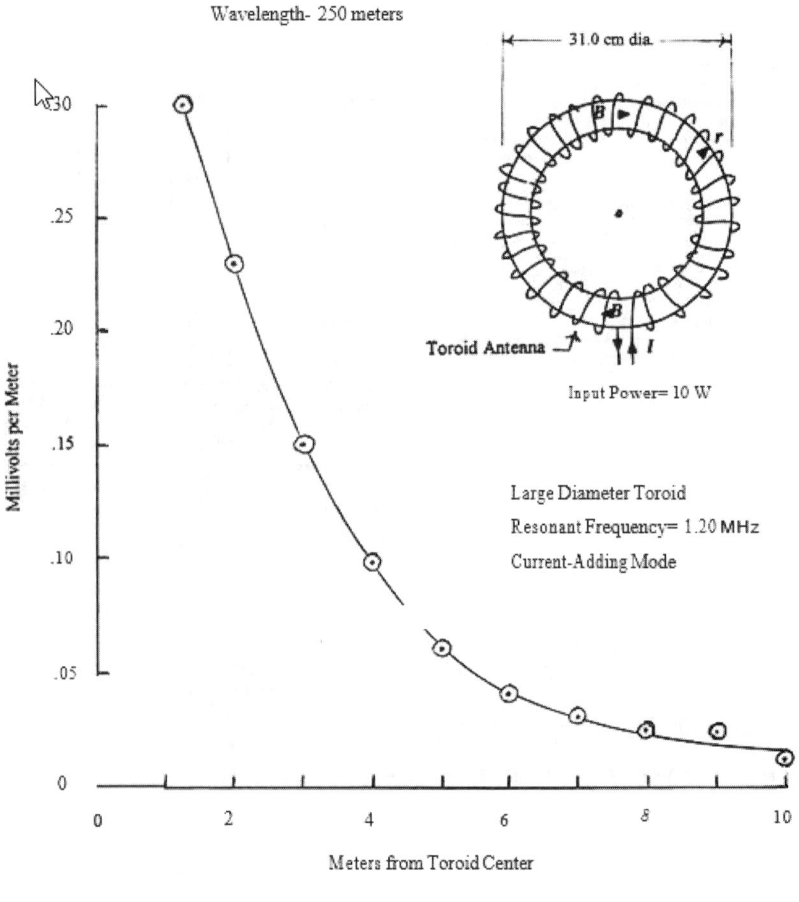

Fig. 12 - Attenuation of Toroid Signal with Range

Search for Gravity Modification

The possibility of gravity modification in the vicinity of transmitting toroid antennas was briefly investigated by use of the Prospector 420 gravitometer and the large diameter toroid radiating up to 0.5 kW of average power at the resonant frequencies associated with current-aiding and current-opposing operation modes. For these powers and operating modes, no discernable gravity modification was detected for the single toroid/gravitometer positioning that time allowed.

As previously mentioned, time limitations required a rapid toroid/gravitometer test set-up which resulted in the gravitational mass being located in a magnetic field region whose intensity was subsequently found to be much less than magnetic field intensity existing in other locations. For the current opposing mode of operation, measured magnetic field intensity at the gravitometer test mass location was only about 15 percent of the maximum intensity measured near the toroids inner diameter. And toroid magnetic field intensity at the gravitometer test mass location was only about 2 percent of the maximum measured magnetic field intensity for the current-adding mode.

One conceivable reason for non-discernable gravity modification is, of course, 5 to 50 times less EM field intensity at the single location probed by the gravitometer, as compared to locations of maximum intensity. But another reason could be dissimilarity in field topologies associated with toroid em emanations and gravity. And still another reason could be enormous possible differences in the frequencies and wavelengths characterizing gravitational fields and those that characterize electromagnetic fields created by transmitting toroid antennas.

SUMMARY AND CONCLUSIONS

Although interesting phenomenon are associated with EM fields created by alternating current flowing at resonant frequencies through toroid coils, no discernable gravity modification (caused by coupling of these fields with those of gravity) was detected. Interesting electromagnetic phenomenon were: (a) standing EM waves along toroid surfaces; (b) EM wave energy focused into more intense beams by asymmetrical toroid shapes; and (c) poss1bly, EM wave propagation through magnetically shielded enclosures There might have been increased probability of detecting a discernable gravity modification if there had been time for gravitometer measurements in regions where toroid field strength was much greater.

The possibility of anomalous wave propagation should be confirmed or refuted by re-testing the large diameter toroid within a magnetically shielded structure that encloses both the toroid and its power leads.

Zero gravity modification within the radio frequency EM fields surrounding transmitting toroid antennas should be confirmed by a gravitometer search throughout the entire vicinity of the large diameter toroid.

ACKNOWLEDGMENTS

This experimental work was motivated by theoretical work by Dr. Terence Barrett with respect to specially conditioned EM fields created by transmitting toroid antennas. Dr. Barrett also contributed useful suggestions as to recommended toroid configurations and test procedures. The most recent experimental work was conducted with the considerable assistance of Mr. Blair Cleveland, who was intimately involved in every aspect of the test preparations conduction and data gathering portions of this work.

REFERENCES

(1) Yang, C.N., "Gauge Theory", McGraw-Hill Encyclopedia of Physics, 2"" Edition, p483, (1993)

(2) Barrett, T.W.,"Electromagnetic Phenomenon not Explained by Maxwell's Equations", Essays on Formal Aspects of Electromagnetic Theory, p6, World Scientific Publ. Co., (1993)

(3) Barrett, T.W., "Toroid Antenna as Conditioner of Electromagnetic Fields into (Low Energy) Gauge Fields", Proceedings of the: Progress in Electromagnetic Research Symposium 1998, (PIERS '98) 13-17 July, Nantes, France (1998

(4) Froning, H.D., Barrett, T.W., "Theoretical and Experimental Investigations of Gravity Modification by Specially Conditioned EM Radiation", Space Technology and Applications International Forum (STAIF) 2000, Editor: Mohamed S. El-Genk, Published by the American Institute of Physics (2000)

Electromagnetic Radiation Experiments with Transmitting, Contra-Wound Toroidal Coils

H. David Froning,[a] George D. Hathaway[b] and Blair Cleveland[b]

[a] *PO Box1211*
Malibu CA, 90262, USA
310-459-5291; froning@infomagic.net

[b]*Hathaway Consulting Services*
1080 19th Sideroad
King City, L7B1KS, Ontario, Canada
ghathaway@ieee.org

Abstract. Except for Quantum Electrodynamics, there has been no real extension of Maxwell's classical electromagnetic (EM) field theory since his electromagnetic EM field equations were developed in 1864. These equations describe the behavior of vector fields of low (U1) Lie group symmetry. In this respect, Terence W. Barrett has used topology, group and gauge theory, to extend Maxwell theory into tensor fields of higher symmetry form: SU (2), SU (3), and higher, that describe the behavior of specially conditioned EM fields. One of Barrett's ways of emitting SU(2) EM fields was driving alternating current through toroidal coils at any of the resonant frequencies that will occur for a specific toroid geometry. Experiments to explore the possibility of achieving such resonant frequencies and SU(2) EM emissions will be described.

Keywords: A Vector Potential, A Fields, Electric Fields, Magnetic Fields, Self Induced Transparency, SU(2), Tensor Fields
PACS: 04.30, 41.20Jb, 41.60Bg

INTRODUCTION

In 1864 Maxwell described a unification of electricity and magnetism with equations that later followers distilled into the four equations now known as the "Maxwell equations" that conform to the laws of electromagnetism (EM) formulated by Gauss, Ampere, Coulomb and Faraday. In the 1950's Feynman and others made Maxwell's classical EM theory compatible with Quantum theory and resulted in Quantum Electrodynamics (QED). QED was consistent with both quantum mechanics and special relativity and precisely predicted interactions between radiation and matter. But, despite its reformulation from quaternionic to vector algebra form, no extension of Maxwell theory has been made during its 146 years of life - despite the inability to accurately explain many observed EM phenomena.

In most cases, EM radiation fields are correctly and adequately described by the classical Maxwell equations which is a theory of U(1) symmetry form. However, in special topologies or situations or boundary conditions, radiation fields are produced that require an extension of Maxwell theory to higher symmetry. Addressing such situations, Barrett (2008) has used topology, group and gauge to derive SU(2) EM radiation fields for those cases of specially conditioned radiation. A specific SU(2) EM radiation field proposed by Barrett is that which is emitted by toroidal coils when alternating current flows through them at any of the resonant frequencies that will occur for a specific toroid geometry. Toroidal coils have been constructed and tested, and. this paper summarizes testing results.

MAXWELL EQUATIONS FOR ORDINARY AND CONDITIONED EM FIELDS

Using group theoretic methods, EM radiation fields of SU(2) symmetry can be created by special conditioning of conventional U(1) EM fields. Table 1 shows Maxwell's (U(1) symmetry) four equations describing electric field strength (*E*), magnetic flux density (*B*) and current density (*J*). The *E* and *B* fields of force can be related to a "magnetic vector potential" (*A*) and "scalar electric potential" (φ). These potentials are unphysical and mere mathematical conveniences in terms of the U(1) field theory. However, in the SU(2) field theory, the potentials *A* and φ do have physicality (Barrett, 2008). Table 2 shows extended Maxwell equations that describe propagation of specially conditioned SU(2) EM fields. These Maxwell Equations are based on tensor, rather than vector field terms, and include *E* and *B* fields as U(1) Maxwell Equations do. But they include additional terms: (i) which can be viewed as the square root of -1 or as an orthogonal rotation occurring in x, y, z, ct spacetime; electron charge (q); and added interactions ***A x E, A x B, A • E***; and ***A• B*** (Barrett, 2008, pp 145-147). These tensors, or matrices function as operators obeying non-commutative, non-Abelian algebra. So, ***A x B*** does not = ***B x A*** for SU(2) fields.

TABLE 1. Maxwell Equations for Ordinary U(1) EM Fields.

Gauss' Law	$\nabla \bullet \boldsymbol{E} = J_0$
Ampere's Law	$\frac{\partial \boldsymbol{E}}{\partial t} - \nabla \times \boldsymbol{B} - \boldsymbol{J} = 0$
Coulomb's Law	$\nabla \bullet \boldsymbol{B} = 0$
Faraday's Law	$\nabla \times \boldsymbol{E} + \frac{\partial \boldsymbol{B}}{\partial t} = 0$

$$E = -\frac{\partial A}{\partial t} - \nabla\phi, \quad B = \nabla \times A$$

TABLE 2. Maxwell Equations for Specially Conditioned SU(2) tensor EM Fields (Barrett 2008).

$\nabla \bullet \boldsymbol{E} = J_0 - iq(\boldsymbol{A} \bullet \boldsymbol{E} - \boldsymbol{E} \bullet \boldsymbol{A})$
$\frac{\partial \boldsymbol{E}}{\partial t} - \nabla \times \boldsymbol{B} - \boldsymbol{J} + iq[\boldsymbol{A}_0, \boldsymbol{E}] - iq(\boldsymbol{A} \times \boldsymbol{B} - \boldsymbol{B} \times \boldsymbol{A}) = 0$
$\nabla \bullet \boldsymbol{B} + iq(\boldsymbol{A} \bullet \boldsymbol{B} - \boldsymbol{B} \bullet \boldsymbol{A}) = 0$
$\nabla \times \boldsymbol{E} + \frac{\partial \boldsymbol{B}}{\partial t} + iq[\boldsymbol{A}_0, \boldsymbol{B}] = iq(\boldsymbol{A} \times \boldsymbol{E} - \boldsymbol{E} \times \boldsymbol{A}) = 0$

Noting that, unlike fields of Table 1, these fields are tensor (matrix operator) fields not vector fields.

The well known Lorentz force (*F*) arises from an electromagnetic interaction that involves *B* and *E* fields and the velocity (v) of charge clusters with charge (e). Table 3 shows force equations for both the U(1) EM vector fields in terms of the magnetic vector potentials and electric scalar potentials that underlie these U(1) vector fields and SU(2) tensor fields in terms of the vector and scalar potentials and it is noted that extra terms are in the SU(2) force equation. Thus, SU(2) field interaction forces can be different in magnitude and direction than U(1) field forces.

TABLE 3. Lorentz Force comparison for U(1) EM and SU(2) EM Fields (Barrett, 2008).

U(1) Lorentz Force	$\mathscr{F} = e\mathrm{E} + ev \times \mathrm{B} = e\left(-\frac{\partial \mathrm{A}}{\partial t} - \nabla\phi\right) + ev \times \left((\nabla \times \mathrm{A})\right)$
SU(2) Lorentz Force	$\mathscr{F} = e\mathrm{E} + ev \times \mathrm{B} = e\left(-(\nabla \times \mathrm{A}) - \frac{\partial \mathrm{A}}{\partial t} - \nabla\phi\right) + ev \times \left((\nabla \times \mathrm{A}) - \frac{\partial \mathrm{A}}{\partial t} - \nabla\phi\right)$

SU(2) EM RADIATION FROM TOROIDAL COILS AT RESONANT FREQUENCIES

Barrett (1998) describes one way of emitting SU(2) EM radiation. It is driving alternating current through toroidal coils at any of the resonant frequencies that will occur for the specific toroid geometry. Figure 1 shows the EM field pattern of a transmitting toroid as a U(1) A potential field overlapping in polarity across the toroid – with a set of equipotential lines of φ_1 pattern and another set of equipotential lines of φ_2 pattern. Also shown is a resonant frequency when the phase is such that (φ_1- φ_2) is maximal, and the various resonant radio frequencies that can occur for a given toroid. At every resonant frequency, the alternating difference in the U(1) potential fields is maximized for the whole toroid. An SU(2) potential field is maximized as well and this results in a maximum transmitted signal.

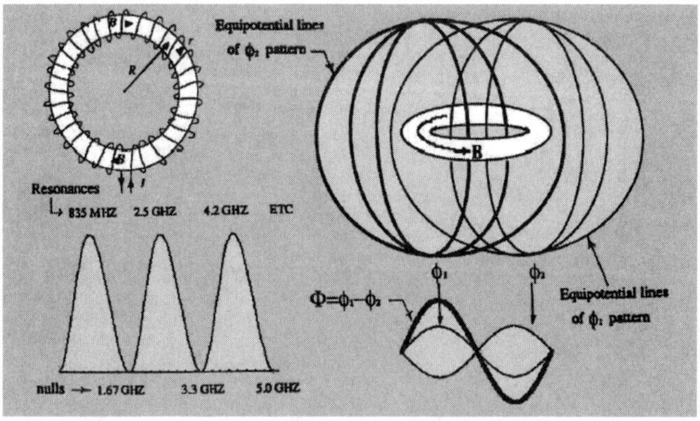

Figure 1. Potential patterns surrounding transmitting toroidal coils.

It was expected that resonant frequencies for radiating toroids would be revealed at the frequencies where the phase of alternating current would reverse and feed-point impedance would peak. Toroidal coils configured to emit SU(2) EM radiation were tested at Hathaway Consulting Services. Frequency sweeps revealed resonances for every tested toroid. Figure 2 shows good agreement between measured resonant frequencies and resonant frequencies predicted by Barrett (1998; 2000) for SU(2) field emission. The sweeps also revealed that EM fields emitted from each toroid would dramatically increase in strength within those narrow frequency bands where the resonances occurred.

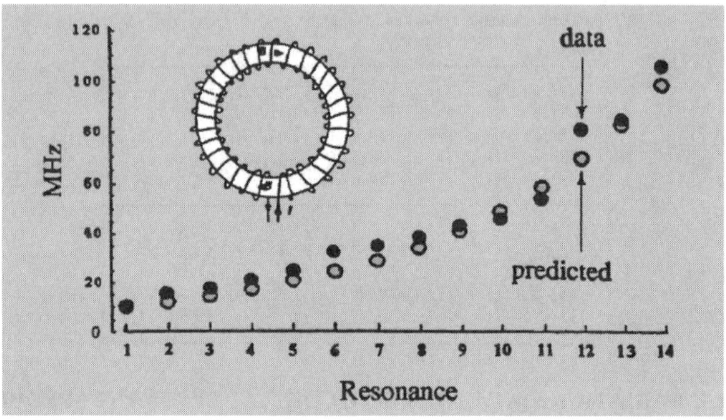

Figure 2. Good agreement between predicted and measured resonant frequencies.

The encouraging initial observations resulted in testing of 4 different toroid configurations shown in Figure 3. They typified variations in toroid major and minor diameter and coil winding density; and included asymmetric shapes. All toroids embodied contra-wound (caduceus) coils whose two interwoven wires make the same pattern as DNA strands. This allowed a current-adding mode, where equal currents flow in the 2 wires in the same direction through the coil; and a current-opposing mode, where equal current flowed in each wire in opposite directions. And, magnetic field probes confirmed that no magnetic field was induced inside each coil for this current-opposing mode.

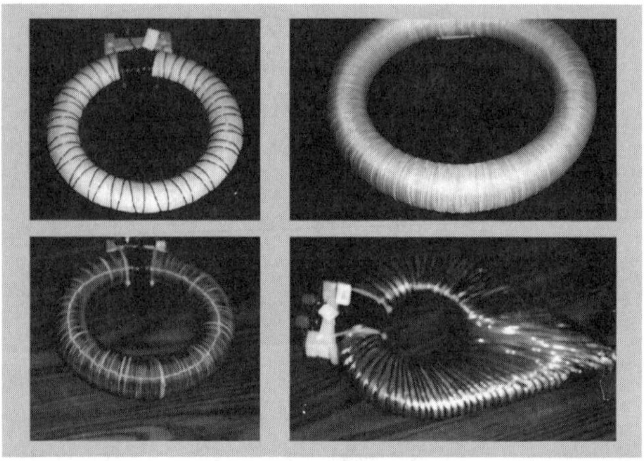

Figure 3. Various contra-wound caduceus toroids that were tested in second test series.

Radio-Frequency sweeping with equipment such as that shown in Figure 4 was conducted to find all resonant frequencies existing in the 400 KHz to 110 MHz range for each toroid. This was done for both current-adding and opposing modes of operation. Then, for at least one resonant frequency for each toroid, magnetic field strength was measured on the surface of (and, in the vicinity of,) each toroid for the current-adding and current-opposing modes.

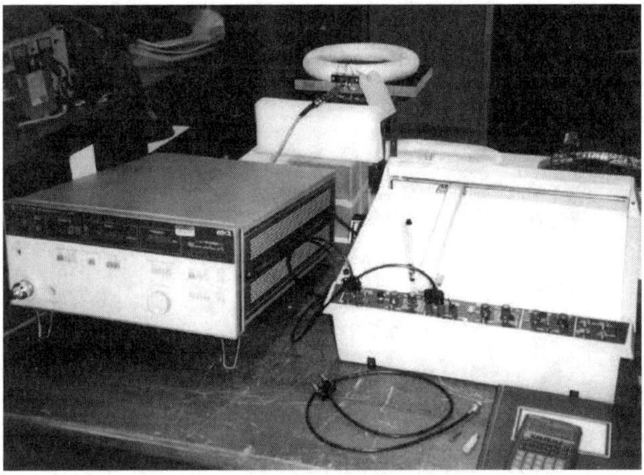

Figure 4. Typical equipment (HP impedance analyzer and plotter) for detecting resonant toroid frequencies.

Figure 5 shows the response of the larger diameter toroid to flowing alternating current in it at frequencies from 400 KHz to 110 MHz for current-adding and current-opposing modes. Shown are many resonances occurring over the frequency spectrum in current-adding mode. Fewer (but much stronger) resonances occur in current-opposing mode.

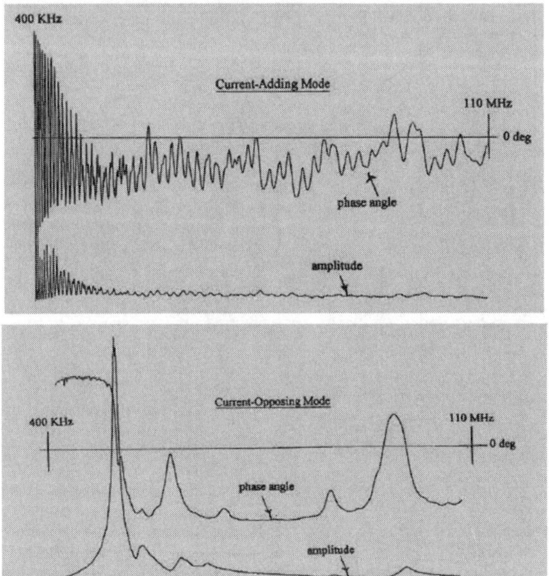

Figure 5. Larger diameter toroid response to frequency in current-adding and current-opposing modes.

Figure 6 shows the magnetic probe used to measure the vertical component of magnetic field intensity at locations on and near the surface of each toroid when the toroid was radiating at resonant frequency. It must be noted that this magnetic probe could not detect the individual undulations of any SU(2) A fields emitted by the toroids. But probe presence would cause symmetry breaking of any SU(2) A fields into U(1) B fields that the U(1) probe could detect.

Figure 6. Magnetic probe used to measure magnetic fields generated by transmitting toroids at resonant frequencies.

Figure 7 shows steep magnetic field gradients forming over tightly-wound toroidal coils that may be emitting SU(2) EM radiation. Shown is higher magnetic field peak for the smaller diameter toroid in current-opposing mode and higher magnetic field peak for the larger diameter toroid in a current-adding mode. This could be due to different toroid diameters or different resonant frequencies or both. Of interest were current-opposing modes of operation for the toroids. For, despite no net current flow or magnetic field inside the coils, magnetic fields formed outside them for this mode. Such field patterns appear impossible for ordinary U(1) EM fields - but they are possible for higher order SU(2) EM fields. Also of interest, was formation of circumferential standing waves of magnetic energy over toroids in a current-adding mode. Eight peaks and nulls occurred for standing waves formed on the smaller diameter toroid surface. Ten peaks and nulls occurred for similar standing waves that formed over the larger toroid's surface.

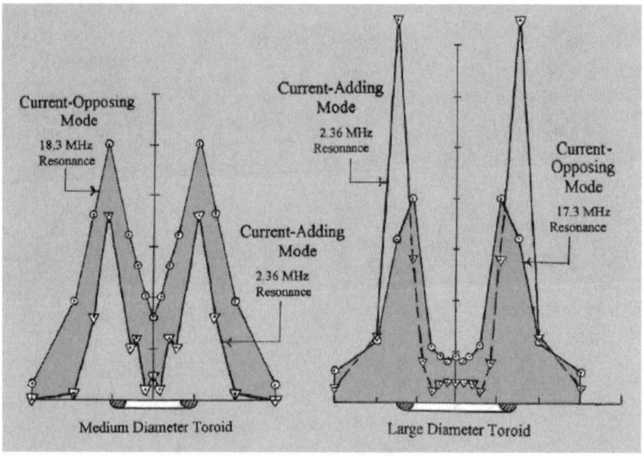

Figure 7. Steep magnetic field gradients forming in the radial direction over tightly-wound toroidal coils.

Figure 8 shows EM magnetic field energy focused into a 'forward direction' from a planar, asymmetric, teardrop-shaped toroidal coil at one of its several resonant frequencies for a current-opposing mode. Such forward focusing might be favorable for applications such as supersonic drag and sonic boom reduction, where work such as Bedin and Mishin (1995) show that thermal and plasma-dynamic effects of EM discharges can reduce aerodynamic drag.

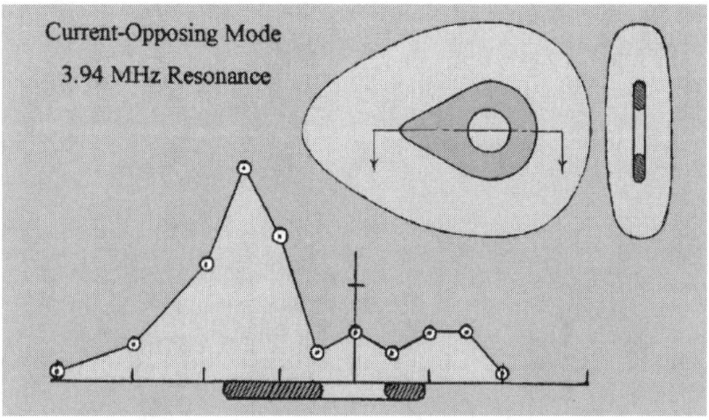

Figure 8. Forward focusing of magnetic field energy from asymmetrical toroids.

As with the small and large diameter circular toroids, standing waves of magnetic field energy formed over the planar, teardrop-shaped toroid when it was radiating in a current opposing mode. However, the standing wave pattern was much more irregularly distributed and both nodes and peaks were difficult to locate with any precision. Magnetic field strength variation with range was measured in one direction out to 10 meters of distance from the large diameter toroid's center in current-adding and current-opposing modes of operation – as is shown in Figure 9. Consistent data was obtained for current-adding mode but less satisfactory data was obtained for current opposing.

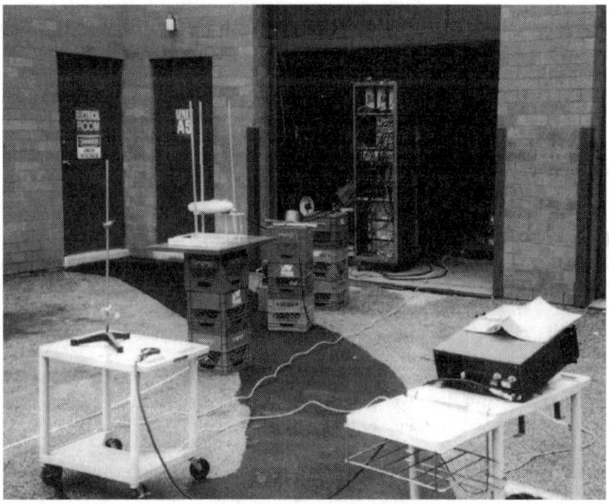

Figure 9. Toroid test setup for measuring magnetic field to 10 meters of range.

Figure 10 shows fall-off in EM field intensity from the larger diameter toroid at: a radiating resonant frequency of 1.20 MHz, and in a current-opposing mode. Measured fall-off in signal strength out to 10 m. from the toroid center deviated somewhat from the $1/R^2$ fall-off in signal strength expected for far-field propagation. But all measurements were well inside the near-field of the ~250 m. wavelength associated with 1.2 MHz resonant frequency. A less rapid fall-off than $1/R^2$ occurred at very short distances and more rapid fall-off than $1/R^2$ occurred at the longer distances.

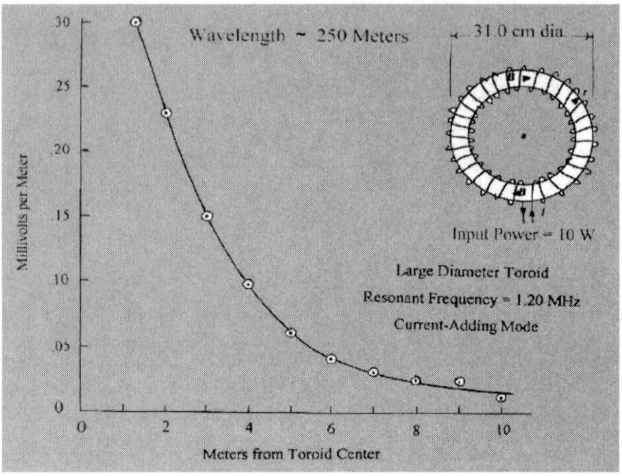

Figure 10. Signal strength fall-off with range for the larger diameter toroid in its current-adding mode.

CONCLUSIONS FROM INITIAL TOROID EXPERIMENAL WORK

Two brief test series have experimentally explored the ideas of Barrett (1998) which suggest that higher order SU(2) EM radiation can be emitted from a toroidal coil when alternating current flows through it at any of the resonant frequencies that can occur for the toroid's geometry. These tests did not prove that SU(2) EM radiation fields were ,indeed, emitted from the tested toroidal coils. But, they revealed interesting and unexpected magnetic phenomena and some appeared consistent with emanation of the SU(2) electromagnetism that was predicted by Barrett (1998).

Our tests could not confirm SU(2) EM field emission from the tested toroidal coils because our ordinary antennas were not capable of detecting the distributed phase patterns of SU(2) A potentials or SU(2) fields. Such detection would have required much more expensive and complex detectors - such as Josephson Junctions - that could detect (but not disturb) the phase patterns associated with states of such fields. Nevertheless, our ordinary U(1) detectors (magnetic probes) detected U(1) magnetic fields outside all tested toroids at resonant frequencies. Such magnetic field detection by our probes would not be expected outside current carrying coils - where only un-detectable U(1) A potentials should exist. But it would be expected if probe presence causes symmetry-breaking of SU(2) fields outside the toroids into U(1) fields. For this would enable detecting the B field components of these U(1) EM fields.

A small number of measurements during the tests indicated significantly stronger signal strength when toroids radiated at resonant frequencies. But there was insufficient test time for detailed exploration of "off-resonance" conditions. This prevented better understanding of what happens electrically inside transmitting toroids and what happens electromagnetically outside them as their resonant frequencies are approached and reached and surpassed. Thus, lack of time for checking or comparing resonance and off-resonance results makes it conceivable that some of our results could ultimately be explained by known U(1) electromagnetism and its U(1) vector and scalar potentials.

However, the most difficult results to explain away are probably those for the current-opposing modes of operation, where counter-flowing alternating currents in contra-wound toroidal coils resulted in no magnetic field whatsoever inside the coils – while, at the very same time, magnetic fields and strong magnetic gradients formed outside them.

FUTURE TOROID TESTING WORK

More testing, involving better field isolation and leakage field considerations, is needed for definitive conclusions on emission of conditioned EM fields from toroidal coils at resonant frequencies and signal benefits they can provide over non-resonant operation. This could be done at places like Hathaway Consulting Services facilities, which now includes an anechoic chamber that will improve the accuracy of any toroid field patterns measured in future testing.

Testing would entail fairly extensive field pattern measurements on only several toroidal coils, with emphasis on an accurate resolution of what happens as resonant frequency is approached and traversed and surpassed. More far-field measurements and determination of the effects of enclosures and magnetic shielding on emitted signal would also be desirable. If results are positive, direct detection of SU(2) fields with Josephson Junctions can also be considered.

ACKNOWLEDGMENT

The Authors wish to acknowledge test consultation and suggestions from Dr. Terence Barrett. And we wish to again acknowledge his authorship of the SU(2) electromagnetism which stimulated our interest in this experimentation.

REFERENCES

Barrett, T. W., "Topological Foundations of Electromagnetism," *World Scientific*, (2008).

Barrett, T. W., "Toroid, Conditioner of Electromagnetic Fields into (Low Energy) Gauge Fields," in the proceedings of the *Progress in Electromagnetic Research Symposium 1998 (PIERS' 98)*, Nantes France, (1998).

Barrett, T. W., "The toroid antenna as a conditioner of electromagnetic fields into (low energy) gauge fields," *Advances in Physics, V. V. Dvnglazov, (ed.)* Nova Science, New York, (2000).

Bedin, A. P. and Mishin, G. I., "Ballistic Studies of the Aerodynamic Drag on a Sphere in Ionized Air," *Pis'ma Ah. Tekh. Fiz*, **21**, (1995), pp. 14-19.

Speculations in Science and Technology **21**, 291–320 (1999)

The toroid antenna as a conditioner of electromagnetic fields into (low energy) gauge fields*

TERENCE W. BARRETT

BSEI, 1453 Beulah Road, Vienna, VA 22182, USA

Treatment of the radiated field from a toroid antenna as two A fields in resonance or a Φ field, and the toroid as a Φ field radiator, results in the prediction of (i) omnidirectional radiation patterns (with small indentations at the poles), and (ii) periodic resonances in the driving conditions. These predictions have been confirmed experimentally, giving validity to the fundamental nature of this topological and group theory understanding of the first order determinants of electromagnetic field dynamics. Resonant gauge (Φ) fields are produced by a toroid radiator as either propagating or standing waves. In the case of a torus with a single winding, an alternating current driver will produce a series of nonmeasurable A vector potential resonances, which overlap and combine into measurable phase factor or gauge field, Φ, waves. Such Φ waves, although generated on a toroidal-solenoidal structure of nonsimple topology, are yet spherical waves – either standing spherical waves, or propagating spherical waves. The topological constraints of electromagnetic fields mapped to a torus driven by single or double wiring are described. It is shown that such mapping results in group symmetries higher than $U(1)$, e.g., $SU(2)$, and that electromagnetic activity is affected by such mapping, being determined by the symmetry forms. Underpinning these symmetry forms are topological conservation laws. The toroid antenna exhibits a series of low and high impedances and permits a $U(1)$ to $SU(2)$ mapping of e.m. fields over a fiber bundle, as well as a mapping of rational and real numbers to complex numbers (in S^3 for the *nonresonant* condition) and quaternions (in S^4 for the *resonant* condition). When in resonance, the singly-wound and the doubly-wound (caduceous) torus emits radiation in $SU(2)/Z_2$ form. The fields emitted for the resonantly driven toroid are alternatively self-dual and anti-self-dual (i.e., instanton solutions to the Maxwell equations of S^4). In resonance, the singly wound and the doubly wound torus produce fields which are both *multiply connected*, and of $SU(2)/Z_2$ form (homeomorphic to S^3), as well as *simply connected*, and of $SU(2)$ from (homeomorphic to S^4). This study has implications far beyond the immediate subject. If the conventional theory of electromagnetism, i.e., 'Maxwell's theory', which is of $U(1)$ symmetry form, is but the simplest local theory of electromagnetism, then those pursuing a unified field theory may wish to consider as a candidate field for unification not only this simple local theory, but other forms of 'conditioned' electromagnetism. As is shown here, other such forms can be either force fields or gauge fields of higher group symmetry, e.g., $SU(2)$ and above.

Introduction

In topology, the torus represents a 2-to-1 mapping of a $U(1)$ field into an $SU(2)$ field. Maxwell's theory addresses local, $U(1)$ symmetry fields, not $SU(2)$ fields [3–7]. Therefore,

---

*Originally presented at the *Progress in Electromagnetics Research Symposium 1998*, (*PIERS' 98*), 13th–17th July, Nantes, France.

it is instructive to consider the fields produced around a toroidally-wound solenoid, the torus being the Cartesian product of two circles, one determining a latitude, the other a longitude. We commence the discussion by considering a linear solenoid (Fig. 1), in which the lines of force (the H field) for adjacent wires annihilate each other except parallel to the surface of the coil on its inside and its outside. It is well known that this field consists of lines of force parallel to the axis of the coil running inside the coil and outside the coil.

If the coil, with N turns, is bent into a circle of radius, R, with the two ends together, the situation changes (Fig. 2). In this torus configuration, no lines of force emerge from the coil but exist continuously inside the coil.

It is useful now to consider the coil and the ring in this torus configuration separately. The current behaves differently in a tightly wound state. Loop currents produce current sheets forming around the ring. With close spacing of the solenoid wires, the current sheet becomes continuous. Whereas the toroidal coil has a sheet current density of

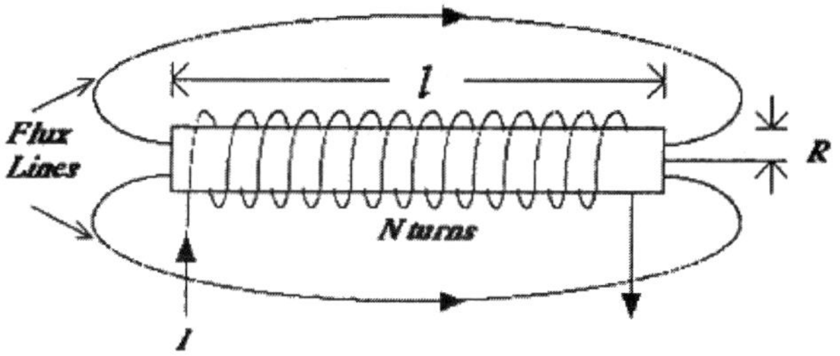

Fig. 1. Linear solenoid.

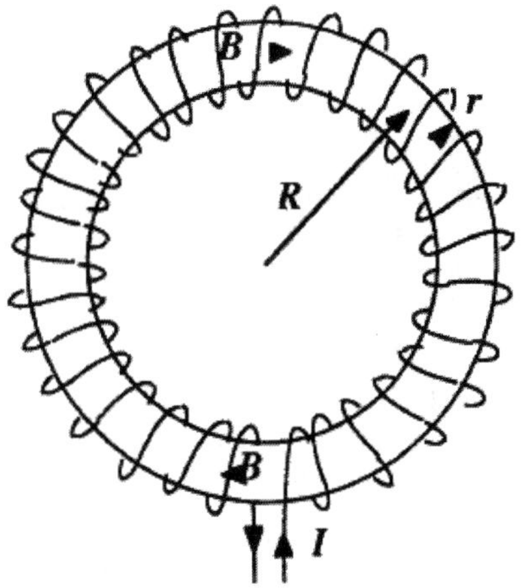

Fig. 2. Toroidal solenoid.

$$K = \frac{NI}{2\pi R} \text{ (A/m)}, \tag{1}$$

the ring has an equivalent current surface:

$$K' = 4K \text{ (A/m)}. \tag{2}$$

Next, the function of the toroidal solenoid depends on the physical composition of the ring. We consider two examples: (1) the ring has no ferromagnetic materials so that the magnetization, $M = 0$; and (2) the ring is composed of ferromagnetic materials, so $M \neq 0$.

In case (1), the magnetic flux density is:

$$B = \frac{\mu_0 NI}{2\pi R} = \mu_0 R \text{ (Wb/m}^2\text{)}, \tag{3}$$

where μ_0 is permeability.

The magnetic field for this case is:

$$H = \frac{B}{\mu_0} = \frac{NI}{2\pi R} = K \text{ (A/m)}, \tag{4}$$

That is, for this case the magnetic field is equal to the sheet current density of the coil winding. For this case also, both B and H are continuous and have the same direction.

The situation changes for case (2), i.e., when the ring is a ferromagnetic material, e.g., an iron ring. The total magnetization is then:

$$M = K' = 4K \text{ (A/m)}, \tag{5}$$

and the magnetic flux density is:

$$B = \mu_0(K + K') = 5\mu_0 K \text{ (Wb/m)}, \tag{6}$$

The magnetic field is:

$$H = \frac{B}{\mu_0} - M \cong K \text{ (A/m)}, \tag{7}$$

and H and B have the same direction.

Thus, the magnetic flux density will vary by a factor of 5 depending on whether the ring of the toroidal solenoid is nonferromagnetic, e.g., Styrofoam, or ferromagnetic, e.g., iron. In the following, we assume that the ring is Styrofoam.

Assuming then a nonferromagnetic ring and using the well known relation between the magnetic flux density and the vector potential, A:

$$B = \nabla \times A, \tag{8}$$

where the vector potential field is defined:

$$A(r) = \frac{1}{c}\iiint \frac{J(r)}{|r|} \mathrm{d}^3 r, \tag{9}$$

and J is the current density, we can plot the isopotential lines around and through the toroidal solenoid (Fig. 3). The E and the B fields within the ring of the solenoid are shown in Fig. 4.

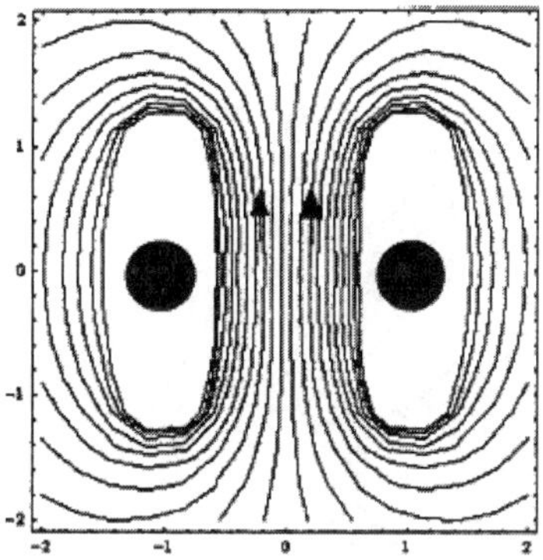

Fig. 3. Plot of a cut through *A* potential field lines surrounding a toroidal solenoid represented as two black dots which join out of the page and into the page.

Discussion

All the above is well known. Addressing, now, the absolute difference of two *A* potentials of opposite polarity generated on a torus, i.e., the *A* field generated on a torus which overlaps one half cycle with itself, some novel observations can be made.

Referring to Fig. 5, the toroidal solenoid is viewed now as a space for mapping two *A* potential alternating patterns ϕ_1 and ϕ_2 onto each other thereby generating a differential phase factor wave Φ. It is this phase factor wave Φ which is either formed around the

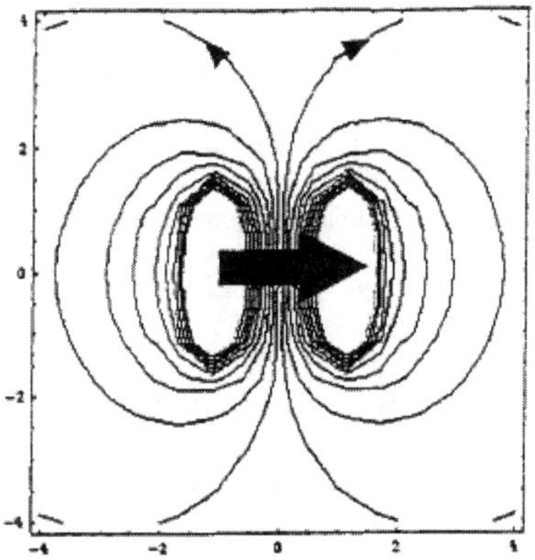

Fig. 4. A second plot of a cut through *A* potential field lines surrounding a toroidal solenoid. The direction of the *B* field in the torus ring is represented by the black arrow.

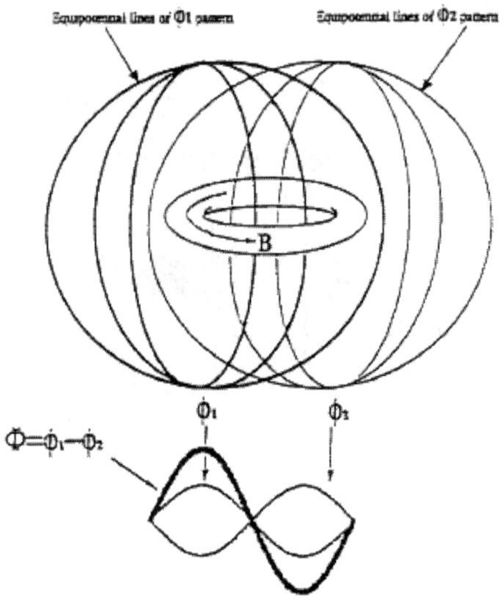

Fig. 5. The toroidal solenoid is viewed as a space for mapping two *A* vector potential alternating patterns ϕ_1 and ϕ_2 onto a differential phase factor wave Φ. In the top figure ϕ_1 and ϕ_2 are shown an exact wavelength out-of-phase on the torus providing maximum generation of Φ (bottom figure). Such out-of-phase condition only occurs for wavelengths of the driving alternating current which are odd multiples of the circumference length of the torus.

toroidal solenoid as standing waves or is transmitted. The question then arises: given an absolute spatial value of the toroidal ring, what is the optimum driving frequency for Φ standing waves and/or propagating waves? Contour lines for the vector potentials, ϕ, and the phase factor, Φ, are shown in the following Fig. 6. Fig. 6D shows that the Φ phase factor waves are spherical waves – either standing spherical waves, or propagating spherical waves.

We suppose a toroidal solenoid with dimensions 12 in. o.d. and 6 in i.d. The average diameter of the current carrying path is 9 in. and the radius is 4.5 in. or 0.1143 m. As the resonance frequencies of interest are related to complex frequencies defined over a torus, the sinusoidal motion is exponential rather than simple harmonic motion. Therefore, with a conductivity $\sigma = 2.5 \times 10^{-8}$ (Ω^{-1}/m), a torus of these dimensions provides a first resonance response at the frequency:

$$\omega_1 = \text{Exp}\left[\xi \times \frac{(2.5 \times 10^{-8}) \times c}{2\pi \times 0.1143}\right], \tag{10}$$

where $c = 2.997925 \times 10^8$, the wavelengths of the resonance conditions are odd multiples of the circumference of the torus, R, and in the above case, $\lambda_{max} = 2\pi R/n$, $n = 1$. ξ is a variable the value of which depends on coil winding. This resonance response is not a maximum gain response but a low impedance resonance. If the wavelength is an even

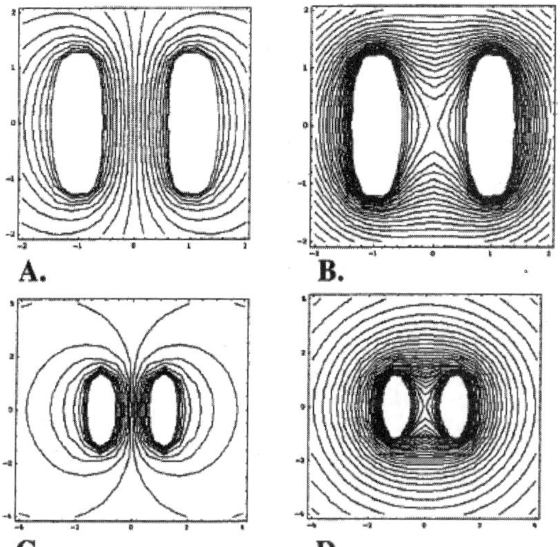

Fig. 6. Contourlines for the ϕ potentials surrounding a torus antenna are shown in (A), and at a greater distance (C). Contourlines for the Φ differential phase factor surrounding a torus antenna are shown in (B). and at a greater distance in (D). (D) indicates that the Φ differential phase factor waves are spherical waves – either standing spherical waves, or propagating spherical waves.

multiple of the circumference of the torus, e.g., $\lambda_{\min} = 2\pi R/m$, $m = 2, 4, 6, \ldots$, there is wave cancellation and a minimum differential phase factor, Φ. Such cancellation provides a minimum gain response, i.e., a maximum impedance response.

In general, therefore, the maximum resonance gain condition (minimum impedance) is:

$$\omega_{\max} = \mathrm{Exp}\left[\xi \times \frac{n \times \sigma \times c}{2\pi \times R}\right], \quad n = 1, 3, 5, \ldots \tag{11}$$

and the minimum resonance gain condition (maximum impedance) is:

$$\omega_{\min} = \mathrm{Exp}\left[\xi \times \frac{m \times \sigma \times c}{2\pi \times R}\right], \quad m = 2, 4, 6, \ldots \tag{12}$$

The resonance maximima and minima for the particular torus of Fig. 2 are shown in Table 1, above. Fig. 7, below, indicates similar information concerning low and high impendance resonances.

Another variable affecting resonance performance is whether the torus is tightly or loosely wound. In the case of the tightly wound torus, the B field is contained on the ring. The field will not be completely contained when the ring is loosely wound. The winding will affect the availability of all possible resonances through the variable ξ. Therefore,

Table 1. Resonance conditions for a toroidal antenna: 12 in. o.d. and 6 in. i.d. (R = 4.5 in. = 0.1143 m) $\xi = 0.2$.

ω_{max} and ω_{min}	$k = l/\lambda = n$ *or* m $l = 2\ \pi R = 0.7182$ meters	State
8.06 Hz	$k = l/\lambda = n = 1$ $\lambda = 0.718$ m	Resonance
65.00 Hz	$k = l/\lambda = m = 2$ $\lambda = 0.359$ m	Null
524 Hz	$k = l/\lambda = n = 3$ $\lambda = 0.240$ m	Resonance
4.23 KHz	$k = l/\lambda = m = 4$ $\lambda = 0.180$ m	Null
34.06 KHz	$k = l/\lambda = n = 5$ $\lambda = 0.143$ m	Resonance
274.6 KHz	$k = l/\lambda = m = 6$ $\lambda = 0.120$ m	Null
2.214 MHz	$k = l/\lambda = n = 7$ $\lambda = 0.106$ m	Resonance
17.85 MHz	$k = l/\lambda = m = 8$ $\lambda = 0.090$ m	Null
143.9 MHz	$k = l/\lambda = n = 9$ $\lambda = 0.080$ m	Resonance
1.160 GHz	$k = l/\lambda = m = 10$ $\lambda = 0.072$ m	Null
9.336 GHz	$k = l/\lambda = n = 11$ $\lambda = 0.065$ m	Resonance
75.42 GHz	$k = l/\lambda = m = 12$ $\lambda = 0.060$ m	Null
.....etc.	.....etc.	.....etc.

a tightly wound torus (Fig. 8A) will exhibit more resonances than a loosely wound torus (Fig. 8B).

The alternating Φ standing waves/transmissions from any physically sized toroidal solenoid, for the wire winding addressed, can be optimized by choice of driving frequencies at the odd resonances with other frequencies nonoptimum. These Φ waves correspond to (low energy) gauge fields or phase factors, and are or $SU(2)$ symmetry form.

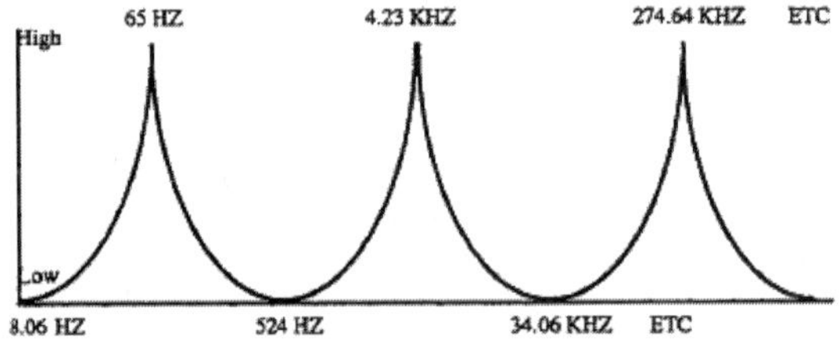

Fig. 7. Representative low and high impedance resonances on the torus of 12 in. o.d. and 6 in. i.d., $x = 0.2$.

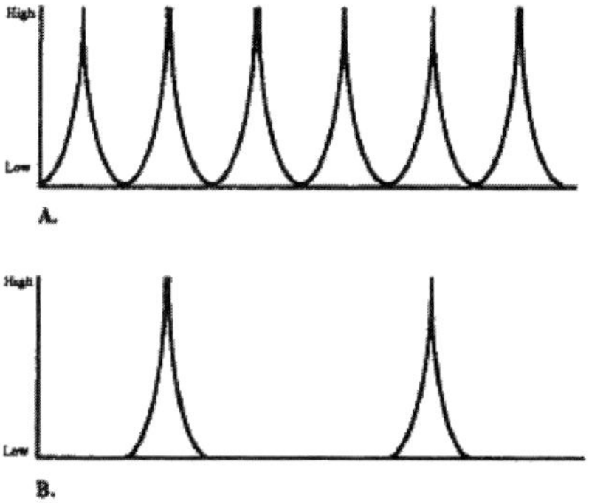

Fig. 8. Representative low and high impedance resonances for tightly wound (A) and loosely wound (B) tori.

The following empirical evidence (Figs. 9 and 10) supports these predictions. Fig. 9 is a plot of range data obtained using a tightly wound torus antenna as a transmit antenna. The plot indicates a $1/r^2$ (one-way) dependence (as opposed to a $1/r$). A less tightly wound torus may exhibit an extended range. There are 15 resonances, 10.1–118 MHz, which are plotted in Fig. 10 as filled circles. These data can be predicted fairly well by:

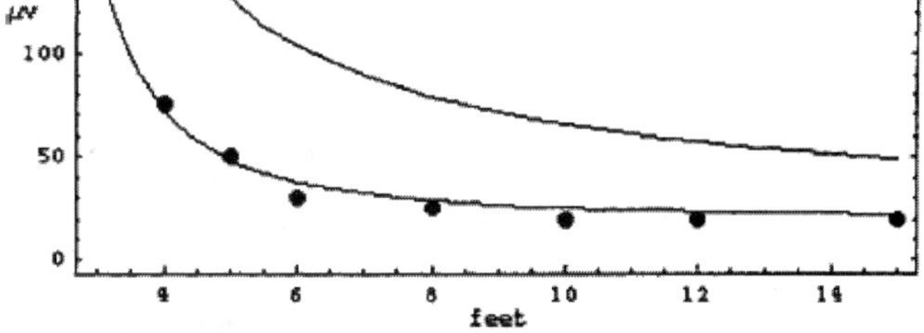

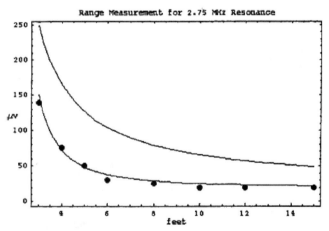

Fig. 9. Upper curve: $400 \times \frac{1}{r - 1.25} + 20$; Lower curve: $400 \times \frac{1}{(r - 1.25)^2} + 20$; Dots: data points. Data courtesy of G. Hathaway and D. Froning.

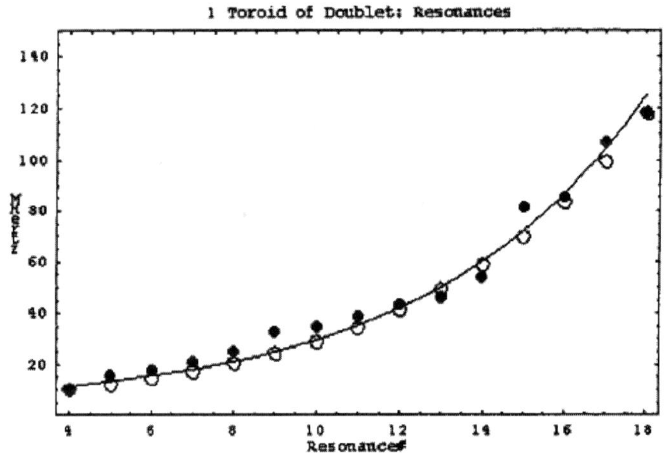

Fig. 10. Some resonances of a singly wound torus. Filled circles: data; unfilled circles: predicted resonances. Data courtesy of G. Hathaway and D. Froning.

$$\text{Resonance} = \text{Exp}\left[\xi \times \frac{n \times \sigma \times c}{2\pi R}\right], \quad n = 1, 3, 5, \ldots \tag{13}$$

where $\sigma = 2.5 \times 10^{-8}$ $(\Omega\,\text{m})^{-1}$ is the conductivity; $c = 3 \times 10^8$ (m/s) is light speed; $R = 0.1651$ m is the torus radius; and $\xi = 0.0122$. The unfilled circles of Fig. 10 show the predicted resonances.

Caduceous coil (double) winding

In the case of the caduceous coil winding (Fig. 11), the resonances depend, firstly, on the driving connections (usually chosen so that the conductance is opposed in the two windings, or 180° phase difference), secondly, on the number of overlaps in the windings, and, thirdly, on the dimensions of the torus.

Topological mappings

Both the single and the double (caduceous) wound torus permit the mapping of $U(1)$ e.m. fields to $SU(2)$ group symmetry form. Fig. 12 is a representation of this mapping. The Φ phase factor differential is represented as an internal degree of freedom and the mapping itself as a space-time to internal space *fiber connection*. The complete space-time to internal space connection is represented as a *fiber bundle*.

If two sets of twin test particles were introduced to the driven torus – the first of the *first* twin set through the outside of the torus and the second of the *first* twin set through the middle the torus, and the first of the *second* twin set through the middle of the torus and

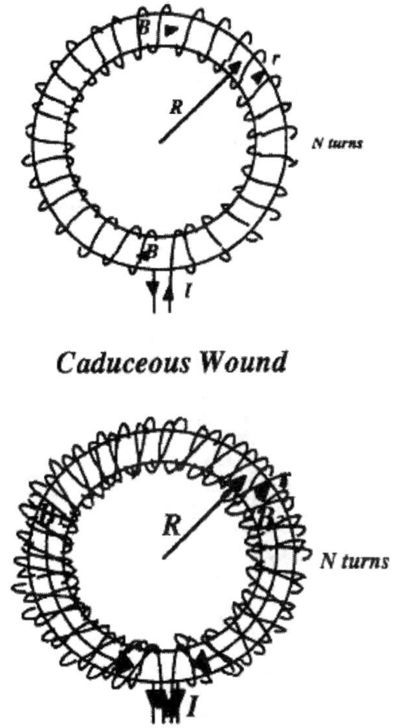

Fig. 11. Single wound torus and caduceous (double wound) torus. In the case of the caduceous winding the driving of the two circuits is 180° out of phase and the winding is left- and right-handed, creating two counterpropagating B fields on the torus.

the second of the *second* twin set through the outside of the torus (Fig. 13) – Aharonov–Bohm effect phases will be detected at overlap locations, $C1$ and $C2$, due to each test particle of each pair being influenced by A fields of opposite polarity. If these phases are compared in a second interferometric level of comparisons, the phase factor field $(C1 - C2)$ will be measured.

Two field mappings are achieved for the resonance condition of the singly wound and the doubly (caduceous) wound torus. A single winding gives a complex number A field representation of the B field (Fig. 14). Either the resonance condition on the single winding or the double (caduceous) winding maps those complex numbers to the sphere S^3.

Details of these mappings are depicted in Fig. 15. Commencing with a half cycle of a sinusoid, $a = b = c = d$, the four half cycles of two monocycles on the torus are formed

SU(2) Group: Linear Transformation of a Complex 2D vector.

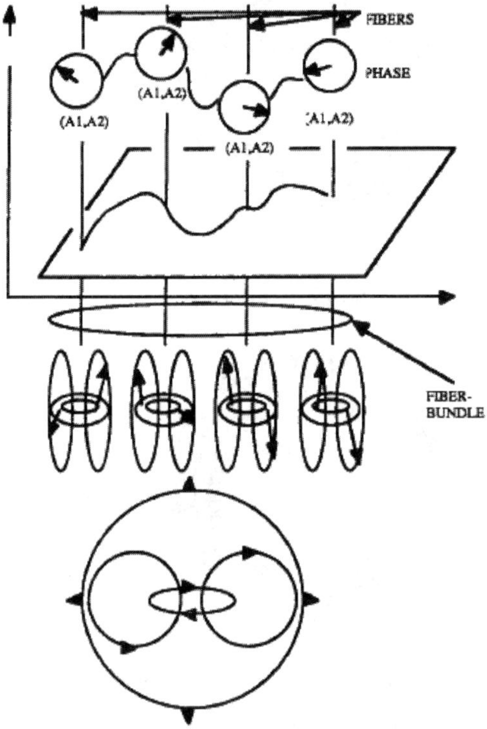

Fig. 12. The topology of a field mapped onto a torus with phase modulation. A fiber connection is shown between the phase in the internal space and the trajectory in space–time. The complete trajectory in internal space is mapped to space–time by a fiber bundle.

(over time), by the I, J, K operations. Thus a complete single quaternion cycle, $H = a + bI + cJ + dK$, is constituted of 2 full sinusoidal monocycles, or 4 half cycles.

The Fig. 15 operation may also be considered a section of a vector bundle over the real numbers of the B field (Fig. 16). At any instant, the driven torus, either singly wound in resonance, or doubly (caduceously) wound, is a section of a vector bundle.

If either the singly wound torus is driven but *not* in resonance, or if the doubly (caduceously) wound torus is driven but *not* 180° out of phase, then the fields on the torus are: (1) closed, (2) exact, (3) of *SU* (2) symmetry, and (4) *simply* connected. But if the singly wound torus is driven *in resonance*, or if the doubly (caduceously) wound torus is driven 180° out of phase, then the fields on the torus are: (1) closed, but (2) *not* exact, (3) of *SU* (2)/Z_2 symmetry form (where Z_2 represents the binary integers), and (4) multiply connected (Fig. 17).

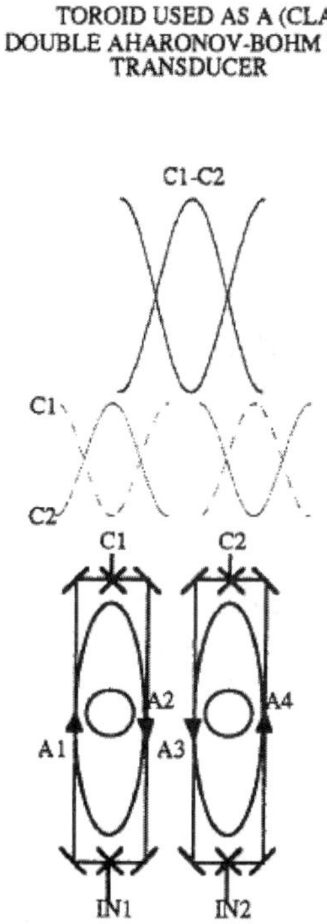

Fig. 13. Opposite fields on a torus used as a double Aharonov–Bohm effect transducer. Two test particles are introduced at *IN*1 and *IN*2 into two interferometers with tracks around opposite sides of a torus antenna. The two arms of the two interferometers interact with the *A* fields *A*1, *A*2, *A*3 and *A*4 which are of alternate directionality. The resulting phase change of the two test particles in the two interferometers are sampled at *C*1 and *C*2. The phase differences *C*1 − *C*2 is shown at top.

The predictions of (i) omnidirectional radiation patterns (with small indentations at the poles), and (ii) periodic resonances in the driving conditions, have been confirmed by experimental tests (see Figs. 9 and 10, above). Much more extensive testing and confirmation has been carried out by the Center for Industrial Research Applications (CIRA) at West Virginia University, Morgantown, WV, USA.

CIRA has conducted studies and tests of a variety of toroidal antennas (cf. [10, 16]). Fig. 18 shows an example of a multilayer printed circuit board of the Contrawound Toroidal Helical Antenna (CTHA). The CTHA exhibits a gain and impedance spectrum with multiple resonances and which is distinctly different from that of, e.g., a standard dipole antenna.

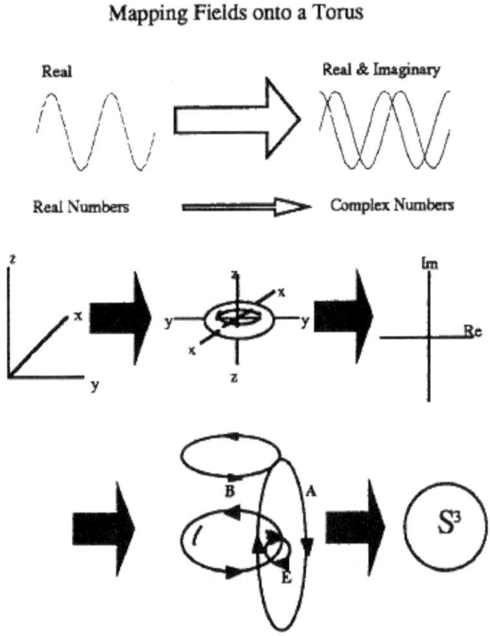

Fig. 14. (Top) Mapping of a real number E field onto a torus antenna results in a complex number representation by the B field due to a first phase representation on a circle. (Bottom) Next, there is a quaternionic representation by the A fields due to two circular representations of the complex number representations by the two A fields on the torus. This results in a phase differential representation. Thus, *in resonance*, the Φ field is the resultant mapping of the E field over S^4.

Figs. 19 and 20 show far field radiations patterns for a typical CTHA. These far field patterns are omnidirectional (with small indentations at the poles) in exact correspondence with the predictions shown in Figs. 3–6 above. The CTHA exhibits other aspects in its radiation pattern which are not accounted for here, but will be addressed in future work. They are:

- Some resonances, e.g., the third, provide a more optimal design than other resonances.
- At various parts of its field pattern, the radiation is present as θ-oriented field energy and at other times as ϕ-oriented field energy (in the spherical polar coordinate system), see Figs. 19 and 20.
- At some resonances, e.g., first and third, the sphere of the far field radiation for a CTHA can produce mixed polarization, i.e., all possibilities of polarization, from left hand circularly polarized, through various degrees of left hand elliptically polarized, to linearly polarized and again through various degrees of right hand elliptical polarization to

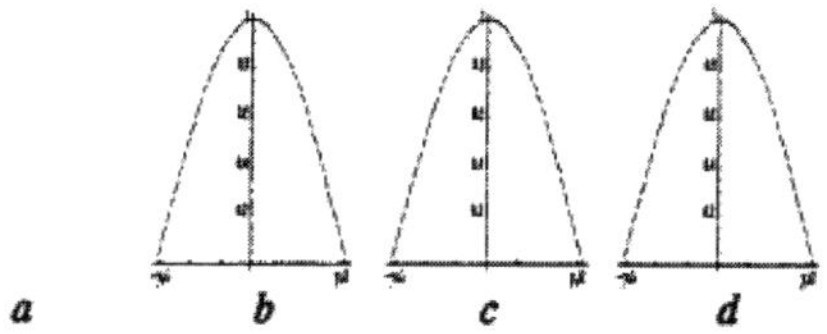

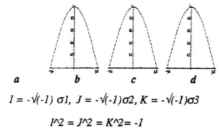

$$I = -\sqrt{(-1)}\,\sigma 1,\ J = -\sqrt{(-1)}\sigma 2, K = -\sqrt{(-1)}\sigma 3$$

$$I\wedge 2 = J\wedge 2 = K\wedge 2 = -1$$

$$H = \{a + bI + cJ + dK:\ a,b,c,d \in \Re\}$$

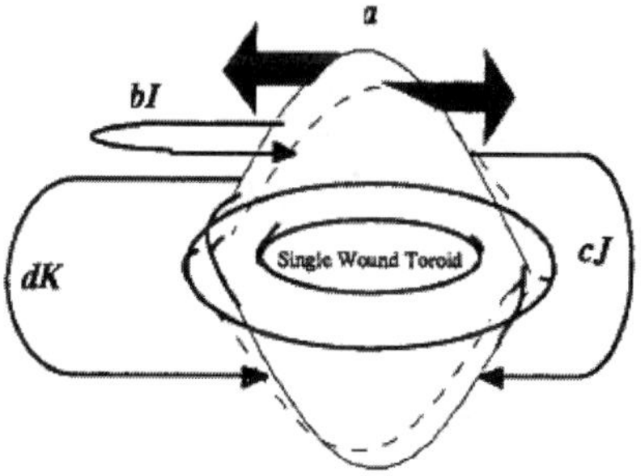

Fig. 15. Representation of a field in quaternionic form on a torus in resonance. The field components are represented as $\pi/2$ sinusoidal base fields: $a = b = c = d$. Mapped onto the torus and with resonance coupling, $a = b = c = d = bI = cJ = dK$. Therefore, a field in resonance on the torus is in quaternionic form: $H = a + bI + CJ + dK$.

right hand circularly polarized. CIRA finds that the typical CTHA regions of circular polarization are 90° either side of the feed, just above and below the horizon. At other resonances, e.g., the second, only linear polarization is obtained.

- Internal interactions of the antenna are the dominant effect on input impedance. Thus the driven antenna is minimally affected by changing environments.

Conclusions

- Treatment of the radiated field of a toroid antenna as two *A* fields in resonance or a Φ field, and the toroid as a Φ field radiator, results in the prediction of (i) omni-directional radiation patterns (with small indentations at the poles), and (ii) periodic resonances in the driving conditions. These predictions have been confirmed experimentally, giving validity to the fundamental nature of this topological and group theory understanding of the first order determinants of electromagnetic field dynamics.

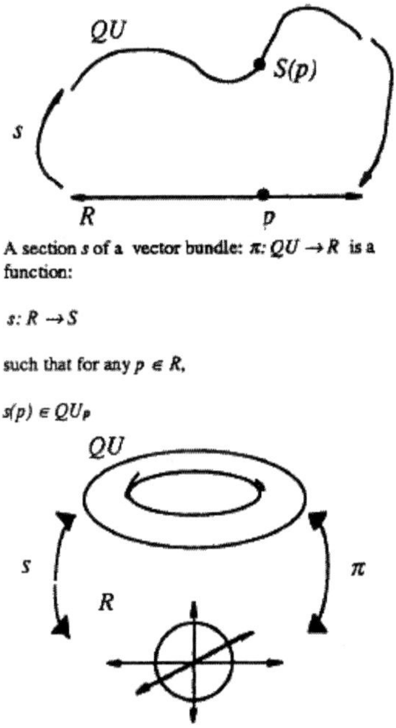

Fig. 16. Section of a vector bundle $\pi : QU \Rightarrow R$. The section assigns to each point in the base space R or the real numbers (i.e., the B field), a quaternion in the fiber over that point. Therefore the quaternionic, QU, space is a bundle over R.

- A field mapped onto a torus by single or caduceous (double) winding maps real numbers of the field into complex numbers at the 2-(differential) form level and quaternions at the 3-(differential) form level.
- $SU(2)$ is the group describing S^3, the unit sphere in the quaternions. However, the radiation of the *in resonance* single winding or double winding mapping on the torus is in quaternionic (or $SU(2)/Z_2$) form, describing S^4, the unit sphere in the quaternions.
- Driven in resonance, e.g., the singly wound toroid driven at wavelengths which are odd multiples of the toroid length and the doubly wound toroid driven at 180° phase lag between the two driving inputs, the radiation is in quaternionic or $SU(2)/Z_2$ form.
- Both (1) the leading and trailing resonance half waves on a singly wound torus and (2) the two mapped fields on a torus in the doubly wound condition are *cohomologous*.
- The field on a torus becomes exact and *simply* connected ($SU(2)$) on S^3 at the 2-form level by mapping two A fields on a torus either singly or doubly wound. Differential forms which are *not exact* are those with integrals equal to zero. Mappings on either the singly wound torus in resonance or the caduceous wound torus *in resonance* results in an

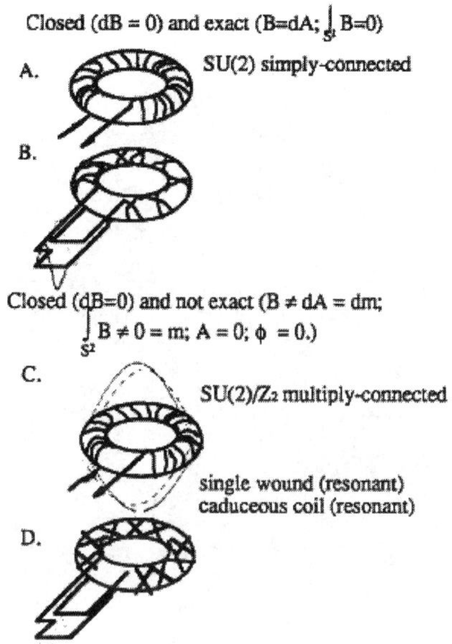

Fig. 17. Fields (*B* fields) on a torus which is either (A) singly or (B) doubly (caduceously) wound, but not in resonance, are: (1) closed (the exterior derivative of the differential form (the *B* field) is zero), (2) exact (the differential form (the *B* field) is the exterior derivative of another differential form (the *A* field), (3) of $SU(2)$ symmetry and (4) *simply* connected. Fields (*B* fields) on a torus which is either (C) singly wound and in resonance or (D) doubly (caduceously) wound and driven 180° out of phase, are (1) closed (the exterior derivative of the differential form (the *B* field) is zero), but (2) not exact (the differential form (the *B* field) is not the exterior derivative of another differential form (the *A* field which is zero), (3) of $SU(2)/Z_2$ symmetry, where Z_2 represents the binary integers and (4) *multiply* connected.

integrated *A* field of zero. Therefore the singly wound *resonance* condition and the doubly wound *resonance* condition result in a *multiply* connected $SU(2)/Z_2$ field on S^4.

- A mapping of a field on a torus which is singly wound in resonance or caduceous wound in resonance is:
  - (A) a homomorphism form $GL(3, C)$ to $SU(2)$, i.e., a representation of C^3 on $SU(2)$.
  - (B) a 3-dimensional complex representation of $SU(2)$ which is a spin-1 representation (quaternionic form).
- A mapping of a field on a torus which is singly wound in resonance or doubly wound in resonance performs an inverse Hodge star operation: $\star\Phi = A_1 \wedge A_2$ and the Φ field is alternatively self-dual and anti-self-dual: $\Phi = \star\Phi$. Self-dual solutions to the Maxwell equations are instantons [13] (Figs. 21 and 23).

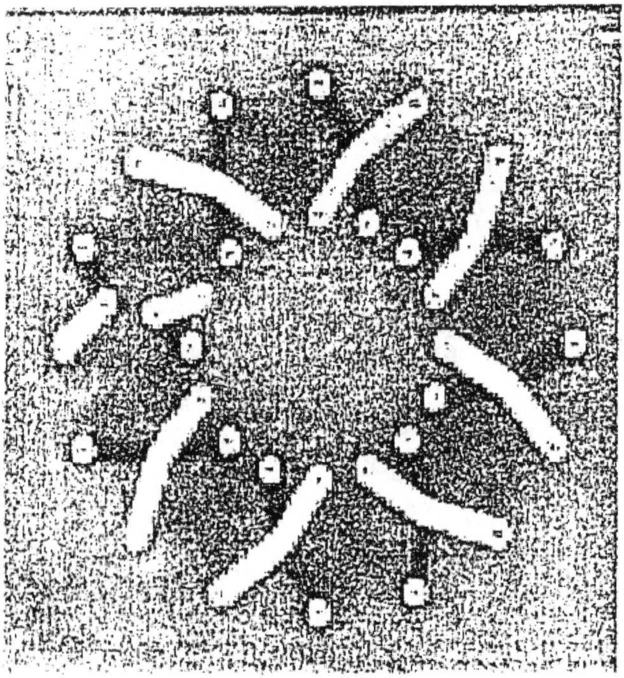

Fig. 18. Multilayer printed circuit board version of the Contrawound Toroidal Helical Antenna (CTHA). From: *Contrawound Toroidal Helical Antenna*, Center for Industrial Research Applications (CIRA), Mechanical and Aerospace Engineering Department, West Virginia University, Morgantown, WV, USA, November 1997, by permission.

• This study has implications far beyond the immediate subject. If the conventional theory of electromagnetism, i.e., 'Maxwell's theory', which is of $U(1)$ symmetry form, is but the simplest local theory of electromagnetism, then those pursuing a unified field theory may wish to consider as a candidate field for unification not only this simple local theory, but other forms of 'conditioned' electromagnetism. As is shown here, other such forms can be either force fields or gauge fields of higher group symmetry, e.g., $SU(2)$ and above.

Appendix

The following definitions are offered in recognition that the concepts of topology and group theory are not well known.

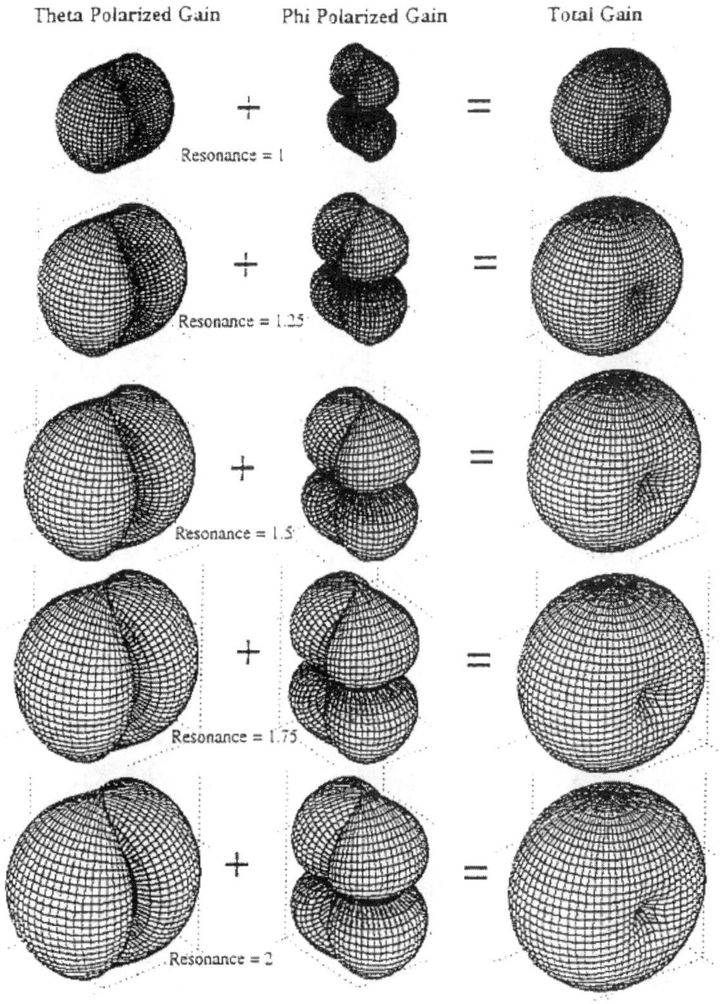

Fig. 19. Far field radiation patterns for typical CTHA at various points of a normalized spectrum. From: *Contrawound Toroidal Helical Antenna*, Center for Industrial Research Applications (CIRA), Mechanical and Aerospace Engineering Department, West Virginia University, Morgantown, WV, USA, November 1997, by permission.

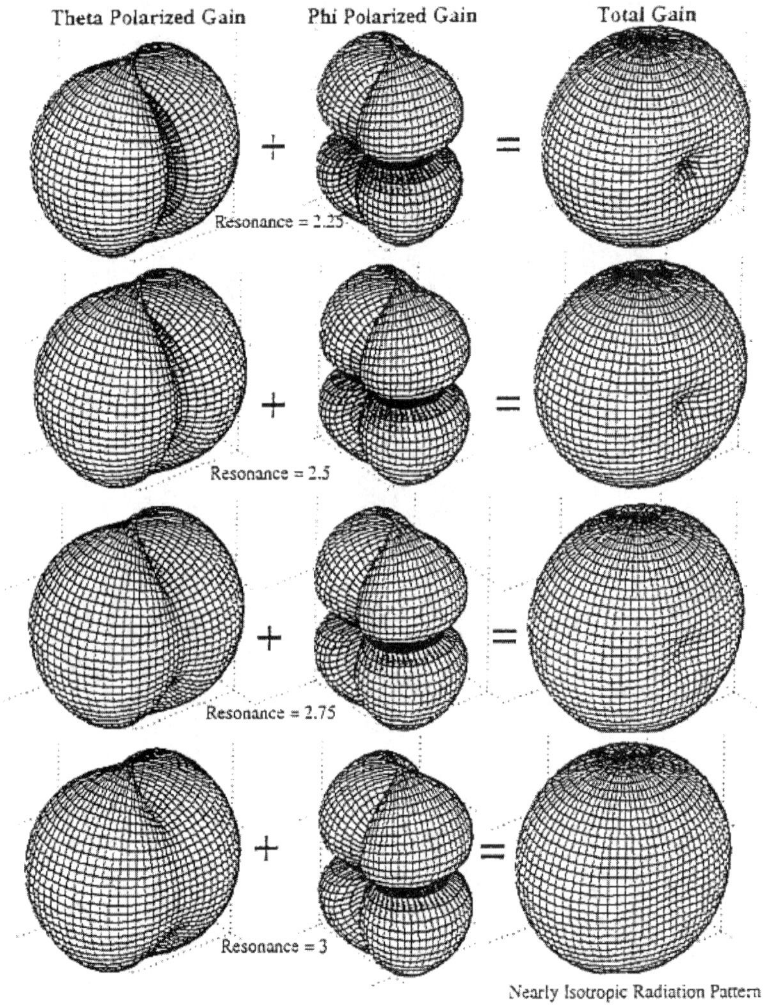

Fig. 20. Far field radiation patterns for typical CTHA at various points of a normalized spectrum. From: *Contrawound Toroidal Helical Antenna*, Center for Industrial Research Applications (CIRA), Mechanical and Aerospace Engineering Department, West Virginia University, Morgantown, WV, USA, November 1997, by permission.

Definitions

U(1) *Maxwell's Equations*

$$\nabla \cdot \vec{B} = 0 \quad \nabla \times \vec{E} + \frac{\partial \vec{B}}{\partial t} = 0$$

$$\nabla \cdot \vec{E} = \rho \quad \nabla \times \vec{B} - \frac{\partial \vec{E}}{\partial t} = \vec{j}$$

can be reduced to the differential forms [2]

$$\mathrm{d}F = 0, \tag{15}$$

which is equivalent to the first pair, because

$$\begin{aligned} \mathrm{d}F &= \mathrm{d}(B + E \wedge \mathrm{d}t) = \mathrm{d}B + \mathrm{d}E \wedge \mathrm{d}t \\ &= \mathrm{d}_s B + \mathrm{d}t \wedge \partial_t B + (\mathrm{d}_s E + \mathrm{d}t \wedge \partial_t E) \wedge \mathrm{d}t \\ &= \mathrm{d}_s B + (\partial_t B + \mathrm{d}_s E) \wedge \mathrm{d}t = 0, \text{ so} \\ &= \mathrm{d}_s B = 0 \text{ and } \partial_t + \mathrm{d}_s E = 0; \end{aligned} \tag{16}$$

and

$$\star \mathrm{d} \star F = j, \tag{17}$$

which is equivalent to the second pair, because

$$\begin{aligned} \star_s \mathrm{d}_s \star_s E &= \rho \\ -\partial_t E + \star_s \mathrm{d}_s \star_s B &= j, \end{aligned} \tag{18}$$

where d is the exterior derivative, $\star$ is the Hodge star operator (defined below) and $\star_s$ is the Hodge star operator on R^3.

It may be obvious by now, but it can be remarked that radiated energy can exhibit higher symmetry forms [15].

Topology defined: A topological space is a set X, together with a family of subsets of X, called open *sets*, satisfying the following conditions:

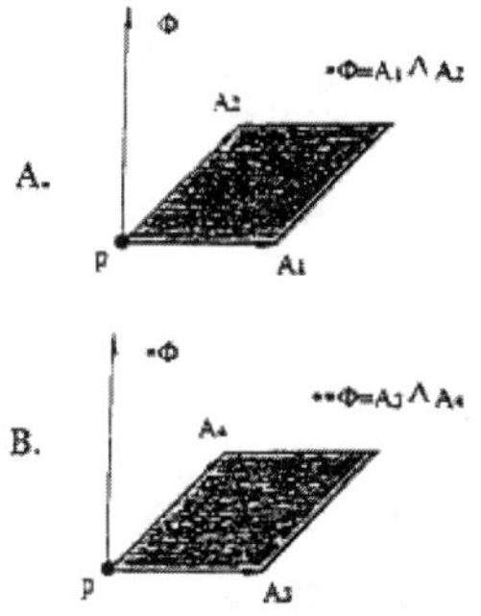

Fig. 21. Self-dual fields, Φ and $\star\Phi$, defined over the 4 half cycles; A_1, A_2, A_3 and A_4.

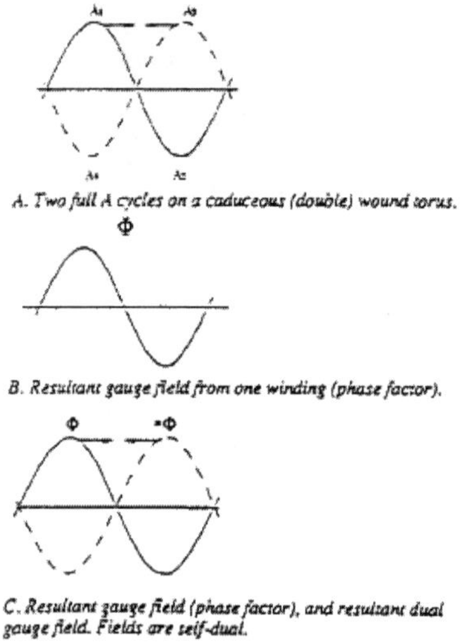

A. Two full A cycles on a caduceous (double) wound torus.
Φ
B. Resultant gauge field from one winding (phase factor).
Φ
*Φ
C. Resultant gauge field (phase factor), and resultant dual gauge field. Fields are self-dual.

Fig. 22. (A) 2 full A field cycles on the double-wound (caduceous) torus. (B) Resultant Φ gauge field from one winding. (C) Resultant alternating self-dual and anti-self-dual Φ gauge fields from the two windings.

(1) The empty set and X itself is open.
(2) If $U, V \subseteq X$ are open, so is $U \cap V$.
(3) If the sets $U_\alpha \subseteq X$ are open, so is the union $\cup U$. The collection of sets taken to be open is called the *topology* of X.

Manifold defined: An n-dimensional manifold, or n-manifold, is a topological space M equipped with charts

$$\varphi_\alpha : U_\alpha \to R^n \tag{19}$$

where U_α are open sets covering M, such that the transition function is smooth where it is defined.

Vector Fields on manifolds defined: A 1-form on any manifold M is a map from $\vec{M}$ to $C(M)$ that is linear over $C(M)$.

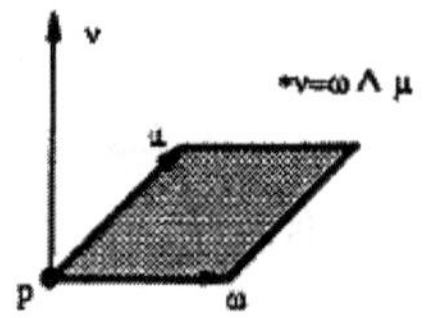

Fig. 23. $\star v = \omega \wedge \mu$

The 1-form df is the differential of f, or the exterior derivative of f.

Just as a vector field on M gives a **tangent vector** at each point of M, a 1-form on M gives a **cotangent vector** at each point of M.

A Cotangent Vector ω at p is defined to be the linear map from the **tangent space** T_pM to R.

Given any vector space, V, the **dual vector space** V^* is the space of all linear functionals $\omega : V \to R$.

Wedge Product ('*Exterior Product*') is defined as a **generalized cross product** preserving antisymmetry (anticommutativity), e.g., the cross product

$$\vec{v} \times \vec{w} = -\vec{w} \times \vec{v} \tag{20}$$

is the wedge product

$$v \wedge w = -w \wedge v \tag{21}$$

plus the application of the Hodge star operator (Fig. 23).

The **Hodge Star Operator** defined:

If $\star\mu$ is the dual of μ, then:

$$\begin{aligned} \star &: dx \wedge dy \mapsto dz \\ \star &: dy \wedge dz \mapsto dx \\ \star &: dz \wedge dx \mapsto dy \end{aligned} \tag{22}$$

The Hodge star operator and the exterior derivative:

$$\star d\omega = (\partial_y\omega_z - \partial_z\omega_y)dy \wedge dz + (\partial_z\omega_x - \partial_x\omega_z)dz \wedge dx + (\partial_x\omega_y - \partial_y\omega_x)dx \wedge dy \tag{23}$$

Let M be an n-dimensional oriented semi-Riemannian manifold. The **inner product** of two p forms ω and μ on M is a function $\langle\omega, \mu\rangle$ on M. The Hodge star operator is defined:

$$\star : \Omega^p(M) \to \Omega^{n-p}(M) \tag{24}$$

and is the unique linear map from p-forms to $(n-p)$-forms such that for all $\omega, \mu \in \Omega^p(M)$,

$$\omega \wedge \star\mu = \langle\omega, \mu\rangle_{\text{vol}} \tag{25}$$

Differential forms on M, denoted $\Omega(M)$, refer to the algebra generated by $\Omega^1(M)$ with the relations $\omega \wedge \mu = -\mu \wedge \omega$ for all $\omega, \mu \in \Omega^p(M)$. $\Omega(M)$ satisfies the rules of an algebra with **functions** substituting for numbers.

Elements that are linear combinations of products of p 1-forms are called p-forms. The space of p-forms on M is $\Omega_p(M)$, and

$$\Omega(M) = \oplus_p\Omega^p(M). \tag{26}$$

Suppose $M = \Re^n$. The 0-forms on $\Re^n$ are functions f. The 1-forms are:

$$\omega_\mu dx^\mu, \tag{27}$$

where ω_μ are functions. 2-forms are:

$$\frac{1}{2}\omega_{\mu\nu}\, dx^\mu \wedge dx^\nu. \tag{28}$$

Therefore:

$$E = E_x\,\mathrm{d}x + E_y\,\mathrm{d}y + E_z\,\mathrm{d}z \quad (1\text{-}form)$$
$$B = B_x\,\mathrm{d}y \wedge \mathrm{d}z + B_y\,\mathrm{d}z \wedge \mathrm{d}x + B_z\,\mathrm{d}x \wedge \mathrm{d}y \quad (2\text{-}form) \tag{29}$$

Similarity of Gradient, Curl and Divergence in terms of exterior derivative and star operators:

$$\begin{aligned} \text{Gradient} \quad & \mathrm{d} : \Omega^0(\Re^3 \rightarrow \Omega^1(\Re^3) \\ \text{Curl} \quad & \mathrm{d} : \Omega^1(\Re^3 \rightarrow \Omega^2(\Re^3) \\ \text{Divergence} \quad & \mathrm{d} : \Omega^2(\Re^3 \rightarrow \Omega^3(\Re^3) \end{aligned} \tag{30}$$

In general:

$$\mathrm{d} : \Omega^p(M) \rightarrow \Omega^{p+1}(M) \tag{31}$$

The electromagnetic field is:

$$\begin{aligned} F &= B + E \wedge \mathrm{d}t \\ \star F &= \star_s E - \star_s B \wedge \mathrm{d}t \\ \mathrm{d} \star F &= \star_s E \partial_t E \wedge \mathrm{d}t + \mathrm{d}_s \star_s E - \mathrm{d}_s \star_s B \wedge \mathrm{d}t \\ \star \mathrm{d} \star F &= -\partial_t E - \star_s \mathrm{d}_s \star_s E \wedge \mathrm{d}t + \star_s \mathrm{d}_s \star_s B \end{aligned} \tag{32}$$

The fundamental importance of the potentials is shown in:

$$\begin{aligned} & \mathrm{d}^2 = 0, \\ & E = -\mathrm{d}\phi, \quad \text{where } \phi \text{ is a function or a 0-form} \\ & B = \mathrm{d}A, \quad \text{where } A \text{ is a 1-form} \\ & F = \mathrm{d}A, \quad \text{where } A \text{ is a 1-form on space-time} \end{aligned} \tag{33}$$

and for any $p \in S$ and any $q \in S$ (but not for the torus, see below),

$$\phi(q) = -\int_\gamma E \tag{34}$$

Homotopy defined: Homotopy refers to the continuous deformation from one topological object to another. Two paths $\gamma_0, \gamma_1 : [0, T]$ from p to q are homotopic if there exists a smooth function: $\gamma : [0, 1] \times [0, T] \rightarrow S$ such that $\gamma(s, \bullet)$ is a path from p to q for each s, and:

$$\begin{aligned} \gamma(0, t) &= \gamma_0(t), \\ \gamma(1, t) &= \gamma_1(t). \end{aligned} \tag{35}$$

The function γ is then said to be a homotopy between γ_0 and γ_1.

It should be noted that the paths around a torus are not homotopic (due to the hole). Therefore for a torus:

$$\mathrm{d}\phi \neq -\int_\gamma E. \tag{36}$$

Holonomy defined [12]: A **connection** is a generalization of the **vector potential**. Integrating the connection around a loop in a certain way obtains $g \in G$ called the '*holonomy*'.

A connection of a **vector bundle** permits the differentiation of sections. A connection also permits '*parallel transport*'.

Homology defined [14]: Homology groups are topological invariants. Objects which are homologous are considered equivalent.

Cohomology defined [9]: Cohomology is the study of holes or obstructions by algebraic methods. Having holes or obstructions is a topological property of space, i.e., a property preserved by all continuous mappings with continuous inverses. Therefore cohomology is a branch of algebraic topology.

De Rham Topology defined: De Rham topology is the study of all closed forms that are *not exact*. Closed 1-forms on a manifold are automatically exact if the manifold is *simply connected*. If a manifold is not simply connected, it has some sort of '*holes*' or '*obstructions*' in it.

The pth de Rham cohomology of a manifold M is a vector space written $H^p(M)$ whose dimension is the number of 'p-holes' in M. If $Z^p(M)$ is the set of *closed* p-forms on M and $B^p(M)$ is the vector space of *exact* p-forms, then those *exact* p-forms are a subspace of the *closed* p-forms:

$$B^p(M) \subseteq Z^p(M) \tag{37}$$

Therefore, the number of *closed* forms that are *not exact* is the quotient:

$$H^p(M) = \frac{B^p(M)}{Z^p(M)}, \tag{38}$$

which is called the pth de Rham cohomology group of M.

An element of $H^p(M)$ is an equivalent class of closed p-forms, where two closed forms, ω, ω' are equivalent if they differ by an exact p-form, i.e., if there is a $(p-1)$-form μ such that $\omega - \omega' = \mathrm{d}\mu$. When ω and ω' are equivalent in this way, they are said to be cohomologous and the equivalent class of ω is its cohomologous class:

$$[\omega] = \{\omega' : \exists\mu \quad \omega - \omega' = \mathrm{d}\mu\}. \tag{39}$$

The singly wound and caduceously wound torus *driven in resonance* tailor the A vector fields into, e.g., ω and ω' forms such that they are cohomologous, and where the phase is, e.g., $\mathrm{d}\mu$. By Stokes theorem:

$$\begin{aligned} &\text{Resonance case, simply connected}: \int_\gamma A = \int_D B = 0, \\ &\text{Nonresonance case, multiply connected}: \int_\gamma A = \int_D B \neq 0, \end{aligned} \tag{40}$$

where D is the boundary.

Therefore, (1) The leading and trailing half cycles of resonance waves on a singly wound torus (180° out-of-phase) are cohomologous; (2) The waves on a caduceously wound torus are cohomologous; and (3) Two trajectories on a stationary interferometer (Aharonov–Bohm, etc.) are cohomologous.

If $H^0(M) = 0$, which is the case of *exact* 0-forms, then M is connected but consists of the zero'th function.

If $H^1(M) = 0$, which is the case of *exact* 1-forms, M is *simply connected*, since then every closed 1-form is *exact*. A *multiply connected* field becomes *simply connected* if two constituent fields of the multiply connected field, e.g., two A fields, are *cohomologous*. This is the situation of a field mapped by the simply wound torus in resonance or the caduceous wound torus in resonance. *Exact* differential forms are those with integrals equal to zero and indicate *simply connected* fields. Mappings of the singly wound torus in resonance and the caduceously wound torus *in resonance* results in a zero integral for the A fields (Equation 40).

These results are represented in the following Table 2.

Homomorphism defined: Given two groups, G and H, we say a function $\rho : G \to H$ is a homomorphism if:

$$\rho(gh) = \rho(g)\rho(h). \tag{41}$$

Isomorphism defined: A homomorphism which is one-to-one and onto is called an isomorphism

Tensor Product defined: Pick a basis $\{e\}$ for V and a basis $\{e'\}$ for V'. Then the tensor product, $V \otimes V'$, is the vector space whose basis is given by all expressions of the form $e_i \otimes e'_i$.

Quaternions and Pauli Matrices: $SU(2)$ group operations can be defined as:

$$SU(2) = \{a + bI + cJ + dK : a, b, c, d \in \Re, a^2 + b^2 + c^2 + d^2 = 1\} \tag{42}$$

For the 3-dimensions, 1,2,3 of the singly wound or caduceous mapping of a field on a torus, (i, j, k) is a cyclic permutation of 1,2,3 and

$$\sigma_i\sigma_j = -\sigma_j\sigma_i = \sqrt{-1}\sigma_k, \tag{43}$$

Table 2. $U(1)$ versus $SU(2)$ Electromagnetics on a Torus.

	Winding: Either caduceously wound or singly wound *in resonance*	Winding: Either caduceously wound or singly wound *out of resonance*
Simply connected (exact differential form)?		
at H^0 level?	Yes, $U(1)$ sym., $H^0 = 0$	Yes, $U(1)$ sym., $H^0 = 0$
at H^1 level?	No, $SU(2)/Z_2 = SO(3)$ sym., $H^1 \neq 0$	No, $SU(2)/Q$ sym., $H^1 \neq 0$
at H^2 level?	Yes, $SU(2)$ sym., $H^2 = 0$	Yes, $SU(2)$ sym., $H^2 = 0$
Multiply connected (inexact differential form)?		
at H^0 level?	No, $U(1)$ sym., $H^0 = 0$	No, $U(1)$ sym., $H^0 = 0$
at H^1 level?	Yes, $SU(2)/Z_2 = SO(3)$ sym., $H^1 \neq 0$	Yes, $SU(2)/Q$ sym., $H^1 \neq 0$
at H^2 level?	No, $SU(2)$ sym., $H^2 = 0$	No, $SU(2)$ sym., $H^2 = 0$

where

$$\sigma_i = \begin{pmatrix} 0 & 1 \\ 1 & 0 \end{pmatrix}, \quad \sigma_j = \begin{pmatrix} 0 & -i \\ i & 0 \end{pmatrix}, \quad \sigma_k = \begin{pmatrix} 1 & 0 \\ 0 & -1 \end{pmatrix} \tag{44}$$

are Pauli matrices.

Defining

$$I = -\sqrt{-1}\sigma_1, \quad J = -\sqrt{-1}\sigma_2, \quad K = -\sqrt{-1}\sigma_3, \tag{45}$$

the identity matrix operating on a maps a or $+A_1$;
I operating on b, maps bI or $+A_2$;
J operating on c, maps cJ or $-A_1$;
K operating on d, maps dK or $-A_2$. Therefore, $\sigma_i^2 = 1$, and

$$a^2 + b^2 + c^2 + d^2 = 1. \tag{46}$$

The field on the singly wound torus and the caduceously wound torus is the unit sphere on H and is of $SU(2)$ symmetry.

Mappings on a torus by (1) single winding driven in resonance, (2) nonresonance, (3) caduceous (double) winding driven in resonance, and (4) nonresonance:

The following mapping:

$$\rho : SU(2) \to GL(3, \Re), \tag{47}$$

is a two-to-one mapping. Mappings of fields on a torus by methods (1), (2), (3) and (4) are described by:

$$\begin{aligned} &(1) \text{ and } (3) \quad \rho_{1,3}^{-1} : GL(3, \Re \;\to\; SU(2)/Z_2, \\ &(2) \text{ and } (4) \quad \rho_{2,4}^{-1} : GL(3, \Re \;\to\; SU(2)/Q, \end{aligned} \tag{48}$$

which are all one-to-two mappings.

As $GL(3, \Re)$ is a subgroup of $GL(3, C), \rho^{-1}$ is also a homomorphism from $GL(3, C)$ to $SU(2)$, or, in other words, a representation of C^3 on $SU(2)$.

Concept of Cover defined: A collection U of open sets covers a topological space X if their union is all of X. The driven torus is a topological space, X, and is covered by the collection, U_α, of scalar and vector potentials which are 0,1 and 2-forms:

- The scalar potential is defined: $E = -d\phi$, where E is a a 1-form and ϕ is a function or 0-form.
- The vector potential for B is defined: $B = dA$, where B is a 2-form and A is a 1-form.
- The vector potential for F is defined: $F = dA$, where F is a 2-form and A is a 1-form on space–time.
- The driven torus, X, thus effects the union of the potentials, U_α, such union covering all of the torus, X.

Closed defined: If the exterior derivative of a differential form is zero, then the differential form is *closed*.

Exact defined: A differential form that is the exterior derivative of some other differential form is *exact*.

Closed 1-forms on a manifold are automatically *exact* if a certain topological condition holds, namely: that the manifold is *simply connected*.

Symplectic topology defined [11]: the study of the *global* phenomena of symplectic geometry, which have no local invariants.

The Action and Magnetic Charge of a Field Mapped onto a Torus by single or double (caduceous) winding: Given a particle's position, λ, and a velocity, λ', then the Lagrangian (the kinetic energy minus the potential energy) is:

$$L = L(\lambda(t), \lambda'(t)). \tag{49}$$

Integrating over time, T, obtains the action:

$$S(\lambda) = \int_0^T L, \tag{50}$$

and the path of a classical and coherent test particle from point a to point b is weighted by a *phase factor*:

$$\exp[iS(\lambda)]. \tag{51}$$

When mapped to a singly wound or doubly wound torus, *in resonance or nonresonance*, the A fields on the driven torus are *closed* and *not exact* at the 1-form level – since the integrals of the A 1-forms threading the hole are *not zero*. Space is *multiply connected* at the 1-form level. However, at the 2-form level the A field on a torus is *closed* and *exact* and the multiply connected space at the 1-form level is mapped to a *simply connected* space at the 2-form level. That is, *at the 2-form level* $\Re \times S^2$ is *simply connected* and every closed 2-form is *exact*. Nevertheless, in the singly and doubly wound case in both resonance and nonresonance, *at the 1-form level* $\Re \times S^2$ has closed forms which are *not exact*, and $H^2\Re \times S^2$ is non-zero and *multiply connected*.

Therefore for the singly and doubly wound tori, both resonant and nonresonant,

$$H^0 = \Re; \quad H^1 = C; \quad H^2 = Q. \tag{52}$$

At the H^1 level (but not at the H^2 level), the field mapped onto a torus has a '*charge*', q, defined:

$$q = \int_{S^2} \star E. \tag{53}$$

In the case of fields on *the singly and doubly wound tori*, both resonant and nonresonant,

- the electric field (a 1-form) is *closed but not exact* – so there is no scalar potential remaining.
- the magnetic field (a 2-form) is *closed but not exact* – so there is no vector potential remaining

In the *exact cases*, (the 0-level $U(1)$ and the 2-level $SU(2)$):

$$\int_{S^2} B = 0, \tag{54}$$

and the fields are *simply connected* at the 0- and 2-form levels.

In the *inexact case*, (the 1-level $U(1)$):

$$\int_{S^2} B = m, \tag{55}$$

where m is the magnetic charge.

$U(1)$ and $SU(2)$ **Groups** defined: $U(1)$ describes 1×1 matrices, which are just numbers. Therefore

$$U(1) = \{\exp[i\theta] : \theta \in \Re\}. \tag{56}$$

$U(1)$ is isomorphic to $SO(2)$, with an isomorphism given by:

$$\rho(\exp[i\theta]) = \begin{pmatrix} \cos\theta & \sin\theta \\ -\sin\theta & \cos\theta \end{pmatrix}, \tag{57}$$

indicating that rotations in the 2-dimensional real vector space $\Re^3$ are the same as rotations of the complex plane, C.

The group $SU(2)$ is the 3-sphere in H (quaternions). There is a two-to-one homomorphism:

$$\rho : SU(2) \to SO(3), \tag{58}$$

where $SO(3)$ is a subgroup of $GL(3, C)$.

Therefore ρ is a homomorphism of $SU(2)$ to $GL(3, C)$, i.e., a 3-dimensional complex representation of $SU(2)$, which is the torus. Mappings of a field on a torus by single or double (caduceous) winding is a 3-dimensional complex representation of $SU(2)$ and a spin-1 representation.

Furthermore, if V is the space of 2×2 matrices, then V can be identified with $\Re^3$. Then ρ is the mapping:

$$\rho : SU(2) \to GL(3, \Re). \tag{59}$$

$SU(2)$ is a two-to-one mapping, or a double cover.

The group $SO(3)$ is not simply connected, but its double cover, $SU(2)$, is, because it is diffeomorphic to S^3, which has no noncontractible loops in it. However, a second resonance can be obtained in both the single and double (caduceous) wound torus, so that pairs of waves on opposite sides of the torus can be identified. In this case, the gauge group is no longer $SU(2)$, but $SU(2)/Z_2$:

$$\frac{SU(2)}{Z_2} \cong SO(3). \tag{60}$$

$SU(2)/Z_2$ is not simply connected.

Bundle defined: A bundle is a structure consisting of a manifold, E, and manifold, M, and an onto map:

$$\pi : E \to M. \tag{61}$$

Fiber defined: For every point $p \in M$ the space $E_p = \{q \in E : \pi(q) = p\}$ is called the fiber over p.

Section defined: A section of a bundle $\pi : E \to M$ is a function $s : M \to E$ such that for any $p \in M, s(p) \in E_p$. Therefore $\pi \circ s$ is the identity map.

A **vector space** defined: A vector space is a vector bundle with a base equal to a single point.

Gauge Theory defined: Gauge theory generalizes electromagnetism by letting the vector potential be not merely a 1-form with components A_μ, but a matrix-valued 1-form with components $A^j_{\mu_i}$.

Curvature defined: Curvatures measures holonomy around '*infinitesimal loops*'.

Maxwell and Yang-Mills Equations defined:
Maxwell's equations are:

$$\star \mathrm{d} \star F = J. \tag{62}$$

Yang–Mills equations are:

$$\star \mathrm{d}_D \star F = J. \tag{63}$$

where d_D is the exterior covariant derivative of differential forms of sections of a bundle.

The variation of the action is zero for all functions f vanishing at $t = 0, T$ if and only if:

$$m\ddot{q}(t) - F(g(t)) = 0, \tag{64}$$

which implies $F = ma$. Therefore $\delta S = 0$ implies that $F = ma$, or that the Euler-Lagrange equations hold:

$$\frac{\partial L}{\partial q^j} = \frac{\mathrm{d}}{\mathrm{d}t}\frac{\partial L}{\partial \dot{q}^j}. \tag{65}$$

The Yang–Mills Lagrangian is:

$$L_{YM} = \frac{1}{2}\mathrm{tr}(F \wedge \star F) = \frac{1}{4}\mathrm{tr}(F_{\mu\nu}F^{\mu\nu})\mathrm{vol}. \tag{66}$$

Integrating the Yang–Mills equations provides the action principle:

$$\delta S_{YM} = 0. \tag{67}$$

Self-duality defined: When A is a self-dual (e.g., $F = \star F$), or anti-self dual, the Bianchi identity

$$\mathrm{d}_A F = 0 \tag{68}$$

automatically implies the Yang–Mills equation:

$$\mathrm{d}_A \star F = 0. \tag{69}$$

Self-dual solutions to the Yang–Mills equations are called **instantons**.

Chern Forms defined: If the space–time M is $2n$-dimensional, then the action is:

$$S(A) = \int_M \mathrm{tr}(F^n), \tag{70}$$

where the nth Chern form is the trace of the n-fold wedge product:

$$\underbrace{\mathrm{tr}(F \wedge \ldots \wedge F)}_{n \text{ factors}} \tag{71}$$

The kth Chern form defines a cohomology class in $H^{2k}(M)$ and the kth Chern class of the vector bundle E over M is the cohomology class of $\mathrm{tr}(F^k)$, where F is the curvature of any connection on E.

References

1. Baez, J. and Muniain, J.P. (1994) *Gauge Fields, Knots and Gravity*, NJ: World Scientific.
2. Baldomir, D. and Hammond, P. (1996) *Geometry of Electromagnetic Systems*, Oxford: Clarendon Press.
3. Barrett, T.W. (1990) Maxwell's theory extended. Part 1. Empirical reasons for questioning the completeness of Maxwell's theory – effects demonstrating the physical significance of the *A* potentials. *Annales de la Fondation Louis de Broglie*, **15,** 143–183.
4. Barrett, T.W. (1990) Maxwell's theory extended. Part 2. Theoretical and pragmatic reasons for questioning the completeness of Maxwell's theory. *Annales de la Fondation Louis de Broglie*, **15,** 253–283.
5. Barrett, T.W. (1993) Electromagnetic phenomena not explained by Maxwell's equations. pp. 6–86, in Lakhtakia, A. (Ed.) *Essays on the Formal Aspects of Maxwell's Theory*, Singapore: World Scientific.
6. Barrett, T.W. (1994) The Ehrenhaft-Mikhailov effect described as the behavior of a low energy density magnetic monopole instanton. *Annales de la Fondation Louis de Broglie*, **19,** 291–301.
7. Barrett, T.W. (1995) Sagnac Effect. pp. 278–313, in Barrett, T.W. and Grimes, D.M., (Ed.s) *Advanced Electromagnetism: Foundations, Theory, Applications*, Singapore: World Scientific.
8. *Contrawound Toroidal Helical Antenna*, Center for Industrial Research Applications (CIRA), Mechanical and Aerospace Engineering Department, West Virginia University, Morgantown, WV, USA, November, 1997.
9. Connes, A. (1994) *Noncommutative Geometry*, New York: Academic Press.
10. Craven, R.P.M., Prinkey, M.T. and Smith, J.E. (1997) *Contrawound Antenna*, United States Patent 5,654,723, dated Aug. 5.
11. McDuff, D. and Salamon, D. (1995) *Introduction to Symplectic Topology*, Oxford: Clarendon Press.
12. Monastyrsky, M. (1993) *Topology of Gauge Fields and Condensed Matter*, New York: Plenum Press.
13. Naber, G.L. (1997) *Topology, Geometry, and Gauge Fields*, New York: Springer.
14. Nakahara, M. (1990) *Geometry, Topology and Physics*, IOP, London.
15. Palais, R.S. (1997) The symmetries of solitons. *Bull. American Mathematical Society* **34,** 339–403.
16. Van Voorhies, K.L. and Smith, J.E. (1995) *Toroidal Antenna*, United States Patent 5,442,369, dated Aug. 15.
17. Ward, R.S and Wells, R.O. (1990) *Twistor Geometry and Field Theory*, Cambridge: Cambridge University Press.

AIAA 95-2368

DRAG REDUCTION, AND POSSIBLY IMPULSION, BY PERTURBING FLUID AND VACUUM FIELDS

H. D. Froning, Jr.
Flight Unlimited
2800 Saddleback Way #31
Flagstaff, Arizona 86004

R. L. Roach
University of Tennessee
Space Institute
Tallahoma, Tennessee 37388

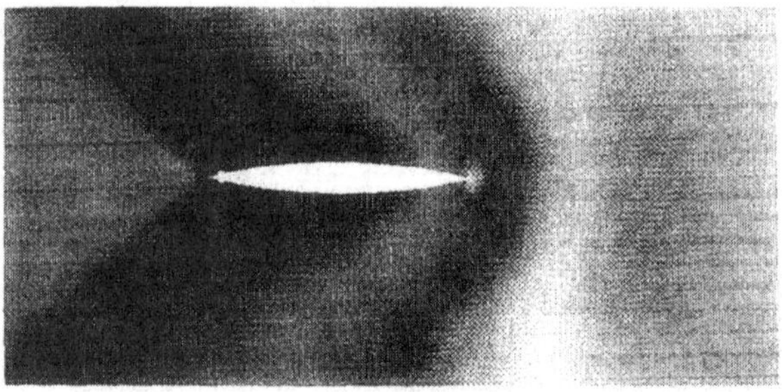

31st AIAA/ASME/SAE/ASEE Joint Propulsion Conference and Exhibit July 10-12, 1995/San Diego, CA

AIAA 95-2368

DRAG REDUCTION, AND POSSIBLY IMPULSION, BY PERTURBING FLUID AND VACUUM FIELDS

H. D. Froning, Jr.
Flight Unlimited
2800 Saddleback Way #31
Flagstaff, Arizona 86004

R. L. Roach
University of Tennessee
Space Institute
Tallahoma, Tennessee 37388

Abstract

This paper explored the possibility of wave drag reduction (beyond that achievable by increasing sharpness and slenderness) by perturbing the γT properties of air at leading and trailing edges of wings. Perturbations that increased γT (and hence local sound speed) in the vicinity of leading or trailing edges decreased drag, while perturbations that decreased the γT of air within these regions, increased drag. And inviscid drag of 2-dimensional shapes was cancelled completely by intense perturbations at both the leading and the trailing edges.

This paper also explored the possibility of reducing the resistance of accelerating bodies by perturbing the $\mu_o \varepsilon_o$ (magnetic permeability and electrical permittivity) characteristics of their surrounding space.

PART I

PERTURBING FLUID FIELDS

Introduction

Thrust and fuel consumption for vehicles that cruise within atmospheres is determined by their drag, which at nearly sonic speed, is mostly due to the "wave" drag caused by coalescence of airflow in forward regions into normal shocks. Thus, significant reduction of such drag would significantly decrease thrust and propellant needs.

Increasing the sharpness and slenderness of wings and bodies is, of course, the current way of decreasing drag at very high speed. And this part of the paper explores the possibility of further drag reduction by perturbing air properties in the vicinity of high-speed craft.

This investigation only considered flow perturbations over two-dimensional airfoil-like shapes. Thus, its results are directly applicable, only, to vehicles with wings of high aspect-ratio and span. Actual perturbations would most likely contain convective movement and be non-circular in shape. However, circular disturbances with no internal movement were assumed for simplicity sake. And, primary study focus was Mach 0.99 because the drag coefficient of most vehicles tends to maximize at near this speed.

Technical Approach

Properties that determine energy storage and dissipation within of a given gas are its: ratio of specific heats (γ). And γ, together with local gravitational acceleration (g), gas temperature (T), and a constant (R) for the given gas, determine the local propagating speed of sound $(gR\gamma T)^{1/2}$ within the gas. Therefore, because the gR of given gases are almost invariant within planetary atmospheres, sound speed can only be changed by varying γT.

Disturbances that locally decrease γT and sound speed within a gas can be from processes that use cold fluids or gases to cool airstreams by transpiration techniques. Disturbances that locally increase γT and sound speed can, of course, be from combustion processes or electrical discharges which cause heating and, disociation/ ionization within a region of air. Shown, as an example of this in Figure 1, is the hot high-speed exhaust of an ion propulsion system (developed by the Rocket Research Company of Redmond Washington) perturbing and increasing the γ and T of nearby air.

Airflow disturbances were centered at either leading or trailing edges, and disturbances at both leading and trailing edges were considered as well. Here, as shown in Figure 2, disturbances extended outwardly in all directions with diminishing intensity until zero intensity was reached at a distance equal to one-half the wing chord (c/2). The disturbance intensity diminished from a maximum value Δ(γT) at a rate roughly proportional to (but somewhat greater than) distance squared. As an example, at distances equal to: c/6, c/4, c/3, disturbance intensity had diminished to values that were approximately 1/2, 1/4 and 1/100 of the maximum value Δ(γT).

Variation in γ cannot be much more than about 10 percent for typical gases, even though enormous T variations can be associated with γ change. But because proper modeling of variations in both γ and T resulted in excessive computational complexity, Δ(γT) values were modeled by Δγ values which were sometimes greater than can actually be achieved in gases. Thus, because large γT changes are not accurately modeled, the results can only be used to indicate trends.

Effect of Airflow Perturbations on Drag

The effect of airflow perturbations on inviscid drag were determined by computational fluid dynamics (CFD) techniques. As shown in Figure 3, drag was increased by disturbances that locally decreased the γT of air at leading or trailing edges, while increasing γT at leading and trailing edges reduced drag. And, leading edge disturbances are seen to decrease drag more effectively than trailing edge disturbances at Mach 0.99 Speed.

Although airflow γT perturbation values investigated in Figure 3 are less than 10 percent of the unperturbed values of γT, much stronger perturbations are possible by more energetic combustion or intense electrical discharge -- which cause significant increase in T. For example, Bussard (1) has shown that "clean" fusion electric power systems can heat air to peak temperatures of the order of 20,000 degrees kelvin - several orders of magnitude greater than ambient air temperature within a Mach 0.99 airstream. Thus, more intense γT perturbations were considered -- for accomplishing greater reductions in drag.

As shown in Figure 4, significant drag reductions were achievable, when γT in the vicinity of wing-leading edges was significantly increased. However, a limiting value of drag reduction was reached, even when very intense γT perturbations of air at the leading edge occurred. However, γT perturbations at both leading and trailing edges reduced drag further. And intense γT perturbations are seen to enable impulsion (rather than resistance) for atmospheric flight.

Pressure coefficient information indicated that favorable leading edge perturbations reduced pressures over the forward portion of the airfoil, while pressures over the aft portion of the airfoil remained almost unchanged. Favorable perturbations at both leading and trailing edges reduced pressures over the forward portion of the airfoil and increased

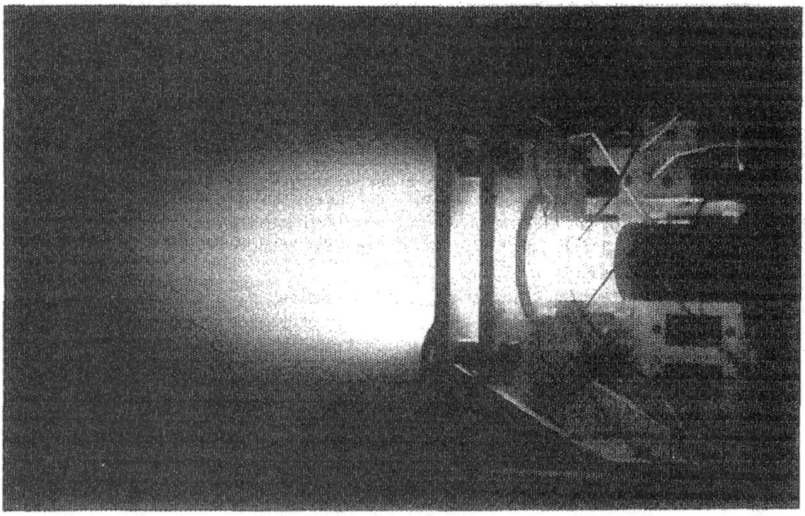

Figure 1 Air Perturbation by an Electric Rocket's Ion Exhaust

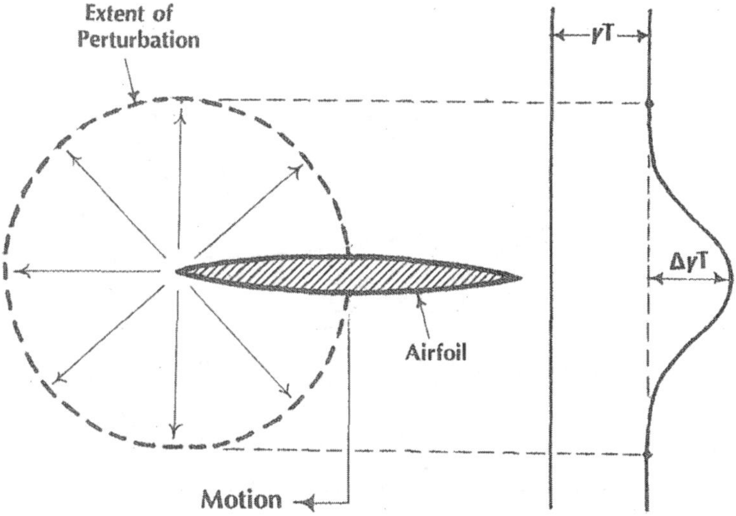

Figure 2 Perturbation Intensity Distribution

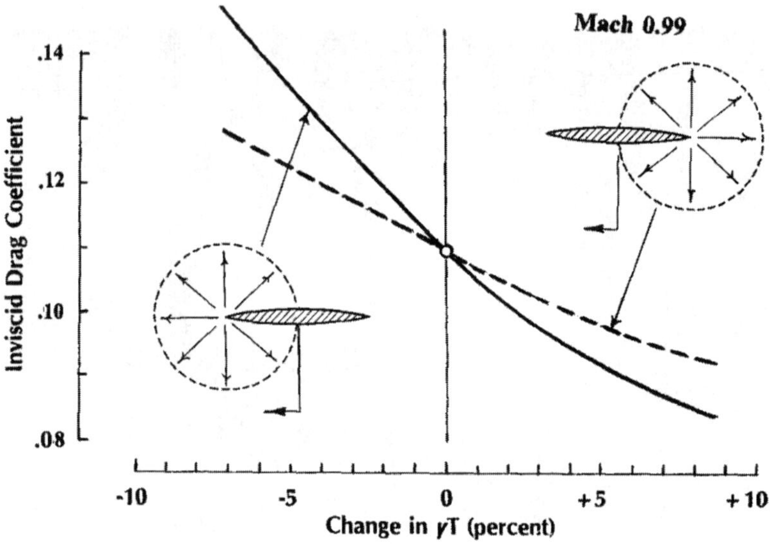

Figure 3 – Influence of Airflow Perturbation

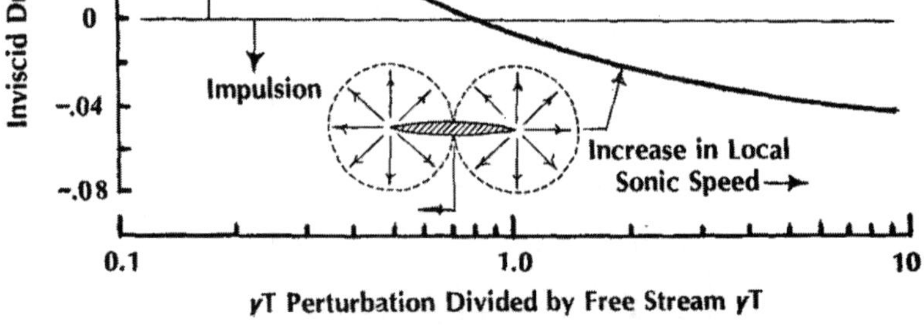

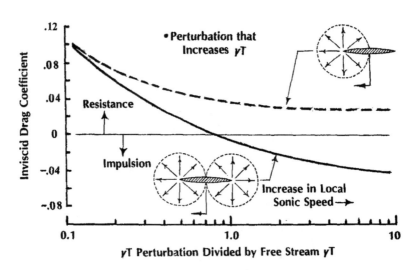

Figure 4 – Influence of Airflow Perturbation

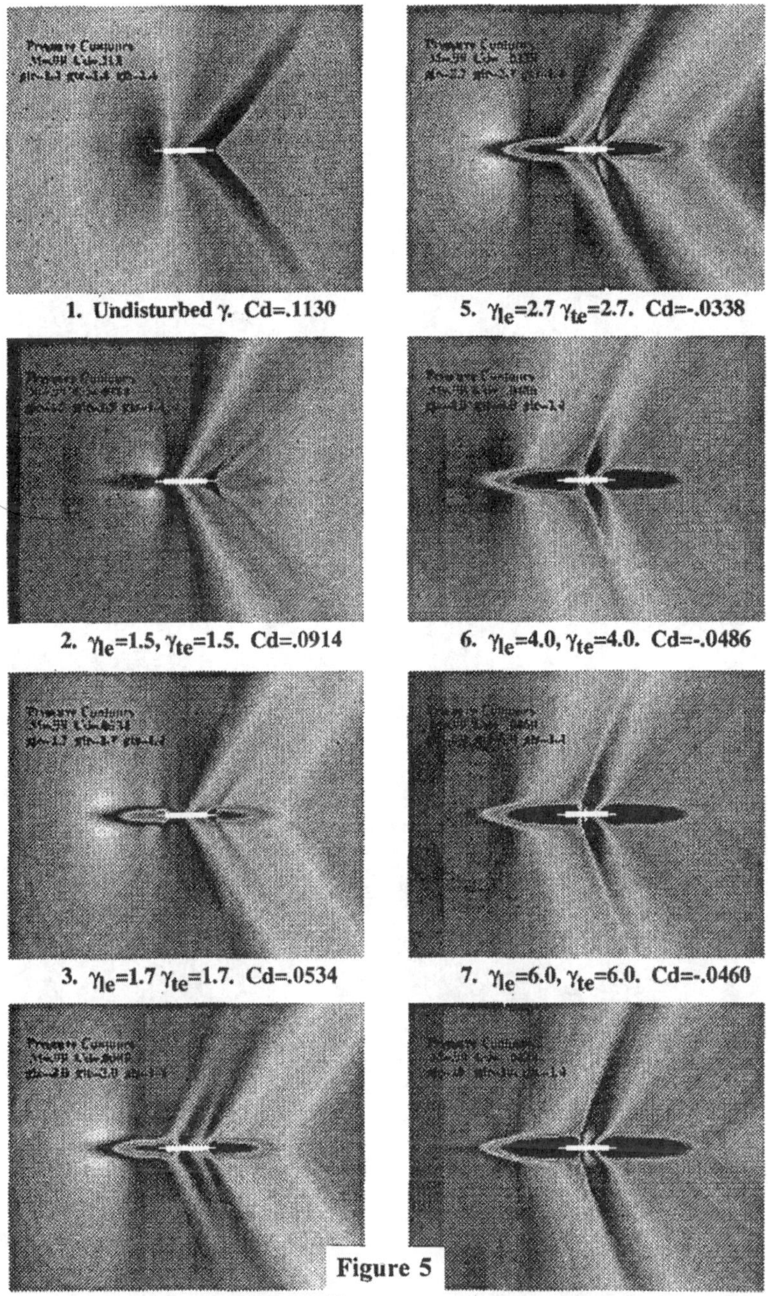

1. Undisturbed γ. Cd=.1130

2. γ_{le}=1.5, γ_{te}=1.5. Cd=.0914

3. γ_{le}=1.7 γ_{te}=1.7. Cd=.0534

4. γ_{le}=2.0, γ_{te}=2.0. Cd=.0085

5. γ_{le}=2.7 γ_{te}=2.7. Cd=-.0338

6. γ_{le}=4.0, γ_{te}=4.0. Cd=-.0486

7. γ_{le}=6.0, γ_{te}=6.0. Cd=-.0460

8. γ_{le}=10., γ_{te}=10. Cd=-.0438

Figure 5

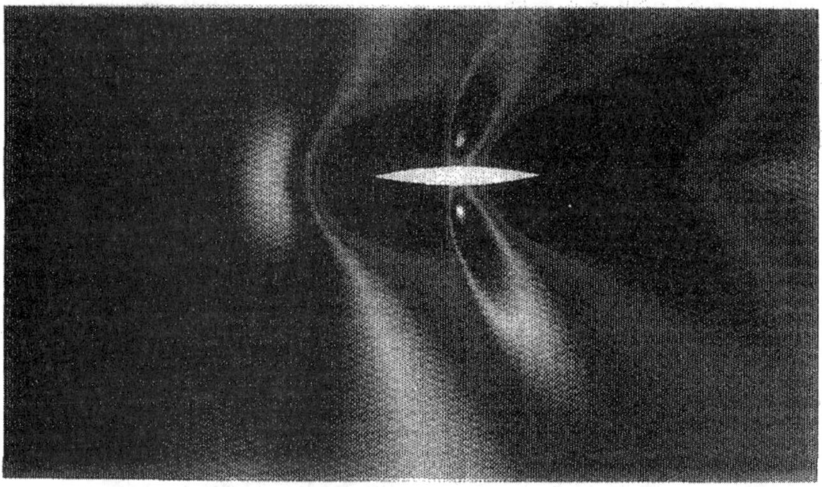

Fig. 6 **Pressure Distribution at Mach 0.99 for a 7-Fold Increase in Leading and Trailing Edge γT**

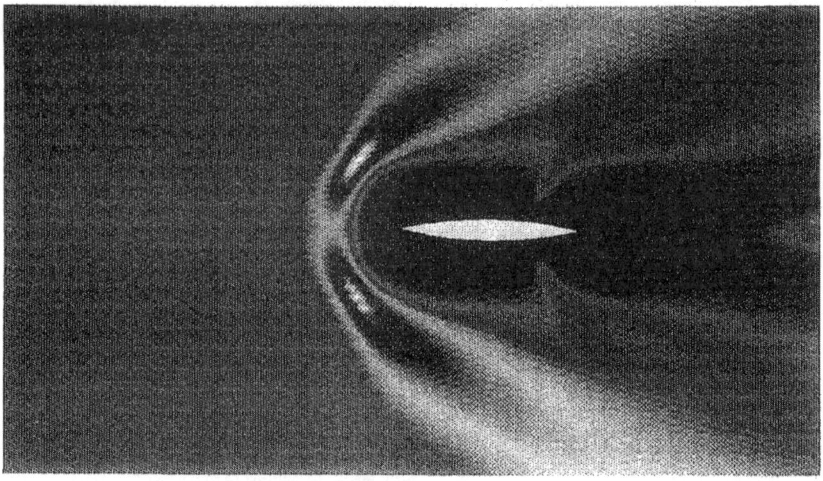

Fig. 7 **Pressure Distribution at Mach 2.50 for a 7-Fold Increase in Leading and Trailing Edge γT**

pressures over its aft portion. Higher aft pressures resulted in further diminishment of airfoil drag and, when leading and trailing edge disturbances were sufficiently intense, enabled impulsion (rather than resistance) at Mach 0.99 speed.

Influence of Disturbance Shape

Changing the circular disturbance shape into an elliptical disturbance shape about twice as long and half as high, was found to increase the amount of drag reduction somewhat. The top left corner of Figure 5 shows pressure distributions about the airfoil and its drag coefficient for unperturbed air at Mach 0.99, and the following pictures show change in pressure distributions and drag coefficient as the intensity of elliptical disturbances are increased at airfoil leading and trailing edges.

Influence of Speed and Disturbance Size and Shape

This investigation emphasized transonic speeds but similar trends were found for supersonic speeds as well. Here, circular perturbations that locally increased air γT in the vicinity of leading and trailing edges decreased inviscid drag coefficient at Mach 2.5. And impulsion --rather than resistance -- was achieved for γT perturbations that were significantly intense. Figures 6 and 7 show distribution of pressure about the airfoil at Mach 0.99 and Mach 2.50 speed, when the γT of perturbed air at leading and trailing edges is about 7 times the γT of unperturbed air. And the favorable gradient, formed by lower forward pressures and higher aft pressures, resulted in impulsion (instead of drag) at Mach 0.99 and 2.50.

Doubling the size of circular disturbance regions about wing leading and trailing edges at Mach 0.99 was found to be counter productive, resulting in increased drag. But, halving the size of disturbance regions resulted in further reduction in drag. This, of course, was encouraging because it indicated that the airflow that must be perturbed may not have to be extremely large.

Conclusions and Future Work

Reduction in wing drag is achievable with disturbances that locally increase the γT properties and sonic-speed of airflow at leading and trailing edges. And sufficiently intense disturbances may actually enable impulsion rather than drag. Additional work, therefore, is needed to ascertain if intense disturbances are achievable with power systems that could be lighter or simpler than current jet propulsion and power systems for high-speed cruise.

It is hoped that future work can examine: flight at angle-of-attack, 3-dimensional vehicle configurations, and disturbances that include heat transfer and real gas effects. Future work would also determine if satisfactory disturbance intensity and drag reduction is achievable with practical power levels and acceptable levels of power system mass.

PART II

PERTURBATION OF VACUUM FIELDS

Introduction

Just as aerodynamic resistance to a vehicle's motion (its drag) may be reducible by favorable perturbation of its surrounding fluid field, so recent developments indicate that inertial resistance to a vehicle's motion (its inertia) might be diminished if its surrounding vacuum field could also be perturbed in a favorable way. This part of the paper will attempt to show that the nature of fluid and vacuum fields -- and their response to vehicle perturbations may be sufficiently similar that certain trends predicted by CFD for

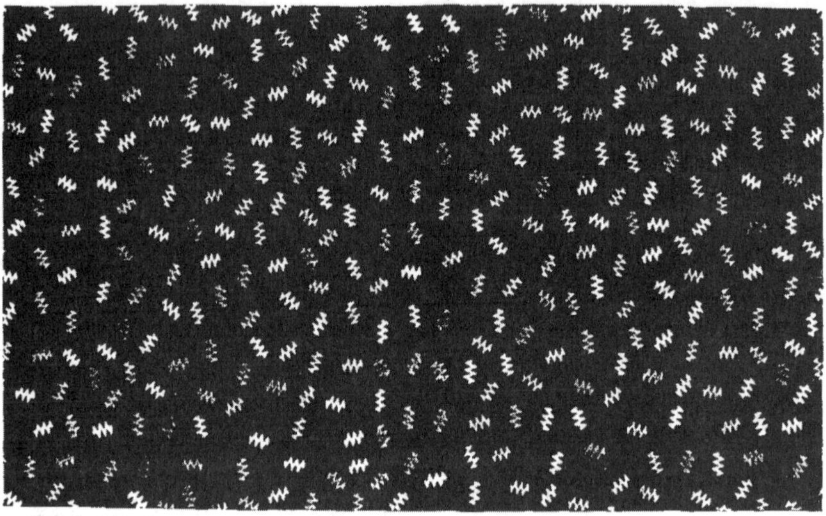

Figure 8 Zero-Point Fluctuations of the Vacuum EM Field

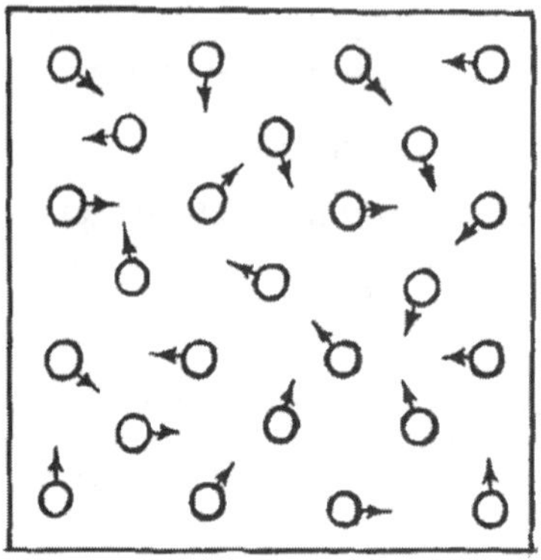

Average Value of Thermal
ρ = Energy Density Within a
Volume of Air at an Altitude (h)

$\rho \cong \rho_0 e^{-kh}$

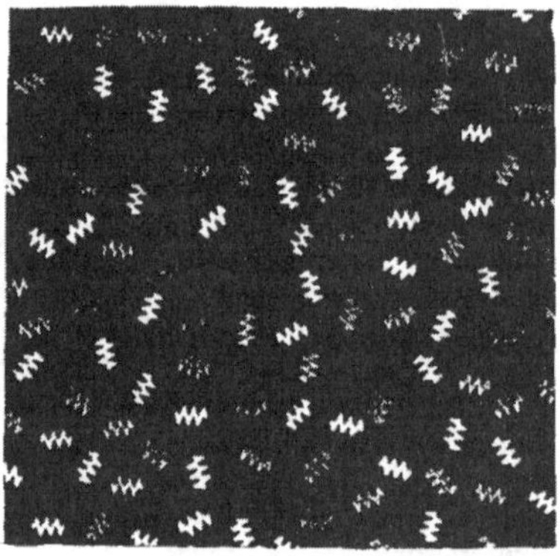

Expectation Value of Zero-Point
$\hat{\rho}$ = Energy Density within a Vacuum
Region of Dimension (L)

$\hat{\rho} \sim \hbar / cL^4$

Figure 9 Density of Air and "Empty" Space

perturbation of fluid fields may be similar to trends for vacuum fields as well.

Vigor and Vitality of "Empty" Space

Although space seems inert and empty, quantum physics (2,3,4) views it as having vigor and vitality over scales of time and space that are too short for the material senses to perceive. And a major contributor to such vigor and vitality are the "zero-point" fluctuations of the vacuum electromagnetic field.

Electromagnetic energy pulsations or waves are believed to be associated with zero-point fluctuations of the vacuum electromagnetic field (sometimes referred to as the zero-point electromagnetic field) and Figure 8 visualizes this electromagnetic vigor and vitality as numerous lightning flashes within a region of space at a given time. And although the vacuum electromagnetic field, and its zero-point energies is not yet completely understood, zero-point electromagnetic energy pulsations are usually modeled as the oscillations of an enormous number of harmonic oscillators, -- with the zero-point energy (E) associated with a given oscillation of angular frequency (ω) and wavelength (λ) being,

$$E = \frac{\hbar\omega}{2} = \frac{\hbar c}{4\pi\lambda} \quad (1)$$

Here, $\hbar$ is plancks constant, c is the speed of light and the total zero-point energy within a vacuum region is the sum of the individual energies of the individual electromagnetic pulsations within it. However, individual pulsations couple in such a way that they tend to cancel each other over regions that are large, compared to their size. Thus, measured zero-point electromagnetic field energy within a large region of the vacuum is less than the sum of the individual zero-point energies of its vibratory modes.

Wheeler (2) has used Heisenberg's Uncertainty Principle to show that the average amplitude of zero-point energy fluctuations will have an "expectation" value of magnitude $\hbar c/L$ ergs within a region of space of dimension (L). Furthermore, since mass is equivalent to energy divided by c^2, expectation values for energy density in terms of equivalent mass is, as shown in Figure 9, $\hbar/cL^4$ gms/cm^3.

Force Development Within Space

Although the vacuum of space contains zero-point fluctuations of many different wave lengths, two plates in very close proximity would exclude those fluctuations of longer wavelength from the narrow gap between them (Figure 10). Casimir (5) predicted that this would result in an attractive force of approximately 0.2 dynes to act upon two plates of one square centimeter area that are 0.5 micro meters apart. And subsequent experiments proved that this force indeed, occurred. And Milonni (6) views the Casimir force as being a consequence of less outward pushing than inward pushing zero-point "radiation pressures" acting upon the plates.

The Resistance of the Vacuum to Accelerated Flight

In 1968 Sakharov made the controversial suggestion that gravity is not a fundamental separate force, but rather, an induced effect caused by the constant interplay between all charged particles throughout the cosmos and all the zero-point fluctuations of the vacuum electromagnetic field. And Puthoff (7) has somewhat substantiated Sakharov's suggestion by showing the possibility of so-called gravitational force being a van der Waals force caused by the long range radiation fields generated by "Zitterbewegung" motion of all charged particles in response to all zero-point fluctuations within the vacuum electromagnetic field.

An investigation by Haisch, Rueda and Puthoff (8) also shows that inertia can be viewed as arising from the vacuum electromagnetic field, acting, as shown in Figure 11, upon every charged sub-elemental entity of an accelerating body in a resisting way. Here, (8) shows that the magnetic component of the Lorentz force arises in accelerated frames as the charged sub-elemental entities interact with the very high frequency zero-point fluctuations to produce an opposing force (F_x) that is perceived as the "inertia" of the body's mass.

In (8) the following equations are derived for the resisting force (F_x) acting on a single elemental charged particle undergoing uniform acceleration (a) and for the inertial rest mass (m_i) associated with the resisting force:

$$F_x = -\frac{1}{2\pi}\Gamma\omega_c\frac{\hbar\omega_c}{c^2}a \qquad (2)$$

$$m_i = \frac{\Gamma\hbar\omega_c^2}{2\pi c^2} \qquad (3)$$

Inertial rest mass is seen to be a consequence of the cut off frequency (ω_c) of zero-point fluctuations, and of the radiation damping constant (Γ) which defines the coupling between the charged particle and zero-point radiation fields.

Interacting with the Vacuum for Flight

The Casimir Effect is generally considered proof that the vacuum can be interacted with (perturbed) and force produced, by the diminishment of zero-point fluctuations of longer wavelength within a given zone. And although this effect, itself, is not useful for thrust, it suggests other possibilities that might. As an example: if energies emitted by a moving ship could interact strongly with the vacuum electromagnetic field and perturb it such that numbers or frequencies of zero-point fluctuations are diminished over its forward part, zero-point radiation pressures would act upon the ship in a thrusting way.

And, Equation (3) implies that diminished ship inertia would be a consequence of diminished zero-point fluctuations or fluctuation frequencies in the forward vicinity of an accelerating ship.

It, of course, is not yet known if energy can ever be crafted with the intensity and modulation required to interact favorably with the vacuum electromagnetic field. However, possibilities include: pulsed or steady-state electromagnetic radiation, electric charge, or magnetic flux.

Similarities in Vehicle Interactions with Fluid and Vacuum Fields

As shown in Figure 9, innumerable collisions of air molecules underlie the energetics of air, while innumerable electromagnetic fluctuations contribute to the vigor and vitality of seemingly empty space. And although the propagating speed $(gR\gamma T)^{1/2}$ of acoustic disturbances in air is about 6 orders of magnitude slower than the speed $(1/\mu_o\varepsilon_o)^{1/2}$ of electromagnetic disturbances in space, both have their speeds determined (as shown in Figure 12) by the energetics of the mediums through which they move.

Figure 13 shows that $\mu_o\varepsilon_o$ and γT perturbations could be smiliar if they cause similar change in disturbance propagation speed within the regions of air and space that they perturb. Geometrical scaling, of course, requires that the respective sizes of vehicles and their perturbations within vacuum and fluid fields be proportional to the disturbance propagating speeds of light and sound.

Similarity in vehicle interactions with fluid and vacuum fields also require that the fields respond similarly to the vehicle disturbances and react back upon them in a similar way. In this respect, just as the aerodynamic recoil of a gaseous medium to moving vehicles with

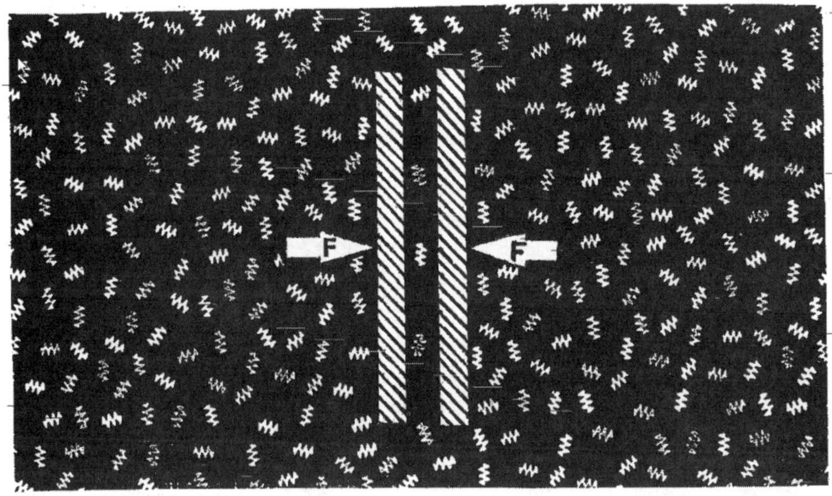

Figure 10 Casmir Force by Perturbing the Vacuum EM Field

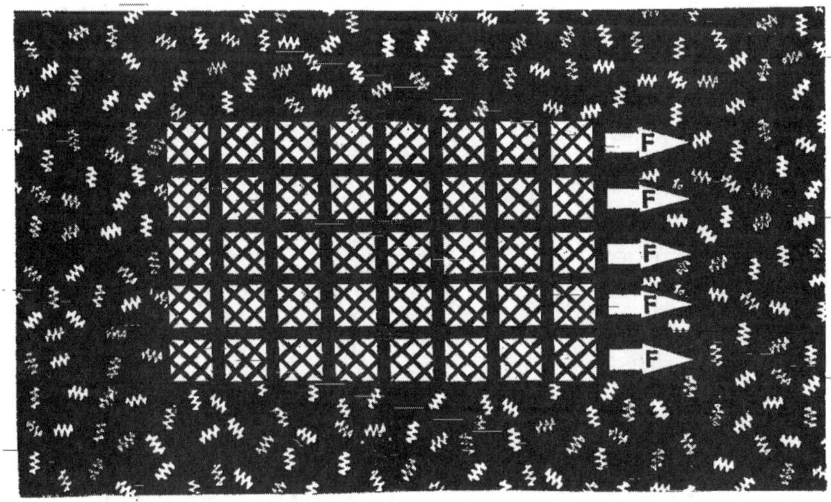

Figure 11 Inertia -- Resistance by the Vacuum EM Field

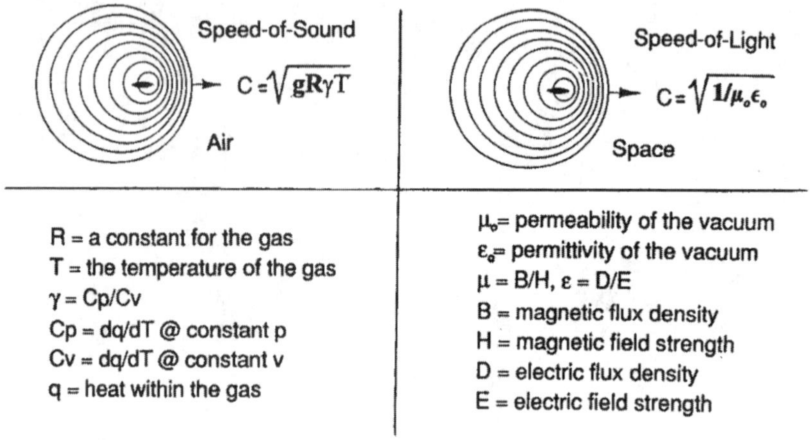

Figure 12 **Speeds of Acoustic and Electromagnetic Wave Fronts in Air and Space**

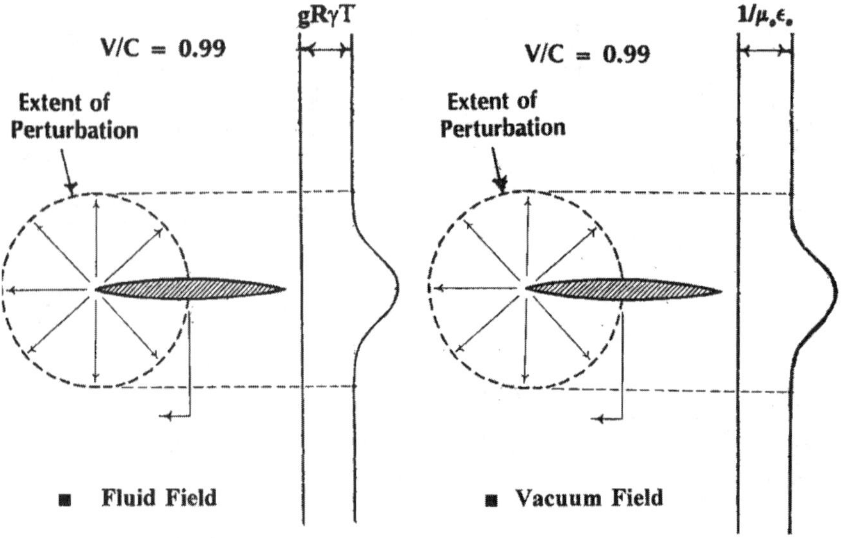

Figure 13 **Perturbations that Might Cause Similar Effects within Fluid and Vacuum Fields**

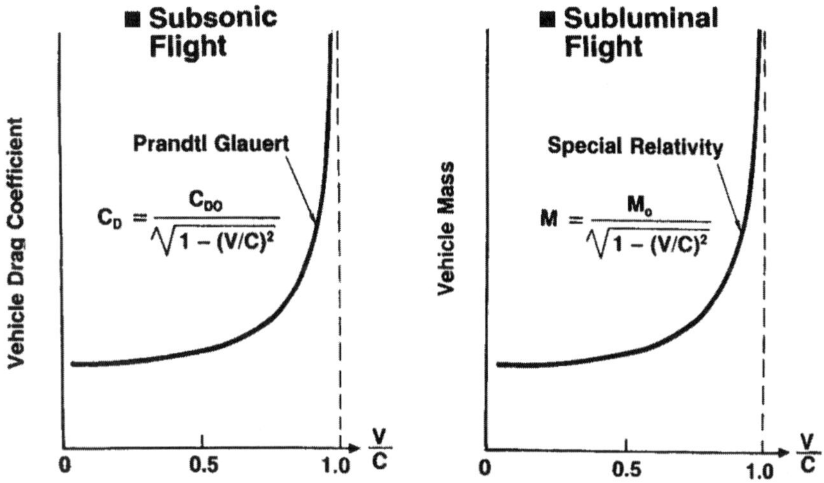

Figure 14 Flight Resistance Increase with Speed

non-zero thickness is perceived as their drag -- so, according to (8), electromagnetic recoil of the vacuum to moving vehicles with non-zero acceleration is perceived as the inertia of their mass. And, because airflow compressibility is analogous to spacetime dilation, Figure 14, shows that simplified theories of fluid dynamics and relativistic dynamics predict the same amplification of drag and inertia as flight speeds approach sound speed in air and light speed in space.

Reducing Resistance of Vacuum Fields

Just as Figure 2 shows fluid field perturbations which increase sound-speed within disturbed regions of a fluid field, so Figure 15 shows perturbations that would increase light-speed within disturbed regions of the vacuum electromagnetic field. And just as vehicle drag is diminished by perturbations that locally increase sound-speed within fluid fields, so Figure 15 shows diminished inertia by perturbations that locally increase light-speed within the vacuum electromagnetic field. Similarly, just as drag reduction is a consequence of a more favorable distribution of atmospheric air pressures about a craft, so inertia reduction could be a consequence of a more favorable distribution of zero-point radiation pressures about it.

Because of inexactness in the analogy, lesser or greater $\mu_o\varepsilon_o$ perturbation than indicated in Figure 15 would be required for a given diminishment in inertial resistance. But, if its predicted trend is correct, there is seen to be the possibility of inertia reduction, and even the possibility of impulsion, rather than resistance, -- if the vacuum electromagnetic field can be strongly interacted with.

****

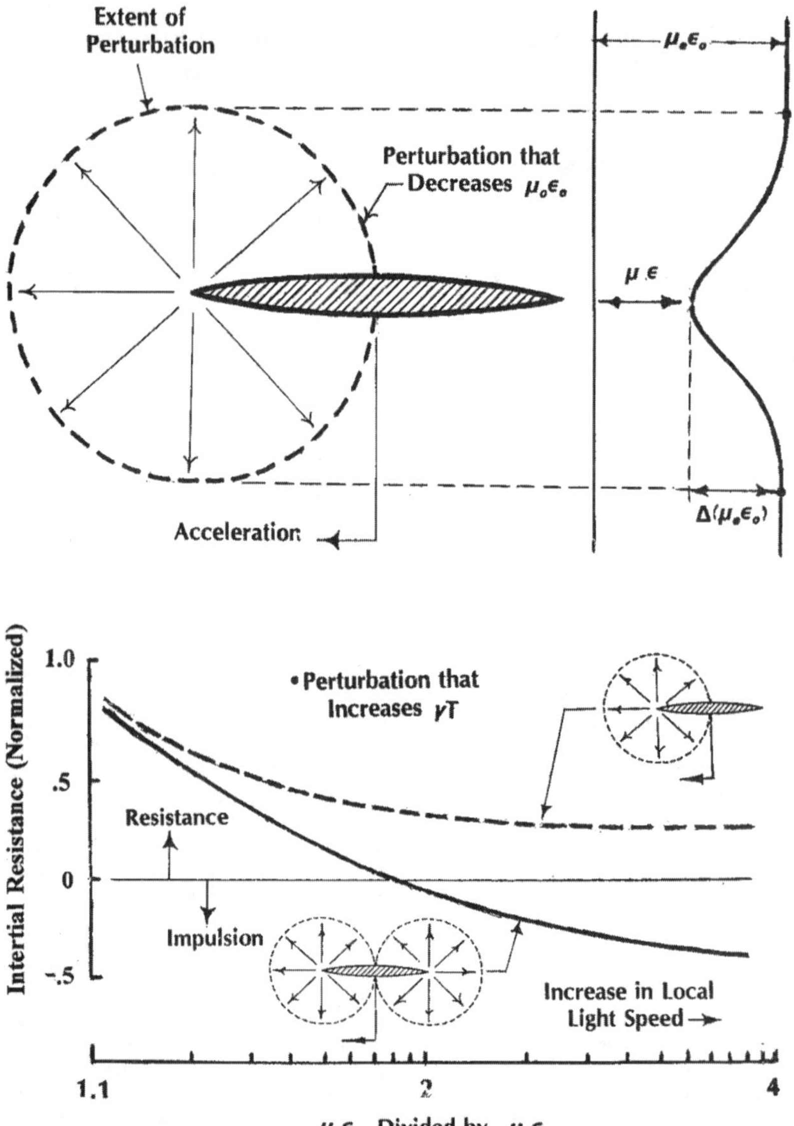

Figure 15 Influence of Perturbation

References

1. Bussard, R.W., the QED Engine System: Direct Electric Fusion - Powered Rocket Propulsion Systems, 10th Space Nuclear Power and Propulsion Conference, Albuquerque, N.M., 1993.

2. Wheeler, T.A., Superspace and the Nature of Quantum Geometrodynamics, Topics in Nonlinear Physics, pp. 615-644. Proceedings of the Physics Section, International School of Nonlinear Mathematics and Physics, Springer Verlag, (1968).

3. Boyer, T.H., Stochastic Electrodynamics, Phys. Rev. D, Vol. 11, No. 4, pp. 790-809, (1975).

4. Atchison, T.J.R., Nothing's Plenty - The Vacuum in Modern Quantum Field Theory, Contemporary Physics, Vol. 26, No. 4 (1985).

5. Casimir, Proc. Ned. Akad. Wet. 51,973 (1948).

6. Milonni, P.W., et al., Radiation Pressure from the Vacuum: Physical Interpretation of the Casimir Force, Phys Rev A, Vol 49, No. 2, pp 678 - 694, (1988).

7. Puthoff, H. E., Gravity as a Zero-Point Fluctuation Force, Physical Review A, Vol. 39, No. 5, pp. 2333-2342 (1988).

8. Haisch, B., et al., Inertia as a Zero-Point Field Lorentz Force, Phys. Rev. A, 1 Feb. (1994).

A97-36363

AIAA 97-3170

INERTIA REDUCTION - AND POSSIBLY IMPULSION - BY CONDITIONING ELECTROMAGNETIC FIELDS

H.D. Froning, Jr
Flight Unlimited
5450 Country Club
Flagstaff, AZ
86004

T.W. Barrett
BSEI
1453 Beulah Road
Vienna, Virginia
22182

33rd AIAA/ASME/SAE/ASEEJoint Propulsion Conference & Exhibit
July 6 - 9, 1997 / Seattle, WA

AIAA 97-3170

INERTIA REDUCTION - AND POSSIBLY IMPULSION - BY CONDITIONING ELECTROMAGNETIC FIELDS

H.D. Froning, Jr
Flight Unlimited
5450 Country Club
Flagstaff, AZ 86004

T.W. Barrett
BSEI
1453 Beulah Road
Vienna, Virginia 22182

Abstract

Although the spacetime metric associated with gravitation can be distorted somewhat by electromagnetic influences (such as magnetic fields of stupendous strength) there is no intimate interaction or coupling between ordinary electromagnetic (em) fields and those of gravitation because they are of a different essence and form. But if the fields associated with ordinary em radiation could be endowed with an essence and form similar to that of gravitation, such radiation might be capable of coupling with gravitation for propulsive use. This paper describes several ways of creating specially conditioned em radiation and, several ways that the fields associated with such radiation might interact propulsively with those associated with gravity or the vacuum's zero-point state.

Introduction

In 1903, the Wright Brothers used energies of chemical combustion and forces of propeller propulsion to achieve the first powered aircraft flight above the earth. Air flight progressed rapidly with propeller propulsion but it was not revolutionized until a new mode of impulsion (jet propulsion) and new technologies (such as microelectronics) enabled swift and economical airline travel approximately 60 years after the Wright Brothers first flight.

Jet propulsion included airbreathing jet propulsion (turbojet engines) for flight within the atmosphere and rocket jet propulsion (rocket engines) for flight within both air and space and in 1957, rocket engines using the energies of chemical combustion and forces of jet propulsion enabled the Sputnik satellite to accomplish the first orbital flight above earth. Spaceflight has progressed rapidly with rocket engines during the first 40 years after Sputnik's flight, but as the efficiency limits of chemical combustion and jet propulsion are being approached, spaceflight has not yet been revolutionized as airflight has. Thus, it is conceivable that spaceflight (like airflight) may not be really revolutionized until a new mode of impulsion is discovered and perfected.

One conceivable possibility for a new mode of impulsion might be field propulsion -- force development by actions and reactions of fields, instead of by the combustion and expulsion of mass. Development of force by actions and reactions of fields has been considered by investigators such as Cox, Engleberger, and Zubrin (1,2 and 3) who have shown that energy emitted by a vehicle in the form of electric charge or magnetic flux can interact propulsively with the ambient electric and magnetic fields of space. But these fields -- caused by flows of charged particles and radiation from stars and rotating celestial bodies -- are relatively mild and the resulting vehicle accelerations are extremely small. Thus, fields and interactions that develop greater vehicle accelerations would be desirable. And such fields and interactions might be associated with gravitation or the quantum electromagnetic ground state of the vacuum -- of empty space itself.

The most desirable type of field interaction would, of course, be that which not only develops thrusting and braking force, but diminishes vehicle inertia -- inertial resistance to its acceleration -- as well. As an example, diminishment of vehicle inertia to almost zero would enable accomplishment of needed flight acceleration with extremely low levels of field propulsion or jet propulsion thrust. Such inertia reduction would also enable enormous vehicle acceleration or deceleration with only modest field or jet thrusting force and with only modest structural stresses and strains exerted upon the vehicle and its inhabitants.

The "Zero-Point" State of Space

Space seems inert and empty but quantum field theory and stochastic electrodynamics views it as possessing vigor and vitality over scales of time and space that are too short for the material senses to perceive. A major contributor to such vigor and vitality are "zero point" energy (ZPE) fluctuations -- innumerable electromagnetic energy pulsations of varying wavelength and frequency which manifest the energetics of the so called "vacuum electromagnetic zero-point field" (ZPF).

Distributions of individual ZPE fluctuations are isotopic throughout undisturbed space and the spectral energy density of the ZPF is Lorentz invariant. Thus, the ZPF acts uniformly over bodies moving at constant speed, causing no net force. But the ZPF is not isotopic and Lorentz invariant in the accelerated frames of accelerating bodies. And investigators

Published by the American Institute of Aeronautics

such as Haisch, Rueda and Puthoff (4) have proposed that the inertia of bodies might be a consequence of spectral distortion of the ZPF in inertial frames of accelerating bodies and of a resisting em Lorentz force that arises from such distortion.

One observed consequence of perturbing the ZPE patterns of so-called "empty" space is the Casimir force. Here, two plates placed closely together, as shown in Figure 1 exclude ZPE fluctuations of longer wavelength from the gap and inward pushing forces act upon the plates, in (5) Milonni associates a "radiation pressure" with each ZPE fluctuation, and views the Casimir force as unbalanced radiation pressures caused by fewer ZPE fluctuations between the plates Scharnhorst (6) calculates that the velocity of light in such a region of reduced ZPE fluctuation density is greater than light velocity in vacuo (unconfined space).

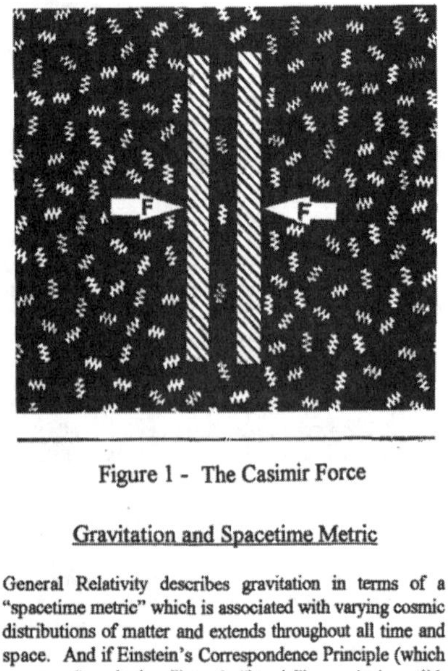

Figure 1 - The Casimir Force

Gravitation and Spacetime Metric

General Relativity describes gravitation in terms of a "spacetime metric" which is associated with varying cosmic distributions of matter and extends throughout all time and space. And if Einstein's Correspondence Principle (which equates "gravitational" and "inertial" mass) is valid, perturbation of spacetime metric within a given region should not only change gravitational influences within the perturbed region -- but the inertias of masses as well.

Gravitation, as the mutual attraction between material masses, has been well understood for many years. But there is not yet scientific unanimity as to its primordial origin or essence. As an example, the preponderance of current scientific opinion is that gravity is one of 4 fundamental forces of nature. But investigators such as Sakharov and Puthoff (7) contend that gravity is not a fundamental force - but one that arises from the continual em interplay between all the charged particles of the universe and all the ZPE fluctuations of the ZPF.

Vehicle Impulsion By Spacetime Warping

Although no means have yet been found for perturbing gravity or the spacetime metric associated with it for propulsive use, Alcubierre (8) has shown a solution to the equations of General Relativity that allows development of enormous flight speed if an accelerating vehicle can "warp" spacetime metric as shown in Figure 2. Expansion of the spacetime metric behind the ship "pushes" it in the desired direction, while contraction of spacetime metric ahead of it adds to impulsion by "pulling" the ship where it wants to go. Spacetime metric throughout the ship is warped into a "flat" configuration in this idealized example. Gravitational influences and inertia are, thus, zero within the ship and it undergoes unlabored motion while rapidly accelerating to (and decelerating from) speeds that can be enormous with respect to earth.

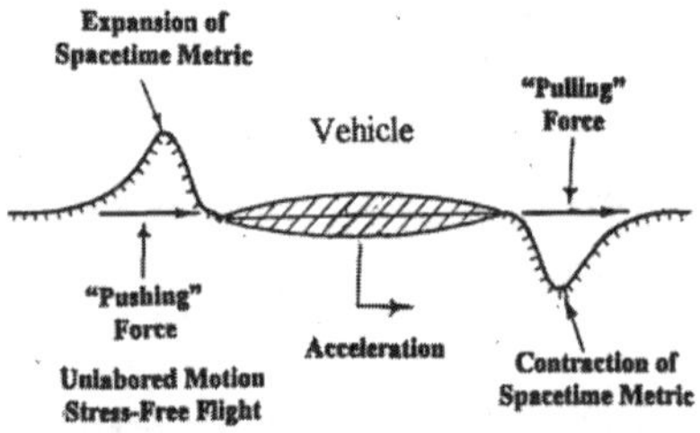

Figure 2 - Warping Spacetime Metric

Alcubierre points out that the favorable spacetime warping shown in Figure 2 would require negative energy densities in the vicinity of the ship, and Froning and Roach (9) have obtained results consistent with the findings of Alcubierre by employing a fluid dynamic analogy and computational fluid dynamic (CFD) techniques. The analogy assumed: a connection between warping of spacetime metric and distortion of the ZPF, and some similarity in the radiation pressures associated with ZPE fluctuations and the atmospheric pressures associated with air particle collisions and recoils.

Perturbation of the ZPF ahead of and behind an accelerating ship was modeled by perturbing of the electrical permittivity and magnetic permeability of the vacuum in the disturbed regions -- as shown in Figure 3. Perturbation intensity was highest at leading and tailing edges, diminishing in a gaussian fashion in the length and breadth dimensions until ambient conditions were reached. Perturbations that decreased permittivity and permeability (and that therefore increased the speed-of-light) in the disturbed regions ahead of and behind the ship were found to decrease ZPF resistance to accelerated flight. Negative energy densities would exist within the perturbed regions, presumably caused by em energy emitted from the ship. And the negative energy densities would be associated with reduced ZPE fluctuation density -- such as that existing between two closely spaced Casimir plates.

Figure 4a shows computed zero-point radiation pressures in the vicinity of a ship that is undergoing acceleration at almost light speed and is emitting no em radiation. Higher-than-ambient pressures act over the ship's forward surface, and those acting over its aft surface are less than ambient -- resulting in an unfavorable pressure gradient that causes resistance to ship acceleration. Figure 4b shows that em emissions that create regions of significant negative energy density ahead of and behind the ship, cause radiation pressures that are much higher than ambient ahead of and near the center of the ship. But the pressures acting upon the front of the ship are less than ambient, while those acting upon the aft are greater -- thereby resulting in a favorable zero-point radiation pressure gradient that causes thrusting (rather than resisting) force to act upon the ship.

Gauge Field Symmetry and Form

Millis (8) notes that electromagnetism is a logical choice for creating an acceleration-inducing force, in that it is related in some degree to spacetime and gravity and is a technology in which we are fairly proficient. And Holt (11) has proposed the possibility of generating coherent patterns of em energy to accomplish field interactions that reduce or amplify gravitation in the vicinity of a hovering or moving ship. Unfortunately, patterns of ordinary em energy do not appear to interact intimately with gravitation. For although the spacetime metric associated with gravitation can be distorted somewhat by em influences (such as magnetic fields of stupendous strength) no intimate interaction or coupling occurs because the fields that underlie gravitation and ordinary electromagnetism are of different essence and form.

In principle, behavior of all matter and radiation can be described in terms of "gauge" fields, with the sources of the fields being conserved quantities. If the essence of the generated gauge field is different from the essence of its source, the field is "abelian". If the essence of the generated field and its source are identical, the field is "nonabelian". Nonabelian fields are, therefore, sometimes viewed as being generated by themselves. In this respect, Gauge fields that describe ordinary em radiation and electrical attraction and repulsion of electrons and protons within atoms and molecules are abelian; while nonabelian fields are associated with processes such as weak and strong interactions within atomic nuclei. Finally, the more intricate configurations of nonabelian fields result in higher internal symmetries. For, the abelian fields associated with ordinary em radiation are of relatively modest U(1) symmetry, while the nonabelian fields associated with weak and strong interactions are of SU(2) and SU(3) symmetry respectively.

Field theorists such as Yang (12) hold that gravitation can be described in terms of a nonabelian field. If so, significant interaction between the nonabelian fields associated with gravitation and the abelian fields associated with ordinary em radiation is not likely. But one of us (Barrett) has shown the possibility of transforming ordinary em fields into specially conditioned em fields of nonabelian form (13). And 13 postulates the possibility of such fields coupling globally with the nonabelian gauge fields that may be associated with spacetime/gravitation through a quantity that may be common to each -- the "A vector potential."

Is the ZPF a Nonabelian Field?

Quantum mechanics and stochastic electrodynamics views the ZPF as a random electromagnetic field that is not intrinsically different from other electromagnetic fields -- especially when in its isotopic, Lorentz invariant configuration in un-accelerated frames. The ZPF can therefore be considered an abelian field of U(1) symmetry. And inertia -- as proposed by Haisch, Rueda and Puthoff (HRP) -- can be viewed as the result of an interaction between two abelian gauge fields (the ZPF and the electric field of the charged substructure of an accelerating mass).

But if a nonabelian field underlies gravitation (as proposed by Yang) and gravitation is a consequence of the ZPF (as proposed by Puthoff) the ZPF, itself, could conceivably contain a nonabelian component of higher symmetry than U(1). And perhaps such a nonabelian field component would be manifested in accelerated frames -- where spectral distortion of the ZPF occurs. If so, there is the possibility of the inertia of accelerated matter being a consequence of an interaction between the abelian and nonabelian gauge fields associated with: (a) the accelerated matter, and (b) the spectrally distorted ZPF in the matter's accelerated frame.

It, of course, is not yet known if nonabelian fields with higher than U(1) symmetry underlie spacetime metric/gravitation or the ZPF. But this is critical to the propulsive efficacy of em radiation whose fields are specially conditioned to achieve higher symmetry and nonabelian form.

Ordinary and Specially Conditioned EM Radiation

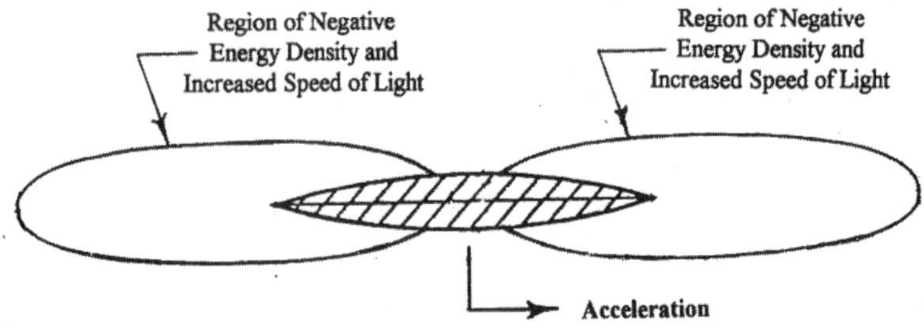

Figure 3 - Regions of Negative Energy Density

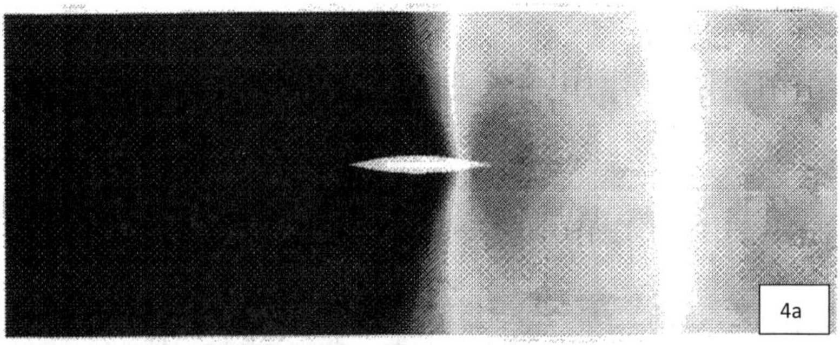

Accelerated Flight at 0.99c No ZPF Perturbation

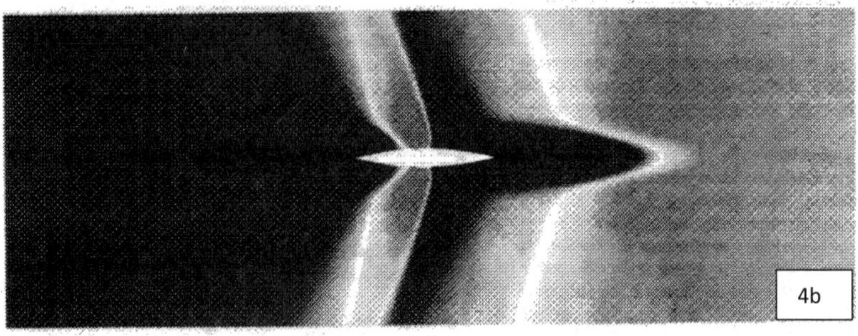

Accelerated Flight at 0.99c ZPF Perturbed

Figure 4 - Zero Point Pressure Distribution

Ed. Note: It helps to read Figure 3 in view of the text where $c^2 = 1/\mu_o\varepsilon_o$ so that an "increased speed of light" in these regions really implies the decrease of the permeability and permittivity of space.

Although even the most complex combinations of frequency and amplitude modulation do not transform ordinary em fields into nonabelian fields of higher symmetry, 13 shows that such a transformation can be accomplished by modulating the polarization of em wave energy radiated from antennas or apertures of RF or laser transmitters. Here, instead of maintaining a fixed linear or circular polarization, the polarization of the emitted waveform is continually rotated through all possible orientations within time intervals that are extremely short. Figure 5 shows that such polarization modulation increases em field symmetry from U(1) to SU(2) and results in a nonabelian gauge field with ability to couple globally with fields of similar form through the action of its vector potential (A). Thus, if nonabelian gauge field configurations with SU(2) components are associated with gravitation, there should be a possibility for modifying gravitational influences within a beam of polarization modulated radiation.

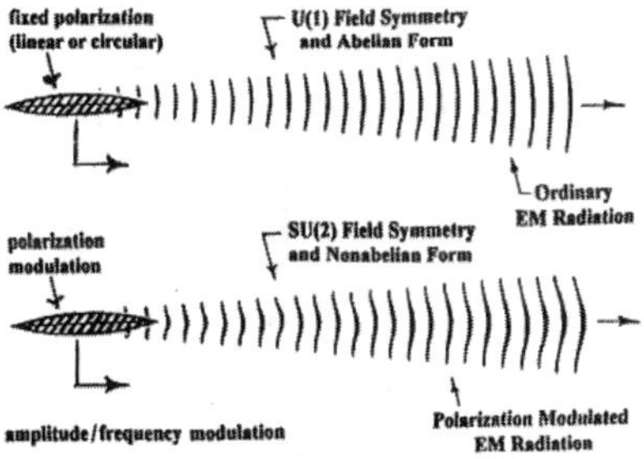

Figure 5 - Electromagnetic Radiation Comparison

The propagating speed (c) of electromagnetic disturbances within a given region of empty space is determined by the electrical permittivity and magnetic permeability of the vacuum through which they move. And such permittivity and permeability is associated with the configuration of spacetime metric throughout that region. Since ordinary em fields do not intimately interact with spacetime metric, electrical permittivity, magnetic permeability, and the speed of light in vacuo remain unchanged and wave fronts within ordinary em beams propagate as planar or spherical disturbances at constant c, as shown in Figure 5. However, if polarization modulated em beams interact with spacetime metric, permittivity, permeability and wave front speed will be changed within the beams -- and the em wave fronts will propagate as non-planar or non-spherical disturbances as shown in Figure 5.

Creation of Polarization Modulated EM Radiation

A means of creating polarization modulated em radiation of SU(2) symmetry and nonabelian form is described by Barrett in (13) and shown in Figure 6. Shown, is: ordinary input em wave energy divided into three fractions. One fraction of the input wave energy is orthogonally polarized (has its polarity rotated 90 degrees) and phase modulated. Another fraction of the input wave energy is expended in accomplishing the phase modulation of the polarization rotated wave energy. The remaining fraction of the input wave energy is then combined with that of the polarized and phase modulated wave energy at a "mixer", and emission of specially conditioned em radiation with continually varying polarization with respect to time results.

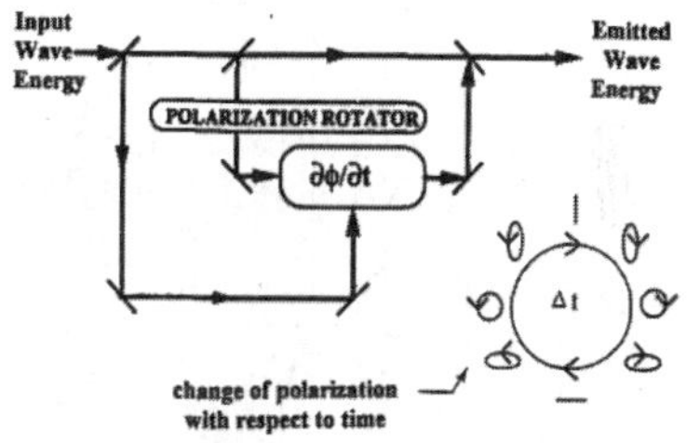

Figure 6 - Polarization Modulation

Since a fraction of input wave energy (beyond that dissipated by circuit resistance and reactance) is expended in accomplishing phase modulation, the electric and magnetic field energy associated with polarization modulated radiation can be less than for ordinary em radiation for a given input power. But energy conservation requires that expended energy not truly vanish, but be transformed into another form. And one possibility would be increased vector potential field intensity within the emitted radiation. If so, significant polarization modulation could significantly diminish electric and magnetic field intensity while increasing A vector potential intensity for coupling with the A vector fields that may be associated with spacetime metric or gravitation.

Figure 7 shows one of the several types of antenna systems that can be configured for emission of polarization modulated RF radiation, together with snapshots of the evolved electric field history after the emitted radiation has traveled 50 and 500 wave cycles from the antenna. The dual-feed crossed antenna system shown in Figure 7 is taken from the polarization modulation patent awarded to Barrett by the U.S. Patent Office, and the particular electric field history shown for this antenna is for a phase modulation frequency that is 0.1 the nominal frequency of the vertical and horizontally polarized waveforms. And it might be noted that the evolved electric field history at 50 and 500 wave cycles would be only a single straight line or a single circle for linear or circular polarization.

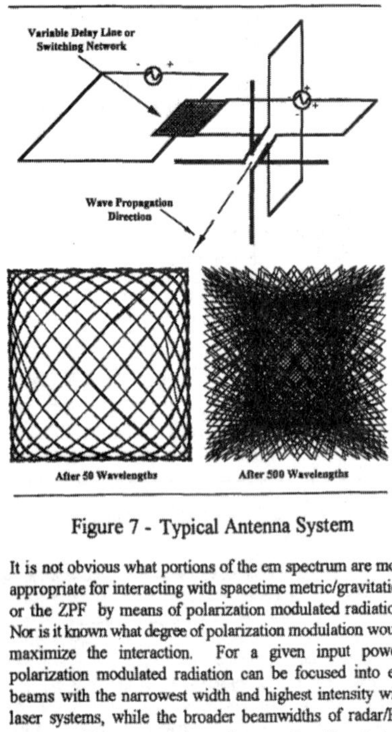

Figure 7 - Typical Antenna System

It is not obvious what portions of the em spectrum are most appropriate for interacting with spacetime metric/gravitation or the ZPF by means of polarization modulated radiation. Nor is it known what degree of polarization modulation would maximize the interaction. For a given input power, polarization modulated radiation can be focused into em beams with the narrowest width and highest intensity with laser systems, while the broader beamwidths of radar/RF systems enable a greater volume of space to be affected. And radars that are masers (that accomplish microwave amplification by stimulated emission of radiation) have waveforms that have the same coherency and monochromatic benefits as lasers.

Polarization modulation of both laser and radar/maser systems should therefore be explored, and increasing amounts of polarization modulation -- up to the highest achievable with modified hardware -- investigated for each. Desired radar and laser system attributes are broad frequency/wavelength bands and provisions for orthogonally polarized waveforms and phase modulation. And radars and lasers that could be most easily modified for polarization modulation would embody multi-frequency bands or wavelengths that can be varied over a very broad range.

Vector Potential Wave Patterns

Specially conditioned em fields which consist almost entirely of a vector potential wave patterns can also be created, as described in (14), by flow of alternating current through a toroid with single windings -- as shown in Figure 8.

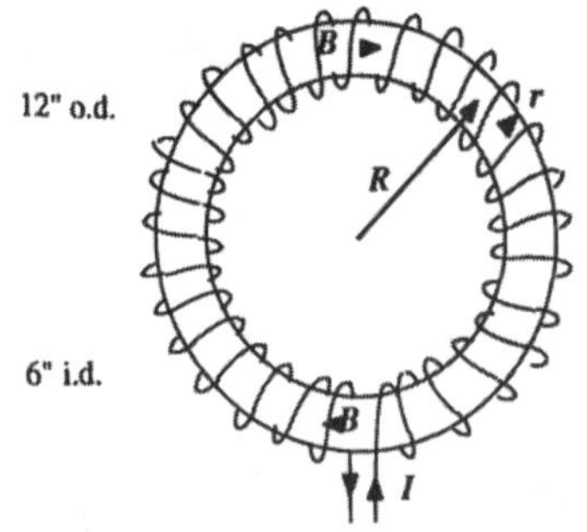

Figure 8 - Toroid Antenna

The resulting magnetic and electric fields to not extend significantly outside the toroid, but its geometry and the alternating current flow produces overlapping A vector potential patterns that combine into "phase factor" waves which extend outward from the toroid over significant distances and represent disturbances in A vector potential. This is shown in Figure 9.

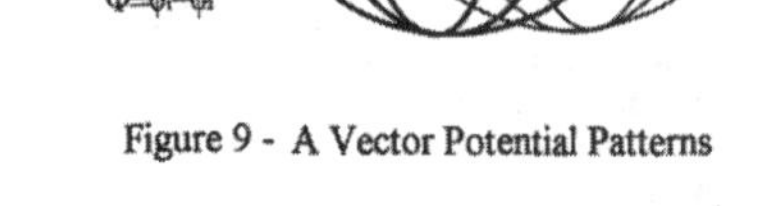

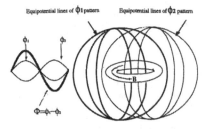

Figure 9 - A Vector Potential Patterns

Phase factor wave patterns -- as shown in Figure 10 -- become almost spherical in shape at distances from the toroid of the order of its diameter or greater. At the present time, it is not known whether the phase factor waves are standing waves or waves that propagate at non-zero speed.

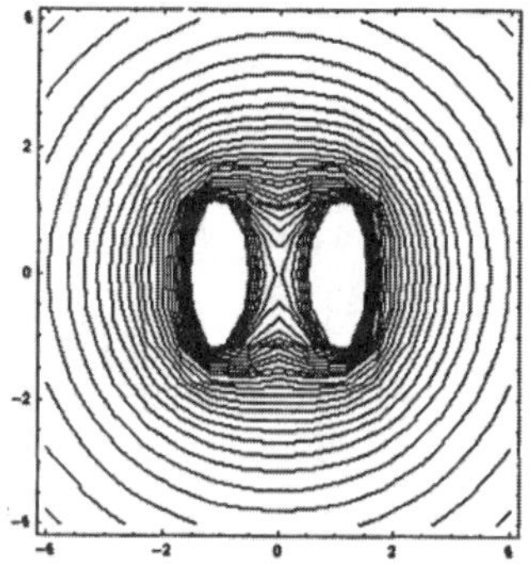

Figure 10 - "Phase Factor" Wave Patterns

The maximum disturbance in A vector potential occurs as phase factor wave intensity peaks at various resonant frequencies. Resonant frequencies for toroids with single windings occur in the RF/Microwave range, being determined by toroid geometry and propagating speed of the alternating current. Figure 11 shows that such resonances occur as a function of frequency for a given toroid geometry and current speed. And an interaction with spacetime metric/gravitation or the ZPF may occur at one or more of the resonant frequencies if the essence of gravitation or the ZPF includes an A vector potential field.

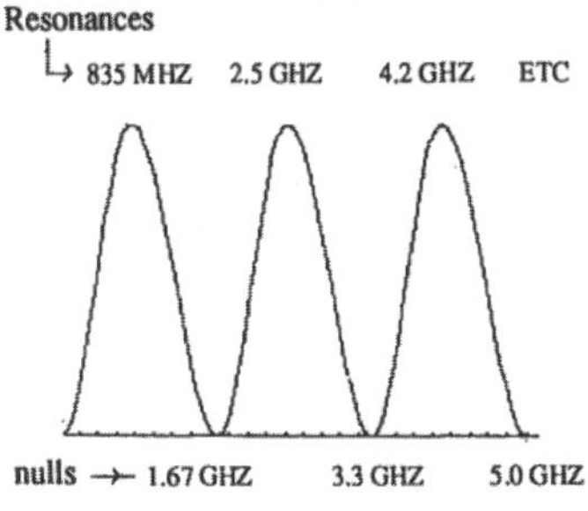

Figure 11 - Resonant Frequencies

Vector potential disturbances of significant intensity may also be achievable at resonant frequencies in the Infra-Red/Visible/Ultra Violet range, using rapid switching by solid-state devices rather than alternating current propagating through wires. But, of course, such devices might not possess the innate simplicity of toroidal coils.

Electrical Power Needs

It must be expected that specially conditioned em radiation of sufficient intensity to significantly influence spacetime metric/gravitation or the ZPF will require significant amounts of electrical power. In this respect, it would be desirable that specially conditioned em beams used in operational field propulsion systems not require significantly greater electrical power than that required by powerful radar and laser systems in use or under development for military aircraft today -- probably somewhere in the 100 to 1000 kilowatt range. But although significant electrical power is likely to be required for generation of specially conditioned em radiation, such power would be invested primarily in creation of intense A vector potential disturbances -- not in generation of intense electric and magnetic fields.

More advanced field propulsion systems that could interact more strongly with gravity or the vacuum's zero-point state could require greater power than is achievable for air flight or spaceflight today. Such power levels, which could be in the 1.0 to 10 gigawatt range, would have to be supplied by systems of extremely high energy density. One example of this type of system might be the aneutronic fusion electric power system proposed by Bussard (15) which emitts no neutrons and causes no radioactivity.

Methods for Detecting Gravitational Changes within Specially Conditioned EM Beams

As mentioned previously, specially conditioned em beams would be very attractive if they would not require significantly more electrical power than that required by high power airborne radar and laser systems in use or in development today. But such power levels are not expected to be available for initial laboratory proof-of-principle tests of modest cost. Thus, one is faced with the challenge of detecting relatively small gravitational or ZPF changes within specially conditioned em beams that are generated by only modest amounts of electrical power.

One means of detecting small changes would be measuring slight changes in weights of objects bathed by specially conditioned em radiation, or changes in forces acting upon them. Another would be use of pendulums or clocks to detect temporal changes due to perturbation of spacetime metric or the ZPF. Another method, shown in Figure 12, would be use of a sensitive ring laser gyro (RLG) -- which is ordinarily used to detect inertial changes due to acceleration. For Barrett (14) has shown that inertial acceleration, as measured by a RLG, can be viewed as a change in the A vector potential field associated with gravitation. Thus any change in A vector potential and gravitation caused by a coupling with specially conditioned em radiation should be detectable by an RLG of sufficient sensitivity.

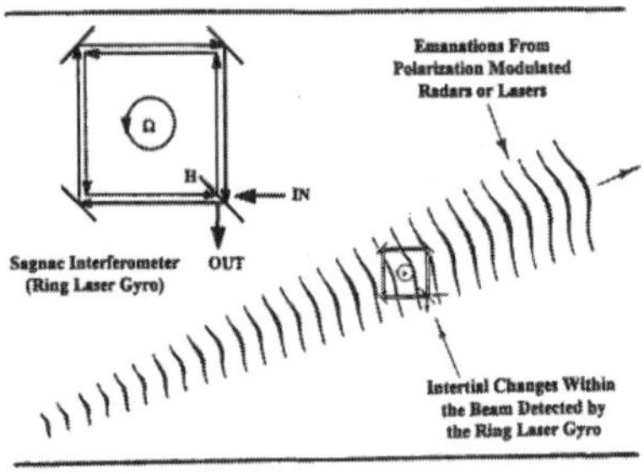

Figure 12 Probing Specially Conditioned Radiation

Still another method would be to detect the influence of specially conditioned em radiation on the measured Casimir force between two closely spaced plates. For any influence that would change ZPE distributions or patterns in the vicinity of the plates would cause a change in measured Casimir force. An experiment which would attempt to do this by bathing the vacuum region surrounding Casimir plates by various types of radiation is shown schematically in Figure 13. More work is needed to determine if such a Casimir force experiment is, indeed, feasible. For incident em radiation that could induce currents and fields within conducting Casimir plates could influence measurements of Casimir force. In this respect, reference measurements, using ordinarily polarized em beams should reveal the magnitude of such influences.

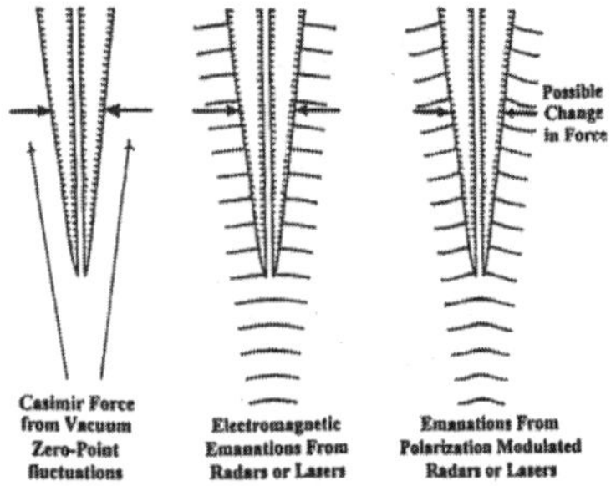

Figure 13 - Casimir Force Experiment

<u>Expanded Maxwell Equations for Specially Conditioned Em Radiation</u>

Figure 14 - Maxwell Equations

	U(1)	SU(2)
Coulomb's Law	$\nabla \cdot E = J_0$	$\nabla \cdot E = J_0 - iq(A \cdot E - E \cdot A)$
Ampère's Law	$\frac{\partial E}{\partial t} - \nabla \times B + J = 0$	$\frac{\partial E}{\partial t} - \nabla \times B + J + iq(A_0 E - E A_0) - iq(A \times B - B \times A) = 0$
Gauss's Law	$\nabla \cdot B = 0$	$\nabla \cdot B + iq(A \cdot B - B \cdot A) = 0$
Faraday's Law	$\nabla \times E + \frac{\partial B}{\partial t} = 0$	$\nabla \times E + \frac{\partial B}{\partial t} + iq(A_0 B - B A_0) + iq(A \times E - E \times A) = 0$

A = Vector Potential **A_0 = Magnitude of Vector Potential**

Ordinary em emanations are solutions to Maxwell's equations. But the U(1) field symmetry associated with them is lower than the SU(2) symmetry of specially conditioned em emanations. Electromagnetic emanations of SU(2) field symmetry must, therefore, be solutions to Maxwell equations of more expanded and symmetrical form. Barrett (13) has derived such expanded Maxwell equations, which are shown in Figure 14.

Maxwell equations of more expanded and more symmetrical form requires additional terms that make the equations more symmetrical with respect to electric and magnetic phenomenon. Expanded Maxwell equations are therefore seen to contain all the terms of Maxwell's equations together with additional terms that involve the coupling of electric and magnetic fields through the action of the A vector potential. And because the dot and cross products within the terms that include the A vector potential obey the commutation relations of nonabelian algebra and quantum mechanics, they are never equal to zero if SU(2) phenomena are present.

The additional terms within the expanded Maxwell equations imply the possibility of em phenomenon that behave as if they possessed a magnetic charge and magnetic current. Because such magnetic phenomenon are not observed in ordinary electromagnetics, their existence is somewhat controversial. But Harmuth (17) shows that magnetic current has a physical basis and is necessary to account for many em transients. And, as will be mentioned next section, there may be some evidence of specially conditioned em phenomenon that can be conditioned to possess SU(2) field symmetry and to behave as if they possessed magnetic charge.

Forces associated with ordinary electromagnetic field interactions are calculated by inputting electric and magnetic field values (E and B) determined by the Maxwell equations into Lorentz force equations that include terms involving cross products of E and B. Lorentz force equations for specially conditioned em fields are more complex because the E and B values for specially conditioned em fields involve additional terms that include cross and dot products of E, B and the A vector potential.

Interactions Involving Magnetic Fields and Ferromagnetic Mass

Experimental work by Mikhailov has provided a degree of confirmation of Barrett's ideas as to the possibility of: (a) em phenomenon with higher field symmetry than the U(1) field symmetry of ordinary em phenomenon, and (b) coupling with such phenomenon with em radiation of comparable field symmetry. Here, experiments by Mikhailov, which used apparatus similar to that shown in Figure 15, has confirmed earlier experimental work by Ehrenhoft, showing that ferromagnetic aerosols in solution behave as if they possess a magnetic charge, reversing their motion with reversal in magnetic field direction -- as shown in Figure 15. Barrett (18) noted that such behavior could be due to spherical boundary (cavity) conditions and global ordering of electron spins that caused the ensemble to react to magnetic influences just as many isolated monopoles would. He also determined that a gauge field of SU(2) symmetry would be associated with magnetically confined ferromagnetic particles in aerosol solution and that such a field should interact with the field associated with polarization modulated em radiation of similar SU(2) form.

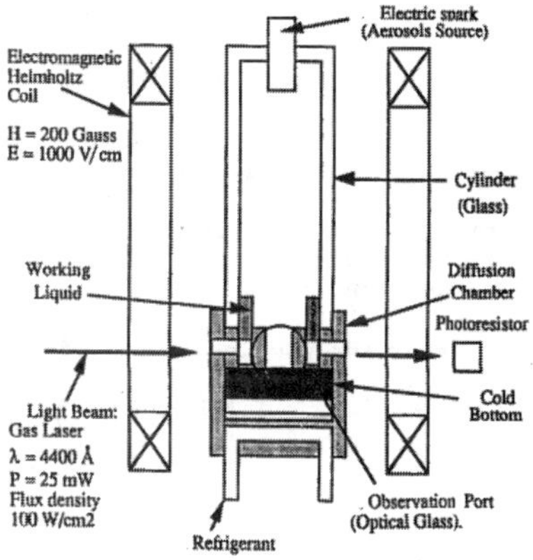

Figure 15 - Ferromagnetic Aerosol Experiment

Taking this suggestion, Mikhailov (19) varied the polarization of the laser beam used to track particle trajectories. He found that fixed polarization of the laser beam caused negligible perturbation of the straight line motion of the particles as they responded to the magnetic field. But bathing the particles with polarization modulated light caused them to oscillate with respect to their motion in the magnetic field direction -- as shown in Figure 16. Unfortunately (19) contains insufficient information for ascertaining what degree of polarization modulation was actually achieved. But Mikahailov reported that oscillations maximized when a given polarization frequency was reached, and oscillations appeared to be in synchrony with the polarization modulation frequency. Mikhailov's experiments therefore provide some substantiation that em phenomenon with higher field symmetry than ordinary em fields exist and can be interacted with by fields of similar symmetry.

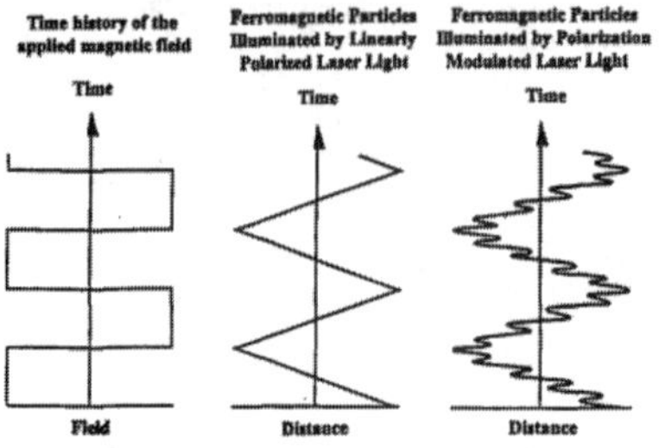

Figure 16 - Ferromagnetic Aerosol Experiment

Summary and Conclusions

- Ordinary em radiation does not interact intimately with gravity/spacetime metric or the ZPF because the fields that underlie them are of different essence and form. Thus, exorbitant expenditure of em field energy appears to be required for an interaction between these fields that causes a significant propulsive effect.

- Ordinary em radiation can be conditioned with the same field essence and form as that which may underlie gravity/spacetime metric or the ZPF. And such specially conditioned radiation may be able to couple propulsively with gravity/spacetime metric or the ZPF through something that may be common to each -- the A vector potential.

- Specially conditioned em radiation can be in the form of elongated beams of significant length, as shown below, or it can be in the form of spherical patterns that emanate in all directions over shorter distances from hovering or maneuvering craft.

- Significant levels of electrical power will probably be required to create specially conditioned em radiation of sufficient intensity for significant influencing of spacetime metric/gravitation or the zero-point state of space.

- But such significant power would be invested, primarily, in the creation of intense A vector potential disturbances -- not in generation of intense electric and magnetic fields.

- The efficacy of specially conditioned em radiation in accomplishing field propulsion depends upon the underlying essence and behavior of gravity, spacetime metric and the ZPF -- none of which is yet completely understood.

- But in the meantime, the propulsive potential of such radiation can be inexpensively proved or dis-proved experimentally by using existing radar, laser and electrical systems modified for the generation of specially conditioned em beams.

* * * * * * * *

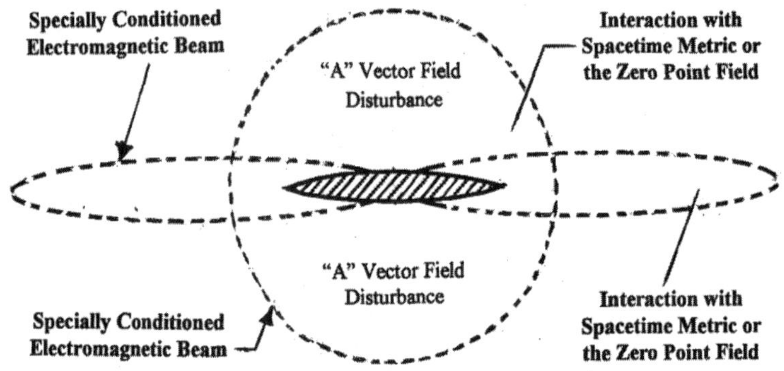

REFERENCES

(1) Cox, J.E., "Electromagnetic Propulsion without Ionization", 16th AIAA Joint Propulsion Conference, AIAA-80-1235, Hartford Conn. (1980)

(2) Engleberger, J., U.S. Patent 3, 504, 868, granted April 7 (1970)

(3) Zubrin, R., "The Use of Magnetic Sails to Escape from Low Earth Orbit", AIAA 27th Joint Propulsion Conference, AIAA-91-2533, Sacramento, CA (1991)

(4) Haisch, B., et al., "Inertia as a Zero-Point Field Lorentz Force", Phys Rev A, 1 Feb (1994)

(5) Milonni, P.W., et al., Radiation Pressure from the Vacuum: Physical Interpretation of the Casimir Force, Phys Rev A, 38, p. 1621 (1988)

(6) Scharnhorst, K., "On Propagation of Light in the Vacuum between Plates", Phys Let B 236, p 354 (1990)

(7) Puthoff, H.E., "Gravity as a Zero Point Fluctuation Force, Phys Rev. A, 39, p 2333; (1988)

(8) Alcubierre, M., "The Warp Drive: hyper-fast travel within general relativity", Classl Quant Grav, Vol 11, IOP Publishing Ltd. (1994)

(9) Froning, H.D., et al., "Drag Reduction and Possibly Impulsion by Perturbing Fluid and Vacuum Fields", AIAA 31st Joint Propulsion Conference, AIAA 95-2368, San Diego, CA (1995)

(10) Millis, M.A., "The Challenge to Create the Space Drive", National Space Society New Roads to the Stars Interstellar Flight Symposium, New York City, NY (1996)

(11) Holt, A.C., "Prospects for a Breakthrough in Field Dependent Propulsion", 16th AIAA Joint Propulsion Conference, AIAA 80-1233, Hartford, Conn. (1980)

(12) Yang, C.N., "Gauge Theory", McGraw-Hill Encyclopedia of Physics, 2nd Edition, p 483, (1993)

(13) Barrett, T.W., "Electromagnetic Phenomenon not Explained by Maxwell's Equations:", Essays on Formal Aspects of Electromagnetic Theory, p 6 World Scientific Publ Co., (1993)

(14) Barrett, T.W., "The Toroid-Solenoid as a Conditioner of Electromagnetic Fields into Gauge Fields", BSEI Report 1-97, obtainable on e-mail from: barrett506@aol.com

(15) Bussard, R.W. "The QED Engine System: Direct-Electric Fusion-Powered Rocket Propulsion Systems," 10th Symposium on Space Nuclear Power and Propulsion, Albuquerque, New Mexico, January, 1993.

(16) Barrett, T.W. & Grimes, D.M., "Advanced Electromagnetism: Foundations, Theory, Applications", p. 278-313, World Scientific Publ Co., (1993)

(17) Harmuth, H.A., "Electromagnetic Transients not Explained by Maxwell's Equations", Essays on the Formal Aspects of Electromagnetic Theory, 87 World Scientific Publ Co., (1993)

(18) Barrett, T.W., "The Ehrenhoft-Mikhailov Effect Described as the Behavior of a Low Energy Density Magnetic Monopole-Instanton", Annales de la Foundation Louis de Broglie, 19, p 291 (1994)

(19) Mikhailov, V.F., "Experimental Detection of Discriminating Magnetic Charge Response to Light of Various Polarization Modulations", Annales de la Foundation Louise de Broglie, 19, p 303 (1994)

AIAA-95-6099
Use of an Almost-Developed Propellant Tank for Flight Demonstration of Spaceplane Art

H. Froning, Jr.
Flight Unlimited
Flagstaff, AZ
M. Little
Rockwell International
Cedar Rapids, IA
P. Czysz
St. Louis University
Cahokia, IL

AIAA SIXTH INTERNATIONAL AEROSPACE PLANES AND HYPERSONICS TECHNOLOGIES CONFERENCE
3-7 APRIL 1995 / CHATTANOOGA, TN

AIAA-95-6099

USE OF AN ALMOST-DEVELOPED PROPELLANT TANK FOR FLIGHT DEMONSTRATION OF SPACEPLANE ART

H. D. Froning, Jr. : Flight Unlimited, Flagstaff, Arizona, USA
M. J. Little: Rockwell International
Cedar Rapids, Iowa
P. A. Czysz: St. Louis University, Parks College
Cahokia, Illinois, USA

Abstract

An advanced aluminum-lithium propellant tank, with provisions for storage and transfer of liquid oxygen and liquid hydrogen and kerosene propellants into rocket engines, is in an advanced state of development in Russia today. This tank, which is being developed for the launcher system "MAKS", is expendible and its non-cylindrical shape is a consequence of the aerodynamic considerations for air launch from an AN225. Nevertheless, it appears capable of being adapted for reuse and its shape packages reasonably well within Reusable Launch Vehicle (RLV) designs. With already developed rocket engines, this tank provides adequate sub-orbital flight performance for an RLV demonstrator that is air-launched. However, adequate performance for ground launch requires, either a low-drag airframe, or use of rocket engines that are under development today.

Preface

Although some argue that sub-orbital flight demonstrators are not a required prelude to development of Reusable Launch Vehicles (RLVs), sub-orbital flights by a small RLV demonstrator (such as the DC-X) address many RLV issues for a fairly modest cost. And some have proposed the next step after this to be sub-orbital flights by an RLV demonstrator that could be as large as two-thirds the size of operational RLVs.

A large RLV demonstrator with only sub-orbital performance would be attractive, only, if its cost and risk could be minimized by having development of all its major components (airframe and TPS materials, rocket engines, propellant tanks) be already completed or almost done. This paper will describe such a large RLV demonstrator – which would embody already developed: airframe and thermal protection materials (TPS) and rocket engines, together with an almost-developed propellant tank.

Introduction

An advanced aluminum-lithium propellant tank with provisions for storage and transfer of liquid oxygen (LO2), liquid hydrogen (LH2), and kerosene into two tri-propellant rocket engines is in an advanced state of development in Russia today. This tank is being developed, for the launcher system "MAKS" by the Molnija Design Bureau. And combustion of its propellants within the rocket engines accelerates a small shuttle-like vehicle to orbital speed.

Figure 1 shows the various elements of the MAKS launch vehicle concept installed upon an AN-225 aircraft, and Figure 2 shows the general configuration and assumed characteristics of the propellant tank element of MAKS. And because the internal configuration of this tank is not yet known, it is shown schematically as being partitioned into 3 different sections for its 3 different propellant types.

NASA studies indicate that tripropellants enable significant reduction in vehicle volume and dry weight and that this may be worth the additional complexity of the additional burning of kerosene. Thus, the tripropellant capabilities of the MAKS tank would be particularly attractive if tripropellants are selected for an RLV demonstrator or an operational RLV.

On the other hand, if non-kerosene burning engines are selected for an RLV demonstrator, the kerosene volume of the MAKS tank would go unused. But kerosene volume is a small fraction of the MAKS tank volume, and it will be shown that with this loss, only a small reduction in demonstrator performance would ensue.

The MAKS propellant tank is not intended for reusable use, and its geometry is a consequence of aerodynamic considerations and air-launch from an AN-225 transport aircraft at subsonic speed. Nevertheless, with minor modifications, it could conceivably, be adapted for reusable use. And its shape packages reasonably well into reusable rocket vehicle designs. Therefore, this paper will examine MAKS propellant tank installations within sub-orbital RLV demonstrators. The paper will also examine the possibility of such demonstrators performing useful work by carrying small upper stages that could place small payloads in orbit about earth.

MAKS Propellant Tank Adaptations for Reusable Vehicle Use

The MAKS propellant tank is in an advanced state of development, with its detailed design completed and full-scale engineering models already undergoing extensive ground test. Thus, the cost to complete its development by the Molnija Design Bureau should be much less than the cost to develop all-new LOX, LH2 and kerosene tanks. Furthermore, since this unique system houses 3 different propellants within a single pressure vessel, it could conceivably be purchased from appropriate organizations in Russia and the Ukraine and be integrated within a reusable airframe for less cost than 3 separate (LOX, LH2, and kerosene) tanks.

However, the following questions are some that would be associated with consideration of the MAKS tank for reusable rocket ship use.

(1) Can the MAKS tank be effectively embodied within reusable rocket vehicles without excessive weight penalty or wasted space?

(2) Can the MAKS tank, when enclosed within an airframe, be properly accessed for repeated: fueling, purging, venting, inspection, and routine maintenance?

(3) Can the insulation of the MAKS tank when enclosed within an airframe, protect against the repeated temperature cycling of many trips to space?

(4) Can the propellant mix in the MAKS tank provide adequate flight performance when used with engines other than the RD-701 MAKS engines and for ascent trajectories that involve launch from either the air or ground.

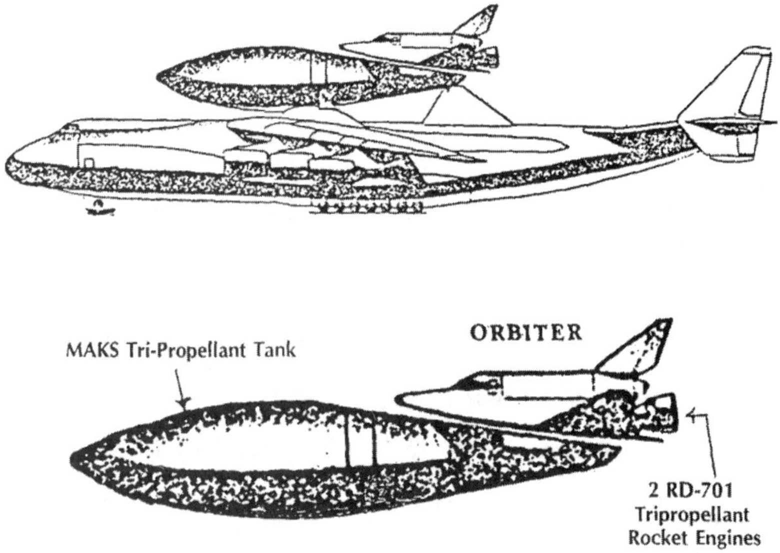

Fig 1 Components of the MAKS System

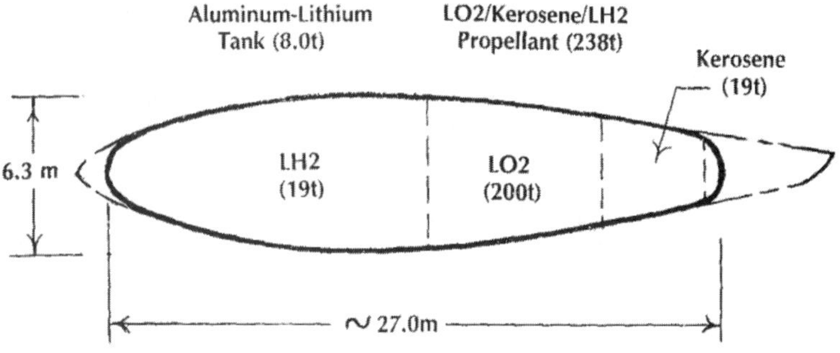

Fig 2 MAKS Tri-Propellant Tank

(5) What would be the cost to complete development of the MAKS tank, including any adaptions needed for reusable vehicle use.

Some of these questions will be addressed in this paper, while others have not yet been addressed at all. Thus, the following sections will discuss each of these questions to some degree.

MAKS Propellant Tank Installations Within Reusable Rocket Ships

Withstanding airloads while housing 3 different propellants within a single and unusual pressure vessel shape probably resulted in the MAKS tank having more structural mass than if it had been installed within the airframe of an RLV. Nevertheless, the MAKS tank is relatively light. And it will be shown that it will be it packages reasonably well within a variety of RLV designs.

When used with the MAKS system, the MAKS tank is oriented such that its least dense propellant (LH2) is the most forward and its most dense propellant (kerosene) is the most aft. However, with some modification, it could conceivably be oriented in the opposite direction if this would be beneficial for reusable vehicle use. And this, of course, would result in a more forward vehicle center of gravity when the tank is full.

Figure 3 shows embodiment of the MAKS tank within three of the different RLV types currently being considered by different contractors for the NASA X-33. One is a wingless vehicle similar in configuration to the Delta Clipper or BETA vehicles (References 3 and 4) which take off and land vertically on the ground. Another is a winged vehicle that could take off vertically from the ground (NASA Reference 5), or be air-launched like Interim HOTOL (Reference 6) and land horizontally on runways. Still another is a vertical takeoff and horizontal landing lifting-body configuration with engines such as the Rocketdyne L2200 "Linear Aerospike".

The structural weight associated with embodiment of the MAKS tank within each of the three different RLV types has not yet been estimated. But, it conceivably could be acceptable for all three.

MAKS Tank Accessibility for Repeated Servicing and Routine Maintenance

The existing MAKS tank is not enclosed within another structure and only requires sufficient inspection, maintenance, and fueling for a single trip to space. It is, therefore, easily accessible for the minimal ground operations associated with its use. However, it is not yet known if the MAKS tank could be satisfactorily accessed when enclosed within an airframe and reused for many trips to space, or if such accessibility and reusability is achievable with minimal modification of this tank.

Although the MAKS tank appears attractive for any of the RLV configurations shown in Figure 3, one corresponding to the winged body configuration was arbitrarily selected for visualizing the accessability and reusability issues associated with use of the MAKS tank. Figure 4 describes this vehicle, which is closely comparable in size and mass to the Interim Hotol vehicle – and which, therefore, could also be air-launched from an AN-225.

As many as 3 oxygen and kerosene burning RS-27A engines plus 4 oxygen and hydrogen burning D-57A engines were considered first for the RLV demonstrator because they are already developed and could make use of all the different

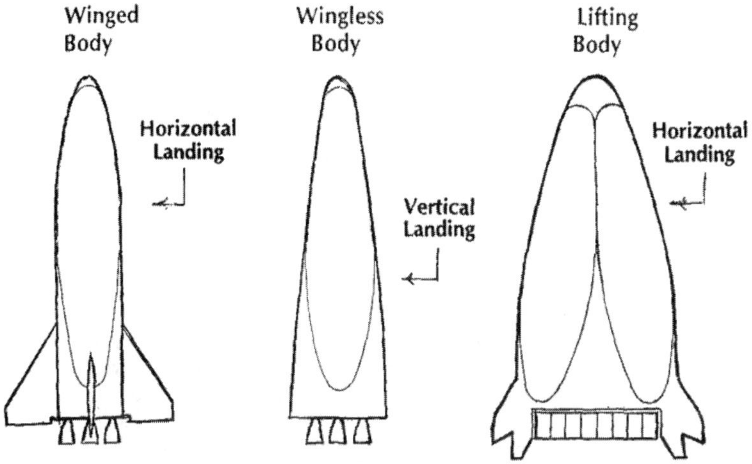

Fig 3 MAKS Propellant Tank within Reusable Flight Demonstrators

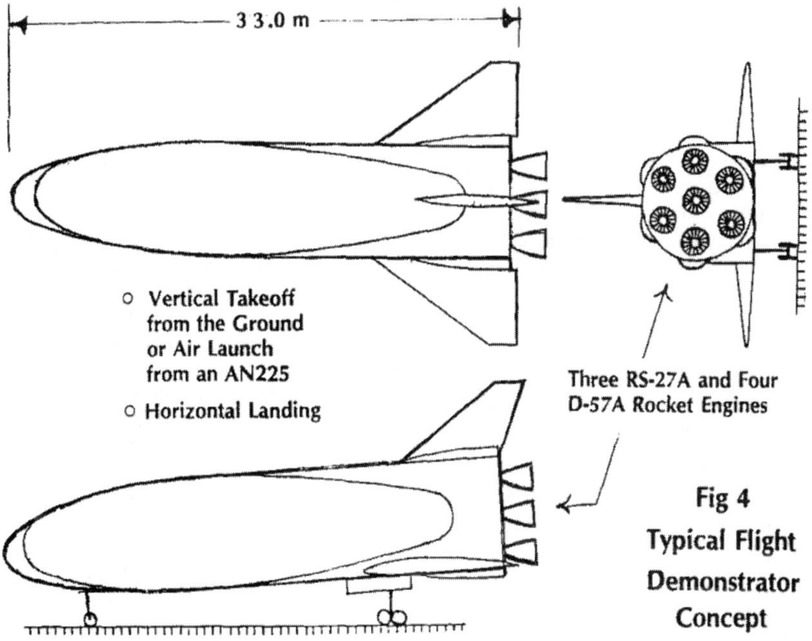

Fig 4
Typical Flight
Demonstrator
Concept

propellants carried within the MAKS tank. These engines utilized all available kerosene and LH2 that can be carried within the MAKS tank, but all available L02 was not used because of the relatively low oxidizer/fuel (O/F) ratios of the RS-27A's.

The higher O/F ratios and effective specific impulse of almost developed rocket engines such as the RD-704 and the RS 2100 (Reference 7) would enable greater RLV demonstrator performance by consuming (at a slower rate) more LO2 propellant from the MAKS tank. And if development of such engines is completed soon, they could be available for demonstrator use.

Thermo structural materials which are considered as "near-term" by NASA in Reference 5, were selected for this RLV. These included: graphite epoxy interstage structure, and the various insulation types shown in Figure 5.

MAKS Propellant Tank Insulation for Many Trips to Space

Insulation on the existing MAKS propellant tank must only protect against the relatively mild heating associated with the ascent phase of flight, and need only survive a single trip to space. By contrast, very high surface temperatures are experienced during the RLV re-entry, and it is not yet known if the MAKS tank insulation can, with minor adaptions, survive the repeated temperature cycling of many sub-orbital trips.

Adequacy of the Present Mix of Propellants within the MAKS Tank

Vehicle synthesis and subsystem mass estimation was accomplished by means of methods developed by P. A. Czysz (Reference 8) and by use of scaling relationships that relate vehicle characteristics to those of already designed craft and although this preliminary investigation has attempted to define the magnitude of demonstrator performance obtainable with propellants from the MAKS tank, much more comprehensive work is needed for proper estimates of the demonstrator's configuration, performance and mass.

Figure 6 shows a preliminary mass estimate for the selected vehicle subsystems, together with flight performance with 3 already developed RS-27A and D-57A engines, and with 3RD-704 engines which are under development today. And in this respect, (8, 9, 10) provided the information used to estimate engine thrust and specific impulse during the vehicle ascent.

The ratios of LO2, LH2, and kerosene within the MAKS tank maximize performance for air-launch and when RD-701 engines are used. Thus, inferior performance is achieved for ground launch and if lower performance engines are used. Here, LOX consumption by the RS-27A engines limited total propellant consumption to only about 158 tons, and demonstrator takeoff weight to about 200 tons. This enabled a maximum speed of approximately Mach 15 for air-launch from an AN-225, and only about Mach 12 speed for vertical launch from the ground.

The higher specific impulses and O/F ratios of tripropellant rocket engines such as the RD-704, which is under development today, enabled slower consumption of most of the available MAKS propellants. As a consequence, speed of more than Mach 17 was achieved for ground launch, with a 265 ton takeoff weight. Bipropellant rocket engines currently under development, such as the RS2100 and RS2200 could not utilize the kerosene carrying capabilities of the

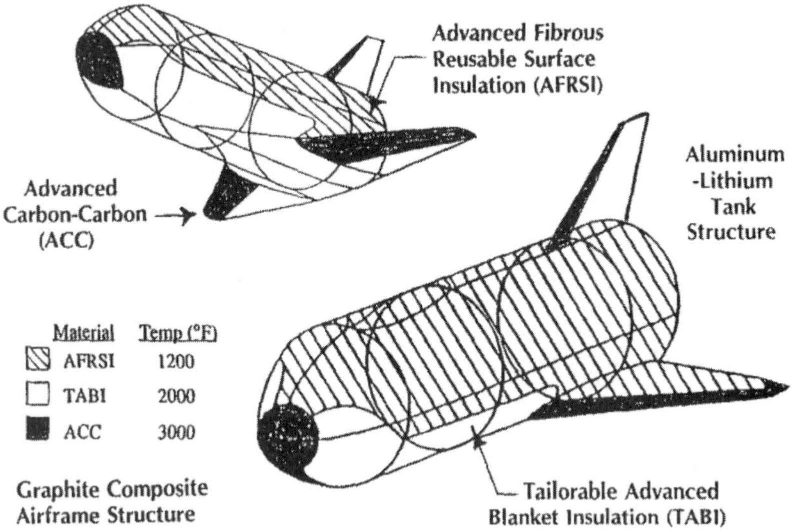

Fig 5 Thermo-Structural Materials for Flight Demonstrator

Demonstrator Component	Three RS-27A and D-57A Rocket Engines		Three RD-704 Rocket Engines	
Propellant Tank	8.0	Maximum Speed for Ground Launch ↓ 3.65 km/sec Mach 12.0	8.0	Maximum Speed for Ground Launch ↓ 5.37 km/sec Mach 17.5
Fuselage	12.7		12.7	
Wings	4.4		4.4	
Thermal Protection	3.3		3.3	
Propulsion	5.3		7.0	
Landing Gear	1.6	Maximum Speed for Air Launch ↓ 4.58 km/sec Mach 15.0	1.6	Maximum Speed for Air Launch ↓ 5.95 km/sec Mach 19.5
Equipment	1.7		1.7	
Margin	5.6		5.7	
Dry Weight	42.6t		44.4t	
Propellant Weight	157.9t		220.7t	
Gross Weight	200.5t		265.1t	

Fig 6 Typical Mass Characteristics of a Flight Demonstrator

MAKS tank. However, they enabled demonstrator speeds that, compared to those of tripropellant engines, were only about 4 percent less.

Cost to Complete Development of the MAKS Propellant Tank

The remaining cost to complete development of the existing MAKS aluminum-lithium propellant tank is not yet known. However, it is surely much less than the cost to develop new aluminum-lithium tanks. The additional cost of modifications of the MAKS tank for reusability is also not yet known. For such cost estimates must await more comprehensive examinations of all the issues that have been described.

Alternate Demonstrator Design

The less-than-desired demonstrator ground-launch performance, with already developed RS-27A and D-57A engines, led to the consideration of ways to increase its speed by reducing its velocity losses due to drag. In this respect, the U.S. Air Force Flight Dynamics Laboratory (FDL) and the McDonnell Douglas Corporation (MDC) evolved a unique low-drag rocket-powered RLV convept during the 1960's – which has been periodically updated and refined.

This concept, shown in Figure 7, involves: a slender low-drag lifting body shape for high ratios of lift-to-drag (L/D) and re-entry cross range , flat panel surfaces for ease of manufacturing, and trapezoidal cross sections to minimize TPS by shielding the vehicle sides and top from the more intense re-entry heat. And the planar lifting body geometry of this configuration, like that of a current Lockheed X-33 lifting body configuration, could be readily adapted for advanced engines such as the Rocketdyne "linear aerospike".Thus, many features of this 30-year-old RLV concept appear attractive for RLV's being studied today.

Figure 7 also shows a typical embodiment of the MAKS tank within an RLV demonstrator of FDL/MDC lifting body shape. The takeoff and dry mass of this configuration was about the same as that of the circular one, with its somewhat heavier fuselage weight being compensated by its having almost no wing weight at all. Here, its major difference was lower drag, which resulted in a very high ratio of L/D.

Because of its low drag losses, the RLV demonstrator shown in Figure 7 was able to achieve (with three RS-27A and D-57 engines) approximately Mach 17 speed when air-launched, and about Mach 14 speed when vertically launched from the ground. And, with three RD-704 engines, it achieved approximately Mach 20 and Mach 18 speed, respectively, when launched from the air and from the ground.

The nose and underside geometry of this type of configuration may also have the potential for providing the pre-compression of airflow needed for demonstrating hypersonic airbreathing flight with an installed ramjet/scramjet engine module (as shown in Figure 8). Considerable work would be required to confirm the viability of this idea. But, it is conceivable that a demonstrator of hypersonic rocket flight might also serve as a testbed for hypersonic airbreathing flight as well.

Demonstrator Growth Potential

Although the primary purpose of a flight demonstrator would be a demonstration of critical technologies for RLV flight, it would be desired that it possess some operational utility as well. In this respect, a

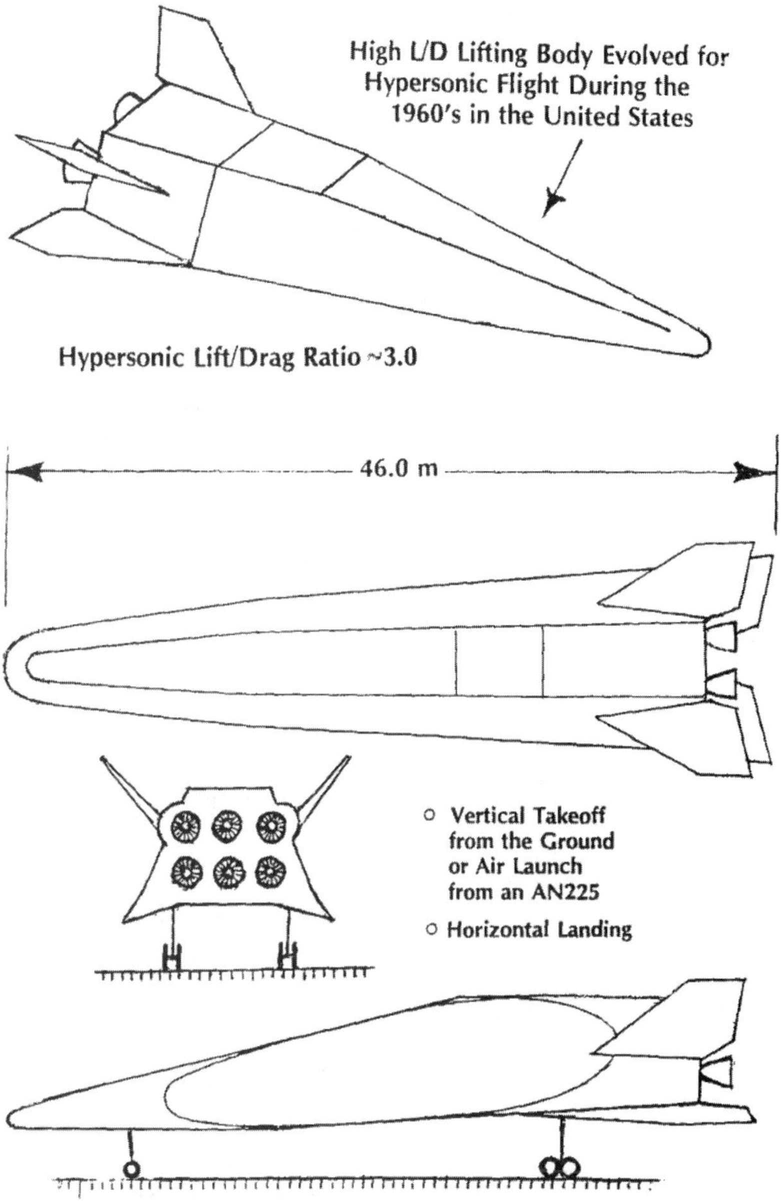

Fig 7 Alternate RLV Flight Demonstrator Configuration

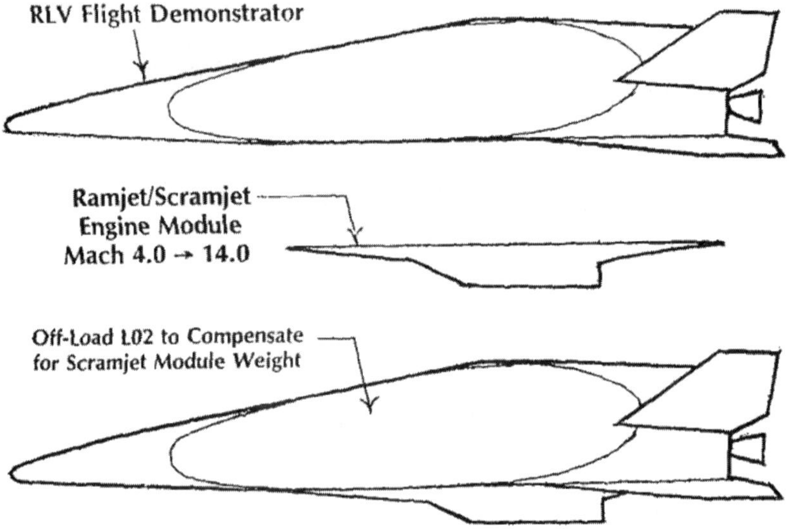

Fig 8 Concept for Demonstrating Rocket and Airbreathing Flight

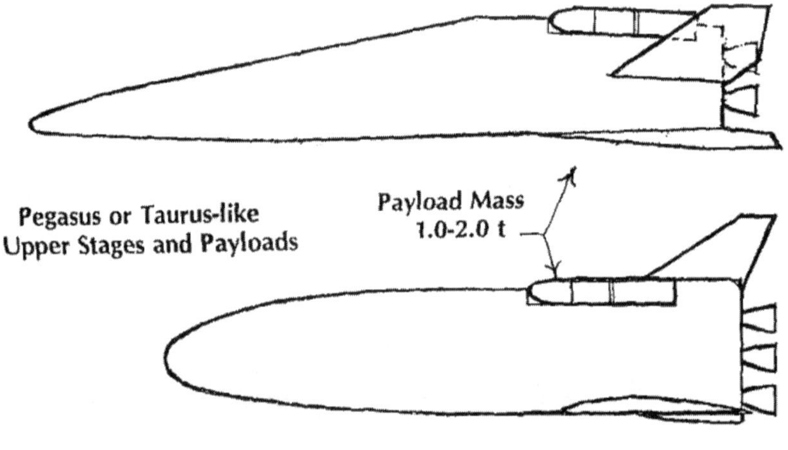

Fig 9 Use of a Flight Demonstrator as a Reusable First Stage

demonstrator with a MAKS propellant tank could carry and release Pegasus or Taurus-like upper stages that could orbit lightweight payloads about earth. As an example, Figure 9 shows installation of a Pegasus upper stage and its 0.3t payload on each side of demonstrators with MAKS tanks. Such vehicles would not necessarily be X-34 RLV candidates for launching lightweight payloads. However, they would involve payload integration and deployment and, therefore, would demonstrate more of the operational issues of reusable vehicle flight.

Conclusions

Preliminary studies indicate that the almost-developed MAKS tank might be effectively embodied within a large sub-orbital demonstrator of RLV flight. Here, significant sub-orbital performance is achievable with existing rocket engines if the demonstrator is air launched. And, adequate ground-launch performance appears achievable with very low drag airframes or with rocket engines that are under development today.

It is, therefore, recommended that more detailed examinations be made of the use of the MAKS tank for sub-orbital demonstrations of RLV flight. This would include participation by the Molnija Design Bureau in determining the feasibility and cost of completing development of this tank for reusable vehicle use.

REFERENCES

1. Lozino-Lozinsky, G. et al., "International Reusable Aerospace System MAKS Present State of the Art and Perspectives", AIAA-93-5165, AIAA/DGLR Fifth Intl. Aerospace Planes Conf., Munich, Germany, 1993

2. Lozino-Lozinsky, G. et al., "Multipurpose Aerospace System MAKS and Its Outlook", IAF-92-0851, 43 Congr. of the IAF, Washington, D.C., USA, 1992

3. Copper, J., "Single Stage Rocket Concept Selection and Design", AIAA 92-1383, AIAA Space Programs and Technologies Conf., Huntsville, Alabama, 1992

4. Kleinau, W., et al., "Development Trend Assessment for Future ETO-Space Transportation Systems", IAF-92-0855, 43rd Congr. of the IAF, Washington, D.C., USA, 1992

5. Stanley, D., et al., "Technology Requirements for Affordable Single-Stage Rocket Launch Vehicles", IAF 93-V.4.627, 44th Congr. of the IAF, Graz, Austria, 1993

6. Parkinson, R., "An-225/HOTOL", AIAA-93-5169, AIAA/DLGR Fifth Intl. Aerospace Planes Conf., Munich Germany, 1993

7. Limerick, C.D., "Rocket Propulsion for Single Stage to Orbit", AIAA 94-3161, AIAA 30th Joint Prop. Conf., Indianapolis, IN, 1994

8. Czysz, P., "Hypersonics Convergence Part 1-5", Textbook published by St. Louis University - Parks College, St. Louis, MO, Copyright: April, 1988

9. Rockwell International/Rocketdyne Division, "Expendable Launch Vehicle Engines", 6633 Canoga Park, CA 91303

10. Fanciullo, T. et al., "High Performance Russian D-57 LO2/LH2 Rocket Engine", AIAA 94-3398, AIAA 30th Joint Prop. Conf., Indianapolis, IN, 1994

AIAA

A00-36681

AIAA 2000-3478
Preliminary Simulations of Vehicle Interactions with the The Zero-Point Vacuum by Fluid Dynamic Approximations

H. David Froning, Jr.
Flight Unlimited
Flagstaff, AZ

R.L. Roach
Ramat Hashron
Israel

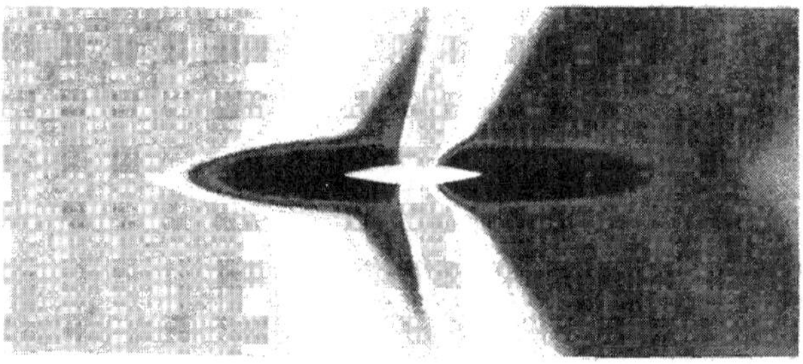

36th AIAA/ASME/SAE/ASEE
Joint Propulsion Conference & Exhibit
17-19 July, 2000/Huntsville, AL

AIAA 2000-3478

PRELIMINARY SIMULATIONS OF VEHICLE INTERACTIONS WITH THE ZERO POINT VACUUM BY FLUID DYNAMIC APPROXIMATIONS

H.D. Froning Jr.
Flight Unlimited
5450 Country Club Drive
Flagstaff, AZ 86004

R.L. Roach
3 Nordau Street
Ramat Hashron 47269
Israel

Abstract

The paper (a) describes the nature of specially conditioned electromagnetic (em) fields that may be capable of coupling favorable with those that underlie gravity and give rise to inertia, and (b) employs computational fluid dynamics techniques to obtain a first-order approximation of the influences of such em fields on reducing or reversing flight resistance at transluminal and superluminal speeds.

INTRODUCTION

Quantum field theory views seemingly empty space as a vigorous and energetic entity over microscopic intervals of time and space. A major contribution to such vigor and vitality are electromagnetic "zero-point" energy (ZPE) fluctuations - innumerable electromagnetic (em) energy pulsations of varying wavelength and frequency which, according to stochastic electrodynamics, pervade all space, as indicated in Figure 1, like an evanescent sea. And a radiation pressure somewhat analogous to hydrodynamic pressure can be associated with each ZPE fluctuation.

Figure 1 - Vigor and Vitality of "Empty" Space

Changed radiation pressures can result in forces, and one observed force caused by such change is the Casimir inward pushing force acting upon two parallel conducting plates placed closely together (1). Milonni (2) shows that one of several possible reasons for this could be - as shown in Figure 2 - diminished outward pushing zero-point radiation pressures caused by exclusion of longer wavelength ZPE fluctuations (and their associated radiation pressures) within the narrow gap between Casimir plates. And Scharnhorst (3) shows that the velocity of light in such a region of diminished radiation pressure should be greater than light velocity in vacuo (in undisturbed space).

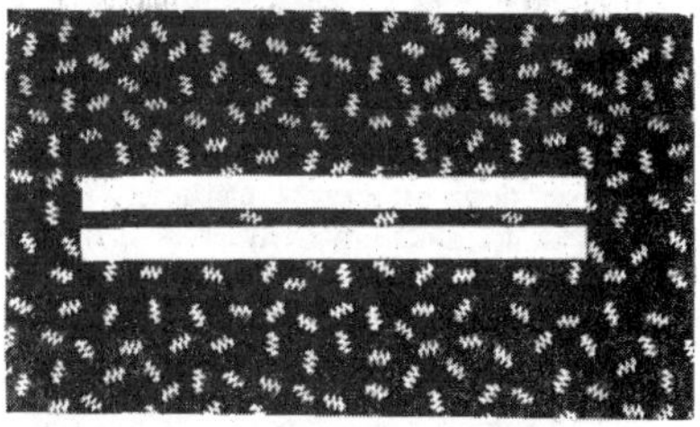

Casimir Force - Unbalanced Zero-Point Radiation Pressures Acting Upon Parallel Conducting Plates

Figure 2 - Unbalanced Zero-Point Radiation Pressures

General relativity describes gravitation in terms of a "spacetime metric". And changed zero-point radiation pressures occur in regions where spacetime metric is "warped" by the moving presence of material objects - just as changed aerodynamic pressures occur in regions where streamlines (paths of air particles) are bent in the vicinity of moving aircraft. In this respect, Alqubierre (4) has shown General Relativity solutions that enable vehicle acceleration by spacetime warping that results in negative energy densities within warped regions. Such densities also occur within regions of diminished zero-point radiation pressure (and increased light propagation speed) caused by fewer ZPE fluctuations between Casimir plates.

Unfortunately, field energy requirements to accomplish Alqubierre's spacetime warping may be prohibitively high. Ford and Roman (5) and Pfenning and Ford (6) suggest that Alqubierre - like warp bubbles enclosing vehicles of 200 meter dimension traveling at 10 times light speed must have a wall thickness no more than 10^{-32} meters and would require more negative energy than the positive energy contained in all the mass of the observable universe.

PART I - SPECIALLY CONDITIONED EM RADIATION

Millis (5) suggests that electromagnetism is a logical choice for creating an acceleration-inducing force, being related in some degree to spacetime and gravity and a technology in which we are fairly proficient. And Holt (6) has proposed generation of coherent patterns of electromagnetic (em) energy for field interactions that reduce or amplify gravitation in the vicinity of a hovering or moving ship. Unfortunately, patterns of ordinary em energy do not appear capable of such interactions. For although the metric associated with gravitation may be "warped" somewhat by em influences such as magnetic fields, investigations such as (7) indicate that stupendous field strengths are required for discernable distortion.

The authors propose that the weak coupling between gravitation and ordinary electromagnetism is a consequence of the entirely different types of fields that underlie them. Here, "abelian" fields that obey the commutation rules of abelian algebra and passes a relatively modest U(1) gauge symmetry are associated with ordinary em radiation, while Yang (8) suggests that a "nonabelian" field that obeys the commutation rules of nonabelian algebra and possesses group symmetry, higher than U(1) underlies gravitation. If so, a very significant interaction between nonabelian gravitational fields and abelian electromagnetic fields would not be expected because of their different essence and form. But, Barrett (9,10) has shown the possibility of transforming ordinary em fields into specially conditioned em fields of nonabelian form and higher symmetry and it is conceivable that such fields might be capable of coupling with the nonabelian fields of similar symmetry that may underlie gravitation and give rise to inertia.

Electromagnetic emanations of higher field symmetry must be solutions to Maxwell equations of more expanded form, and Barrett (9) has derived expanded Maxwell equations for nonabelian em fields of SU(2) symmetry. These equations include existing Maxwell terms plus additional terms that involve the coupling of electric and magnetic fields through the action of A vector and scalar potentials.

Examples of Specially Conditioned Radiation

Although the most complex combinations of frequency and amplitude modulation do not transform ordinary U(1) em fields into nonabelian fields of SU(2) symmetry, Barrett (9, 10) and (11) shows several ways of accomplishing such a transformation. One way is modulating the polarization of the em wave energy emitted from microwave or laser transmitters. The other way is driving alternating current through toroidal coils at resonant frequencies. For either type of modulation, specially conditioned em radiation of a SU(2) or higher symmetry is attained depending upoln the degree of polarization modulation or amount of toroid transmitter sophistication used. Preliminary polarization modulated laser tests in Russia and toroidal coil tests in Canada have confirmed many of our theoretical predictions (11). New experiments with more powerful polarization modulated lasers are now planned for the future, while experiments with more powerful toroidal coils of circular and non-circular configuration will soon be conducted.

Polarization modulation of em wave energy is illustrated in Figure 3. One fraction of the input wave energy has its waveform rotated 90 degrees with respect to that of the other fraction, while wavelengths of the two waveforms are made to differ by a small amount of phase modulation. Both fractions of wave energy are combined at a "mixer" and specially conditioned em radiation, whose electric and magnetic fields rotate through all possible polarizations within very short intervals of time - is then emitted.

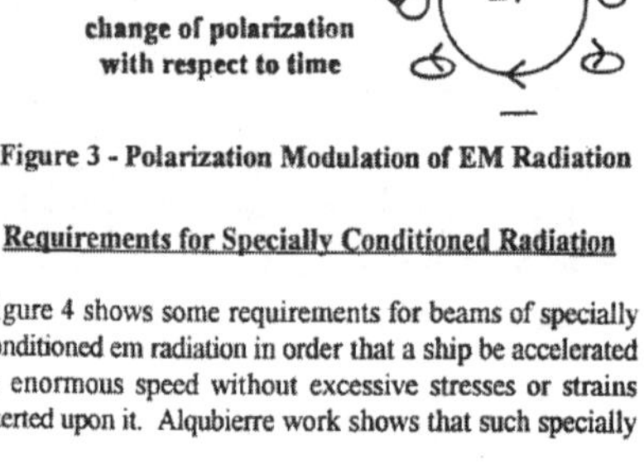

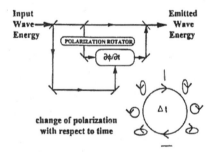

Figure 3 - Polarization Modulation of EM Radiation

Requirements for Specially Conditioned Radiation

Figure 4 shows some requirements for beams of specially conditioned em radiation in order that a ship be accelerated to enormous speed without excessive stresses or strains exerted upon it. Alqubierre work shows that such specially conditioned em beams must result in favorable warping of spacetime metric and increased em propagation velocity, and, our work indicates that favorable warping must be accompanied by a favorable re-distribution of zero-point radiation pressures in the vicinity of the moving ship.

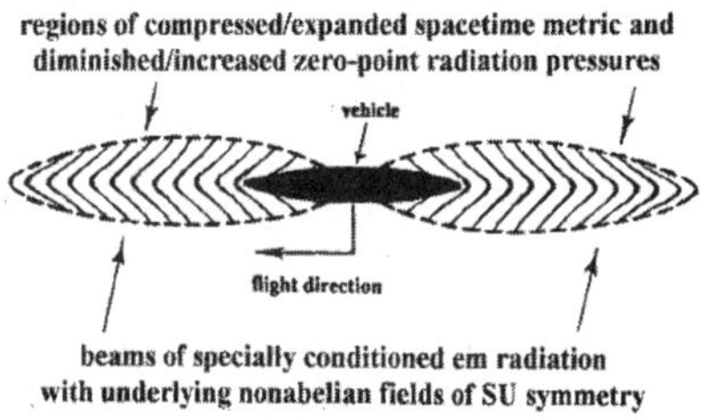

Figure 4 -Specially Conditioned EM Radiation Needs

The propagating speed (c) of electromagnetic disturbances within a given region of empty space is determined by the electrical permittivity (ϵ_o) and magnetic permeability (μ_o) of the vacuum through which they move, and since ordinary em fields to not intimately interact with spacetime metric, ϵ_o and μ_o the propagating speed of em waves in vacuo remain unchanged. But, if specially conditioned em beams interact with spacetime metric such that ϵ_o and/or μ_o is reduced, em wave propagation speed greater than in undisturbed vacuo would occur within such beams, and em wave fronts would propagate as shown in Figure 4.

PART II - SIMILARITIES IN FLIGHT THROUGH AIR AND SPACE

Froning and Roach (12) suggest that vehicle interactions associated with the fluid (gas) dynamics of air and the quantum dynamics of space are sufficiently similar that computational fluid dynamics (CFD) techniques can crudely simulate vehicle interactions that warp spacetime metric and perturb the ZPF. Here, Figure 5 shows: that innumerable collisions and recoils of air molecules comprise the energetics of air, while innumerable zero-point electromagnetic fluctuations comprise to the vigor and vitality of space. In this respect, unbalanced zero-point radiation pressures acting upon parallel conducting plates separated by a 14 angstrom gap, would exert the same pushing pressure as sea-level atmospheric pressure exerts.

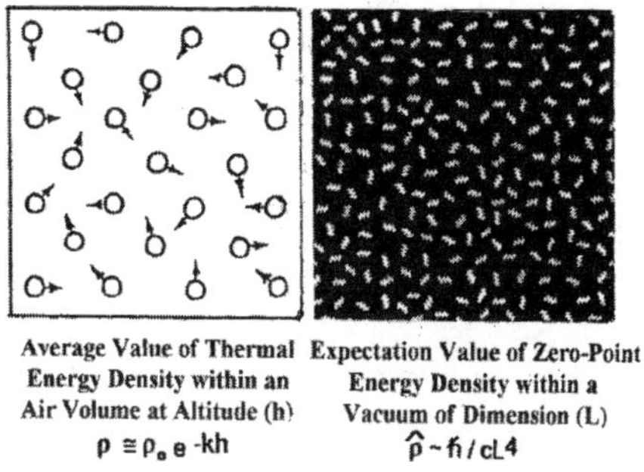

Average Value of Thermal Energy Density within an Air Volume at Altitude (h)
$\rho \cong \rho_o e^{-kh}$
Expectation Value of Zero-Point Energy Density within a Vacuum of Dimension (L)
$\hat{\rho} \sim \hbar / cL^4$

Figure 5 - Density of Air and "Empty" Space

Resistance to Flight in Air and Space

Special relativity predicts that inertia of accelerating vehicles is amplified by a factor $[1-v/c)^2]^{-1/2}$ as vehicle speed (v) approaches the propagating speed of light, while linear aerodynamic theory predicts an identical amplification in aerodynamic resistance (vehicle drag) as vehicle speed (v) approaches the propagating speed of sound. Thus, as shown in Figure 6, simplified theories of compressible aerodynamics and relativistic mechanics predict the same rate of flight resistance increase as vehicle flight speeds approach sound speed in air and light speed in space.

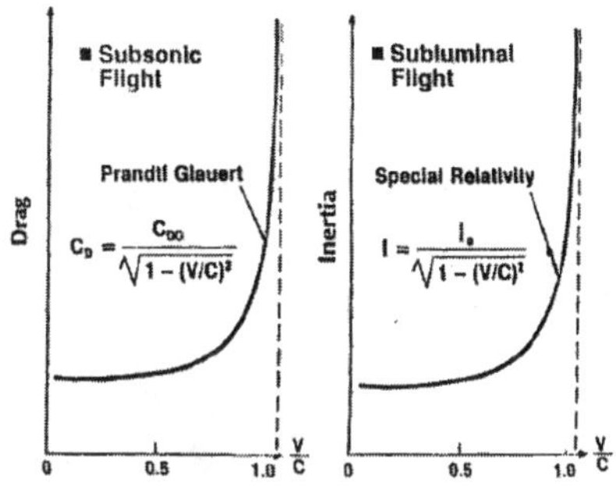

Figure 6 - Flight Resistance Increase with Speed

Just as air disturbed by a moving body exerts a resisting aerodynamic force on the body which is perceived as its drag, so Haisch, Rueda, Puthoff (13) propose that the equilibrium state of the electromagnetic zero-point vacuum is disturbed by an accelerating body and exerts a resisting

Ed. Note: The variable "gamma" $\Upsilon = (1-(v/c)^2)^{-1/2}$ in vacuum engineering and relativity theory, not to be confused with the specific heat of air (Fig. 9). It is also labeled ϒle and ϒte (in AIAA 95-2368) as well as "gle" and "gte" respectively in the landscape simulations at the beginning of this Appendix.

electromagnetic force on the body which is perceived as its inertia. Such electromagnetic resistance can be viewed as the Lorentz force caused by interaction of the magnetic component of the electromagnetic zero-point field (ZPF) with each charged particle that comprises the electrical substructure of the accelerating body. And more recent work by Haisch and Rueda (14, 15) derives the same results by an analysis of the momentum flux of the zero-point radiation associated with the ZPF impinging on accelerated objects (15) conjectures that transferred linear momentum and translational kinetic energy is radiated outward from accelerating objects by the propagation of wave-like disturbances within spatially decaying evanescent fields.

Such electromagnetic resistance to flight is somewhat similar to aerodynamic resistance to flight. Here, air friction at the aircraft surface results in the "viscous" component of the aircraft's drag (including pressure drag from flow separation). And outward propagation of disturbances from the aircraft, in the form of spatially decaying pressure waves, form a pressure gradient about the moving aircraft which determines its "wave" drag. Nearly symmetrical pressure gradients form at speeds much less than sound speed, and wave drag is very slight. But as speed approaches sound speed, compressibility effects cause a non-symmetrical pressure distribution and the resulting wave drag becomes the dominant contributor to total drag.

Figure 7, generated by computational fluid dynamics (CFD), illustrates a pressure distribution for a two-dimensional airfoil-like shape moving at 99 percent of the speed-of-sound. Incident airflow is compressed while passing over the craft's forward contour and higher-than-ambient air pressure causes a resisting "push". Airflow expands over the rearward curvature and lower-than-ambient pressures causes additional wave drag in the form of a resisting "pull". Then, if this air pressure gradient is somewhat similar to a zero-point radiation pressure gradient, Figure 7 can also be viewed as a crude first order approximation of adverse zero-point radiation pressures acting upon a much larger craft accelerating at 0.99c.

Figure 7 - Aerodynamic Pressure Distribution at 99 Percent of Sound Speed

If unbalanced zero-point radiation pressures act upon a body, the resulting force would be somewhat similar to the Casimir force acting upon closely spaced plates. And total inertial resistance to vehicle acceleration (like aerodynamic resistance) partitioned into two components as shown in Figure 8. One, somewhat analogous to aerodynamic viscous drag, a consequence of interactions along and within the boundaries of a moving body. The other one, somewhat similar to aerodynamic wave drag, would be a consequence of the ZPF/vehicle interaction being propagated as a disturbance beyond the body into the surrounding ZPF. As a consequence, unbalanced zero-point radiation pressures would act upon the vehicle exterior in the form of a Casimir-like resisting force.

Aerodynamic Resistance	Vacuum Field Resistance
Viscous Drag Exerted on the Charged Particle Substructure of a Vehicle's Surface by Moving Air	Lorentz Force Exerted on a Vehicle's Charged Particle Substructure by the Zero-Point Field
Wave-Drag Exerted on a Vehicle's Exterior by Unbalanced Aerodynamic Pressures	Casimir-like Force Exerted on a Vehicle's Exterior by Unbalanced Zero-Point Radiation Pressures

Figure 8 - Contributors to Flight Resistance

Just as wave drag of moving aircraft increase significantly as they approach the propagating speed of acoustic disturbances, so similar increase in inertia occurs as the propagating speed of electromagnetic disturbances is approached by accelerating craft. The propagation modes of acoustic and em disturbances are, of course, not the same. But Figure 9 indicates that acoustic and em wave front speeds are solutions of similar wave equations and, for a given temperature and altitude, are determined by the physical structure energy storage/dissipation characteristics of the mediums through which they move.

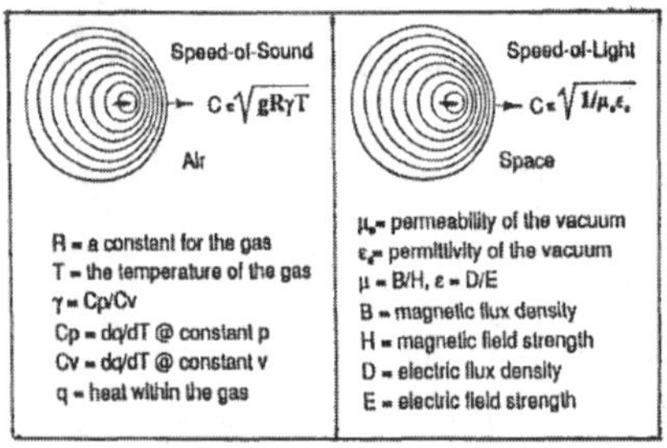

Figure 9 - Acoustic and Electromagnetic Wave Speed

Disturbance Propagation Through Air and Space

Just as disturbances that perturb the gaseous structure (R) and specific heat characteristics (γ) of air will affect sound propagation speed and vehicle drag, so those that would perturb ϵ_o and μ_o within a region of space would affect em propagation speed within the perturbed region and, possibly, vehicle inertia. It was therefore assumed that perturbations of spacetime metric would be accompanied by distortion of the zero-point vacuum - in terms of its magnetic permeability μ_o and electric permittivity ϵ_o. And as shown in Figure 10, the first-order effect of perturbing the ϵ_o and μ_o of the zero-point vacuum was simulated by perturbing the R and γ characteristics of atmospheric air.

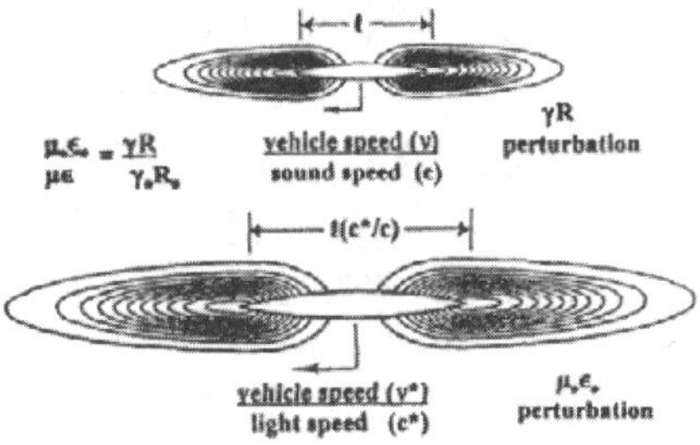

Figure 10 - Scaling of Acoustic and EM Disturbances

As shown in Figure 10, disturbances caused by specially conditioned em radiation extended outwardly with decreasing intensity from leading and/or trailing edges of wing-like vehicles until zero perturbation intensity was reached. As indicated, vacuum field perturbations were simulated by fluid field perturbations that resulted in the same percentage change in disturbance propagation speed within the region of perturbation. Ratio of vehicle speed to disturbance propagation speed were held equal, while size of vehicle and field perturbation region scaled, of course, with the ratio of light speed to sound propagation speed.

Because ability to approach or exceed sound-speed is determined by wave drag, computational effort was simplified by solving only the inviscid Euler equations of fluid dynamics for wave drag, and by considering only 2-dimensional airfoil-like shapes. However, as indicated in Figure 11, the resulting disturbance topology is 3-dimensional-with disturbance intensity being maximum at vehicle leading and/or trailing edges, and diminishing in a gaussian fashion in parallel and perpendicular directions.

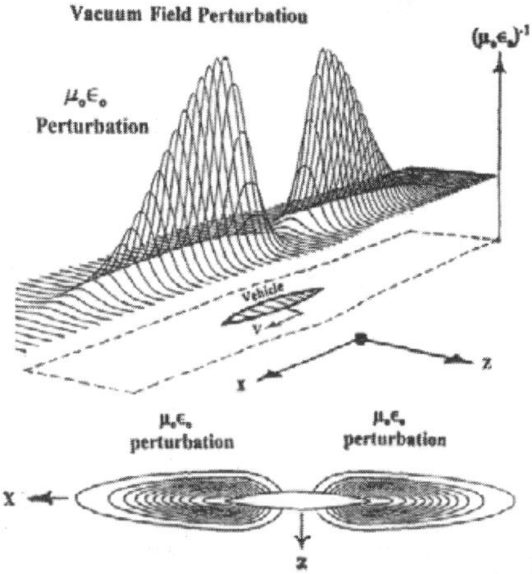

Figure 11 - Topology of Vacuum Field Disturbance

PART III - DISTURBANCE EFFECT ON FLIGHT RESISTANCE AND IMPULSION

Our first-order simulation is based upon what we believe to be certain similarities in flight through the atmosphere of earth and vacuum of space. But because these similarities are by no means exact, our results are useful, only, for indicating general trends. Also, the CFD computations only addresses the effects of external pressures acting upon vehicles. They, thus, considered only the flight resistance (inertia) associated with redistribution of zero-point radiation pressures over vehicle exteriors - not the Lorentz force (described by Haisch, etal) acting upon the charged particle substructure that comprises vehicle interiors.

Influence of Disturbance Shape, Intensity, Location

Early studies by Froning and Roach (13) considered omnidirectional beams whose circular disturbances were generated at vehicle leading or trailing edges, while more recent studies considered elongated beams with increased length and decreased width. As might be expected, narrower and more elongated beams were more effective in reducing flight resistance. Figure 12 shows the influence of the location and intensity of elongated beams on flight resistance reduction at 99 percent of the speed of light.

Ed. Note: $c^2 = 1/\mu_o\varepsilon_o$ which in modern physics designates the values for the unperturbed vacuum of space. Figures 10 and 12 honor that definition by indicating the value of $\mu\varepsilon$ is for a perturbed vacuum, specifically for permeability and permittivity of space. Thus $\mu_o\varepsilon_o$ nor c are not really constant.

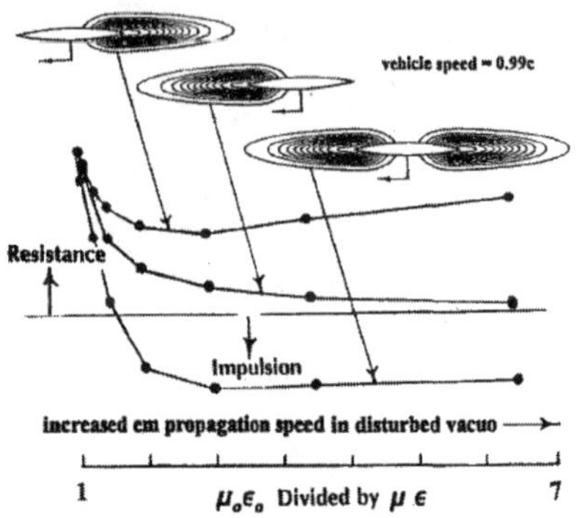

Figure 12 - Influence of Disturbance Location and Intensity at 99 percent of Light-Speed

Diminished vehicle resistance to acceleration (inertia) was achieved by specially conditioned em beams that diminished $\mu_o \epsilon_o$ and increased light-speed within the vacuum regions they perturbed. Emission of radiation ahead of the vehicle was more effective in reducing resistance than emission of radiation behind it, but maximum reduction was achieved for emission of radiation both ahead of and behind the vehicle. The zero-point radiation pressure distributions shown in Figure 13 and 14 result in about a 25 percent resisting force reduction by specially conditioned em radiation emitted ahead of the vehicle, and about a 50 percent force reduction for radiation emitted both ahead of and behind the vehicle.

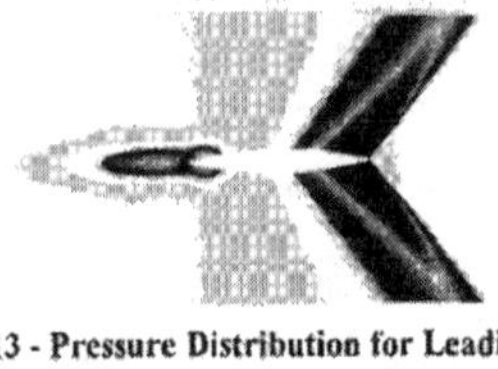

Figure 13 - Pressure Distribution for Leading Edge Disturbance at 99 percent of Light-Speed

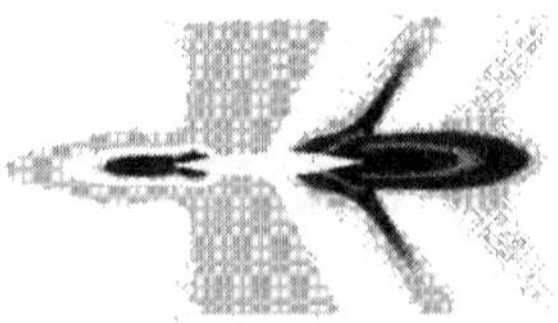

Figure 14 - Pressure Distribution for Leading and Trailing Edge Disturbance at 99 percent of Light-Speed

Vacuum ZPF disturbances that significantly diminish $\mu_o \epsilon_o$ ahead of and behind the vehicle - such as those shown in Figure 15 - enable impulsion (rather than resistance) at almost light speed. This is because much lower than ambient zero-point radiation pressures exist over the forward portion of the vehicle for a net "pulling" effect.

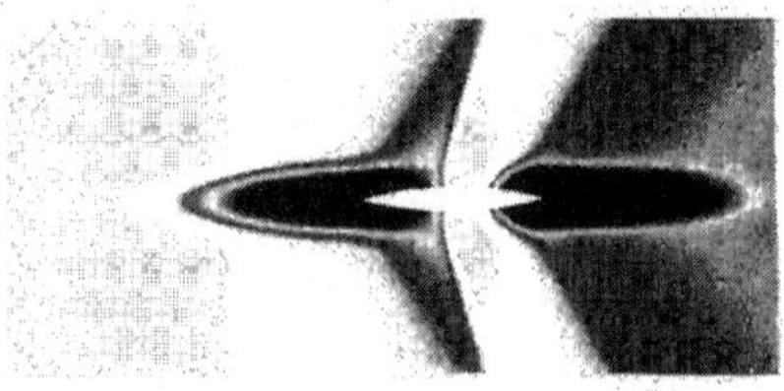

Figure 15 - Pressure Distribution that Causes Impulsion at 99 percent of Light-Speed

Influence of Disturbances at Superlaminal Speeds

For the spacetime warping considered by Alqubierre (4) light propagation speed increases within the region of warping such that vehicle speed is always less than light speed within this region, even when its speed with respect to earth is greater than the propagating speed of light in un-warped space. Figure 16 shows such a condition, with specially conditioned em radiation diminishing $\mu_o \epsilon_o$ to values that result in very high light propagation speed in the vehicle vicinity. This enables slower-than-light speed with respect to adjacent space even though vehicle speed is 2.5 times light speed in undisturbed vacuum. And the resulting zero-point radiation pressure distribution is so favorable; that a net thrusting force acts upon the vehicle.

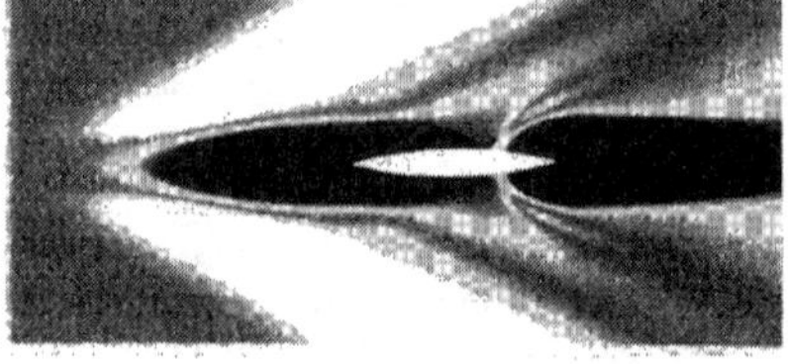

Figure 16 - Pressure Distribution that Causes Impulsion at 2.5 times Light-Speed

Since significant spacetime warping and zero-point field distortion would be occurring within the perturbed region surrounding a ship moving faster-than-light with respect to undisturbed space, it would probably be invisible in observer frames that are at rest with respect to earth. However, its presence might be indirectly detected if Chrenkov radiation emanates from density gradients that cause sudden decreases in light propagation speed.

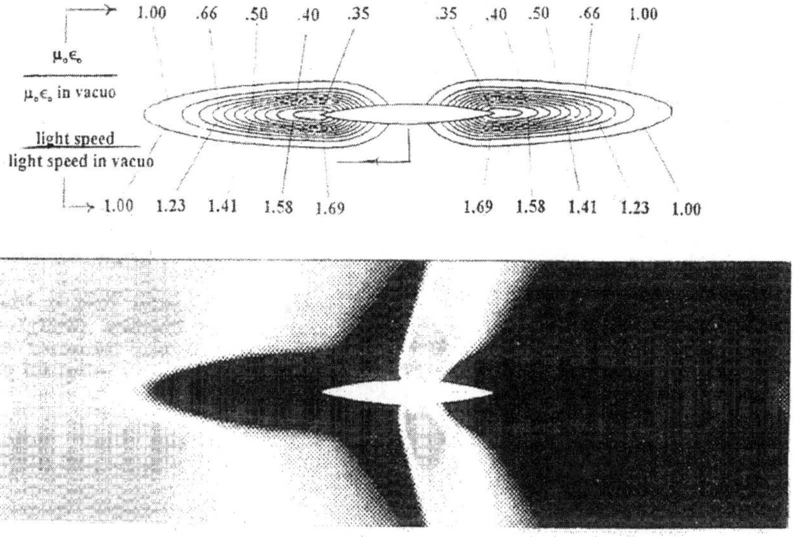

Figure 17 - Perturbation Pattern and Pressure Distribution that Causes Impulsion at V = .99C

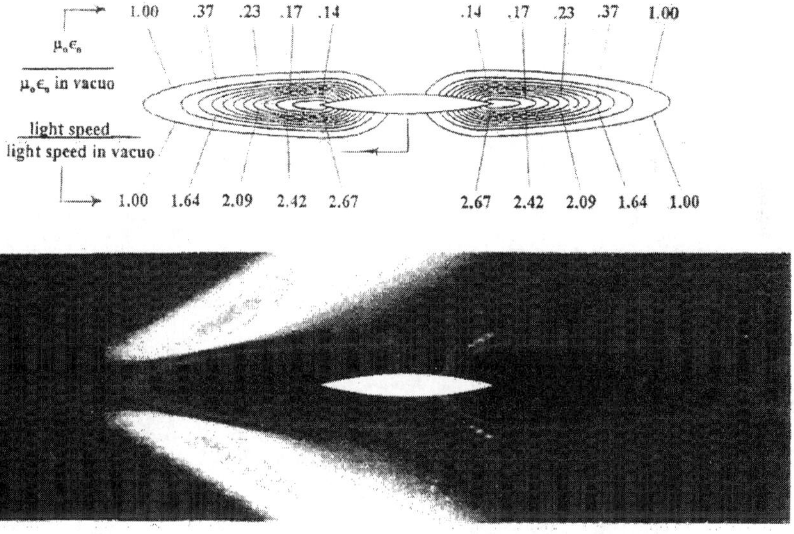

Figure 18 - Perturbation Pattern and Pressure Distribution that Causes Impulsion at V = 2.5C

Requirements for Acceleration and Deceleration

Enhancement of vehicle acceleration or deceleration is achievable. Conditioned em radiation that decreases $\mu_o \epsilon_o$ in the forward and rearward vehicle vicinity at high speeds aids acceleration while increased $\mu_o \epsilon_o$ aids deceleration. Decreased $\mu_o \epsilon_o$ in the forward vehicle vicinity and increased $\mu_o \epsilon_o$ in the rearward aids acceleration at low speed, while increased forward $\mu_o \epsilon_o$ and decreased rearward $\mu_o \epsilon_o$ aids deceleration.

SUMMARY AND CONCLUSIONS

If a virtual sea of zero-point em energy fluctuations pervade spacetime - as proposed by stochastic electrodynamics - and if accelerating vehicles are bathed by an asymmetric distribution of such fluctuations, the vehicles are acted upon by zero-point radiation pressure gradients that resist or assist vehicle motion. Thus, in addition to the Lorentz force acting upon the charged particle interiors of accelerating vehicles (as suggested by Haisch etal) there may also be a Casimir-like force from un-balanced zero-point radiation pressures acting upon their exteriors.

If specially conditioned em radiation can significantly diminish vacuum electric permittivity (ϵ_o) and magnetic permeability (μ_o) in the vicinity of an accelerated ship, the effects of gravity and inertia can be significantly changed at subluminal, transluminal and superluminal speed.

Increase in light propagation speed will occur within regions where μ_o and/or ϵ_o, are diminished. Thus, if sufficient light propagation speed increase is achieved within such regions, slower-than-light speed with respect to disturbed space will occur - even if vehicle speed is much greater than light speed in undisturbed vacuo.

REFERENCES

(1) Casimir, Proc. Ned. Akad. Wet. Vol 51, p793 (1948)

(2) Milonni, P.W., et al., Phys Rev A, 38, p1621-1630 (1988)

(3) Scharnhorst, K. Phys Let B 236, p354-359 (1990)

(4) Alqubierre, M. Class Quant Grav 11, L73-L77 (1994)

(5) Ford, L.H. and Roman, T.A., "Quantum Field Theory Constrains Traversable Wormhole Geometries", Phys. Rev. D, Vol 53, No. 10, p5496 (1996)

(6) Pfenning, M.J. and Ford, L.H., "The Unphysical Nature of Warp Drive", Class. And Quant. Gravity, Vol 14, No. 7, p1743 (1997)

(7) Millis, M.A, "The Challenge to Create the Space Drive", National Space Society New Roads to the Stars Interstellar Flight Symposium, New York City, NY (1996)

(8) Holt, A.C., "Prospects for a Breakthrough in Field Dependent Propulsion", 16th AIAA Joint Propulsion Conference, AIAA 80-1233, Hartford, Conn (1980)

(9) Davis, E.W., "Interstellar Travel by Means of Wormhole Induction Propulsion (WHIP)," Space Technology and Applications International Forum" (STAIF-98), AIP Conference Proceedings 420, p.1502-1508, Albuquerque, N.M., (1998)

(10) Yang, C.N., "Gauge Theory", McGraw-Hill Encyclopedia of Physics, 2nd Edition, p 483, (1993)

(11) Barrett, T.W., "Electromagnetic Phenomenon not Explained by Maxwell's Equations", Essays on Formal Aspects of Electromagnetic Theory, p 6, World Scientific Publ Co., (1993)

(12) Barrett, T.W. & Grimes, D.M., "Advanced Electromagnetism: Foundations, Theory Applications", p. 278-313, World Scientific Publ Co., (1993)

(13) Froning, H.D., Barrett, T.W., Hathaway G., "Experiments Involving Specially Conditioned EM Radiation, Gravitation and Matter," 34th AIAA Joint Propulsion Conference, AIAA 98-3138, Cleveland, OH (1998)

(14) Froning, H.D. and Roach, R.L., "Drag Reduction and Possibility Impulsion by Perturbing Fluid and Vacuum Fields," 31st AIAA Joint Propulsion Conference, AIAA 95-2368, San Diego, CA (1995)

(15) Haisch, B., Rueda, A., Puthoff, H., "Inertia as a zero-point field Lorentz Force," Phys. Rev. A, Vol. 49, p678 (1994)

(16) Rueda, A., Haisch, B., "Contribution to Inertial Mass by Reaction of the Vacuum to Accelerated Motion", Phys, Rev. A, Vol. 39, p2333 (1998)

(17) Rueda, A., Haisch, B., "Contribution to Inertial Mass by Reaction of the Vacuum to Accelerated Motion", Found. of Physics, Vol.28, p 1057 (1998)

Technical Notes:

The Computational Fluid Dynamics (CFD) simulations described in this paper assumed instant development of perturbation patterns in the vicinity of an accelerating vehicle. Geomeries of vehicles, perturbation patterns, and pressure contours correspond to what would be seen in moving inertial frames that are accelerating such that they remain at rest with respect to the inertial frame of the accelerating vehicle. At relativistic vehicle speeds, such geometries (and associated kinematics) would, of course, be different in other inertial frames - such as those of observers at rest with respect to earth.

The results presented in this paper are consistent with some of those of Alqubierre and others (4, 5, 6) - in that regions of negative energy density (regions containing less than ambient zero-point radiation pressure) were associated with zones where favorable ZPF distortion occurred at very high flight speeds with respect to earth. And the greatest propulsive effect was achieved when metrical warping and ZPF distortion occurred in both the forward and the rearward vicinity of accelerating ships. But, in our analysis, favorable metrical warping and ZPF distorting was accomplished by much milder field energy gradients than those required in (5, 6) and field energy was deposited in regions much closer to the ship.

Acknowledgments:

The concept of specially conditioned em radiation described in this paper originated with Dr. Terence Barrett of BSEI. 1453 Beulah Road, Vienna, Virginia 22182. The CFD work described in this paper was performed by Dr. Roach while he was affiliated with the University of Tennessee Space Institute. Dr. Roach's work was sponsored by Flight Unlimited with internal company funds.

AIAA

A01-34344

AIAA 2001–3658
Specially Conditioned EM Radiation Research with Transmitting Toroid Antennas

H.D. Froning Jr.
Flight Unlimited
Flagstaff, AZ, USA

G.W. Hathaway
Hathaway Consulting Services
Toronto, ON, Canada

37th AIAA/ASME/SAE/ASEE Joint Propulsion Conference and Exhibit
8-11 July, 2001
Salt Lake City, UTAH

AIAA 2001-3658

SPECIALLY CONDITIONED EM RADIATION RESEARCH WITH TRANSMITTING TOROID ANTENNAS

H. David Froning
Flight Unlimited
5450 Country Club Dr.
Flagstaff, AZ 86004, U.S.A.

George Hathaway
Hathaway Consulting Services
39 Kendal Ave.
Toronto, ON/M5R IL5, Canada

Abstract

Experimental work to: (a) determine em field characteristics associated with em radiation created by alternating current flowing through toroidal coils at resonant frequencies, and (b) determine if the specially conditioned em fields associated with such radiation could cause a discernable gravity modification, is described. This experimental work was the result of collaboration between Flight Unlimited (FU) and Hathaway Consulting Services (HCS), performed at the laboratories of HCS in Toronto, Canada during a test period in 1998 and during a test period in 2000. Tested toroid configurations included: circular toroids with differing diameters and winding densities; and asymmetrical toroids for focusing em radiation into narrower and more intense beams. The toroid configurations and the AC power and instrumentation systems available at HCS limited the experimental work to the relatively low radio frequencies (400 kHz to 110 MHZ) of the electromagnetic spectrum.

INTRODUCTION

Just as airflight was not revolutionized until propeller propulsion was superceded by a new mode of impulsion (jet propulsion), so spaceflight may not be revolutionized until jet propulsion is superceded by a new mode of impulsion (field propulsion). Field propulsion would develop thrust by actions and reactions of fields instead of by combustion and expulsion of mass. And field actions and reactions that would greatly reduce propellant (the major portion of rocketship mass) and engine thrust requirements would be those that would reduce the resistance of gravity and inertia to ship acceleration.

One conceivable way of reducing the resistance of gravity and inertia is by accomplishment of a favorable coupling between those fields which underlie electromagnetism and gravity. But no significant coupling of ordinary em fields with those that give rise to gravity may be achievable because their essence is completely dissimilar. Yang (1) notes that "nonabelian" fields which probably give rise to gravity are of more intricate topology and higher internal symmetry than the "abelian U(1)" fields that underlie ordinary electromagnetism. In this respect, Barrett (2, 3) has identified two ways of transforming ordinary em fields into specially conditioned em fields of nonabelian form and higher than U(1) symmetry. One identified way of creating such fields is modulating the polarization of em wave energy emitted from microwave or laser transmitters. Such polarization modulation creates em fields of nonabelian form and SU(2) symmetry within beams of radiated power that can be focused into very narrow beams of very high energy density. Thus an experiment to detect possible gravity modifications within narrow polarization modulated laser beams has been submitted to the NASA Breakthrough Propulsion Physics (BPP) program. This experiment is described in (4).

Another way of transforming ordinary em fields into specially conditioned em fields of nonabelian form and SU(2) symmetry is with toroidal coils through which alternating current is flowing at resonant frequencies. Barrett (3) shows that such specially conditioned em radiation includes - not only electric and magnetic field energy - but A Vector potential field energy as well. Barrett predicts that A Vector field intensity maximizes at discreet resonant frequencies. Thus, if an A Vector potential field underlies the essence of gravitation, gravity modification might be possible in the vicinity of toroids transmitting at such frequencies.

Fabrication and testing costs were significant for polarization modulated laser beams. However, they were found to be relatively modest for toroidal coils configured for operation in the lower (radio-frequency) range of the em spectrum. Thus a cooperation between Hathaway Consulting Services (HCS) and Flight Unlimited (FU) was established to: test Barrett's hypotheses as to specially conditioned em radiation emitted from toroidal coils; and to determine if gravity modification could occur within such radiation. Probability of gravity modification by radio frequency radiation from inexpensive toroids was deemed to be very low. But it was hoped that the tests would reveal interesting electromagnetic phenomenon and extend our knowledge of electromagnetics.

TRANSMITTING TOROID ANTENNAS

EM wave propagation by transmitting toroid antennas has been examined by various investigators for more than a decade. Examples are U.S. Patent No. 4,751,515 awarded to Corum for an "electrically small, efficient electromagnetic structure that may be used as an antenna or waveguide probe" and U.S. Patent No. 5,442,369 awarded to Van Voorhies et al. for an antenna that "has windings that are contra wound in segments on a toroid form and that have opposed currents on selected segments". In this respect, Barrett (4) has shown that specially conditioned em fields of SU(2) symmetry and nonabelian form can be created by transmitting toroid antennas - as in Figure 1.

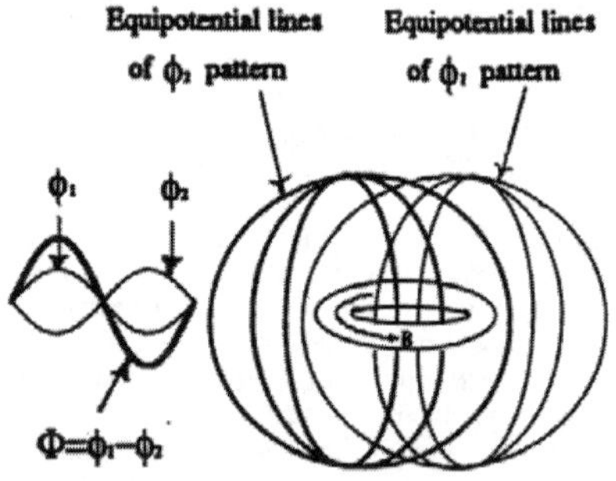

Fig. 1 – A Vector Potential Patterns

The magnetic and electric fields which encompass a transmitting toroid are accompanied by A Vector potential fields, and the alternating current flow produces overlapping A Vector potential patterns which encircle the toroid ring, as shown in Figure 1. These A Vector patterns combine into "phase factor" waves which represent disturbances in A Vector potential. The maximum disturbance in A Vector potential occurs as phase factor wave intensity peaks at the resonant frequencies where A Vector potential patterns are exactly out-of-phase, and a predicted pattern of these disturbances is shown in Figure 2.

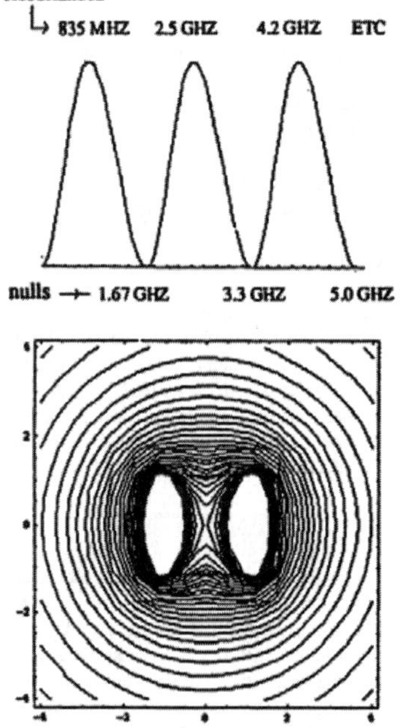

Fig. 2 – "Phase Factor" Wave Patterns

Resonant frequencies are determined by the shape and dimensions of the toroid, and by the propagating direction and speed of the alternating electric current thru its windings. And, if an A Vector potential field underlies the essence of gravitation, the probability of gravity modification in the toroid vicinity would be highest at resonant frequencies.

INITIAL TOROID EXPERIMENTS

Initial experimental work involved: (a) fabrication of transmitting toroid antennas that, according to (4), should emanate specially conditioned em radiation; and testing of the toroids at low power levels at the laboratories of HCS in Toronto. The general goals of this initial work, which was performed on March 6 and 7 during 1998, were: perfection of techniques for fabricating toroidal coils, detection of resonant phenomenon indicative of A Vector potential resonances with such coils; and identification of problems associated with operating toroidal coils over wide frequency ranges and at significant power levels.

Most of the goals of the initial work were achieved. Toroid antennas with conventional and caduceus windings were successfully fabricated, and although no instrumentation (such as Josephson Junction arrays) were available for directly detecting A Vector fields, measured resonances (reversals in phase and amplification of signal strength) were in good agreement, as indicated in Figure 3, with Barrett's predictions predicated on occurrence of A Vector fields. Heat generated by current flow within the relatively thin windings of the toroids and their relatively fragile styrofoam interiors limited input power to less than 100 watts in the initial experiments. This identified the need for thicker wires and stronger structures for higher toroid power and temperature.

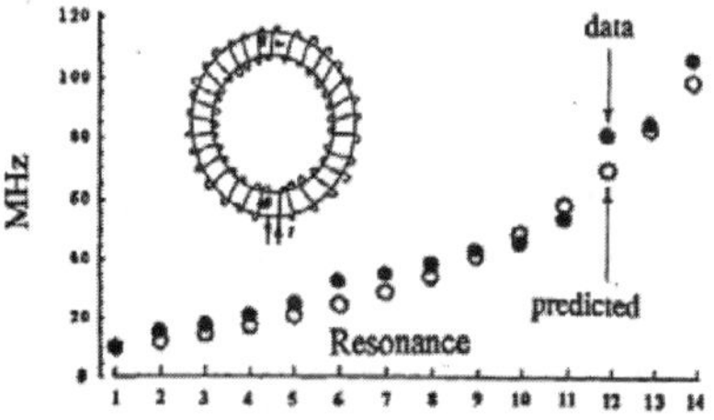

Fig. 3 – Correlation of Theory and Experiment

FOLLOW-ON TOROID EXPERIMENTS

Because results of the initial toroid experiments were somewhat encouraging, it was decided to have a follow-on experimental program, which included toroids configured for much higher power levels, at HCS between June 9 and June 15, 2000. It included: (a) signal phase/amplitude tests to precisely determine the resonant frequency characteristics of each different transmitting toroid configuration; (b) magnetic field measurements to map em field intensity in the vicinity of each toroid; (c) propagation characteristics of toroid radiation; and (d) limited gravitometer testing to search for a gravitational disturbance at a one location near one of the transmitting toroids.

Toroid Configurations Tested

To our knowledge, transmitting toroid antennas built and tested by most other investigators have been designed for communication purposes - with wires loosely wound (widely separated) around the toroid's ring in order to maximize far field intensity and range. By contrast, our tested toroids were "tightly wound" to maximize near-field intensity for possible gravity modification - not far field range for communication. Our tested toroids were "contra-wound" in a caduceus pattern to allow 2 types of modulation. One, in which current flowed in opposite directions in crossing wires, resulted in an "opposing" or "bucking" mode which caused opposing magnetic fields that cancel themselves along the toroid ring centerline. The other, in which current flowed in the same direction - resulted in an "adding" mode. Figure 4 shows the 4 different toroid configurations that were tested during the follow-on experimental program.

The loosely-wound toroid (upper left) was built for comparing its near-field intensity with that of the tightly wound toroid (lower left). Both toroids had similar cross sections (approximately 4.0 cm) and the same outer diameters (21cm) and wire size (No. 20). The greater winding density of the tightly wound toroid (350/333 inner/outer turns vs 26/25 inner/outer turns) resulted in greater near-field intensity for a given input power. The toroid in the upper right was configured with a larger outer diameter (31 cm.) than the lower left one and No. 20 wire size but its cross-section is the same. The larger diameter resulted in more windings (398/384 inner/outer turns of No. 14 wire). And the "tear drop" shaped toroid (lower right) was configured to focus radiation into more intense and elongated beams. Its length, breadth and thickness was 26.5, 18.0, 2.5 cm. It had a hole diameter of 7.3 cm; and 95/88 inner/outer turns of No. 14 wire. And, because of their stronger structure (hard maple wood) and larger wire diameter, the tear drop and larger diameter toroids could withstand the heating associated with 1.0 KW of radiated em power.

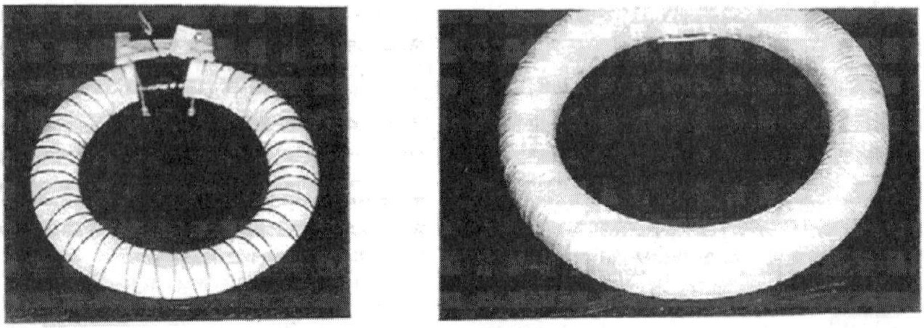

Fig. 4 – Toroid Configurations Tested

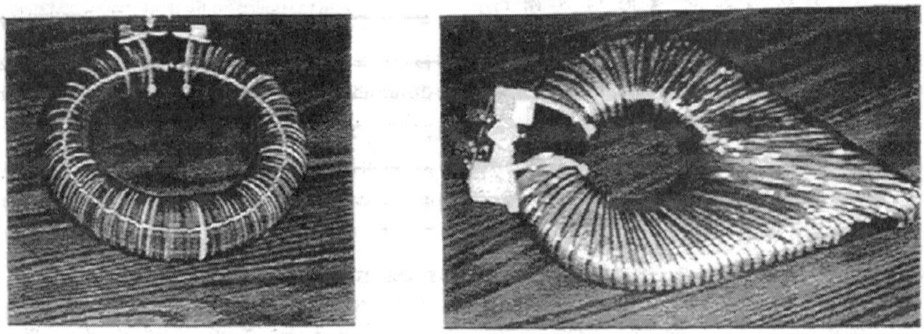

As in the first test series, resonant conditions (revealed by reversal in signal phase and rise in signal amplitude) were searched for at all ac frequencies between 400 KHZ and 110 MHZ. This was done for each toroid configuration for current opposing and current adding modes of operation. Equipment used for the resonance sweeps was an HP 4193 vector impedance analyzer. Figure 5 shows part of the test set-up for detecting resonant modes for each toroid configuration and each operating mode.

Although resonances were detected throughout almost the entire 400 KHZ to 110 MHZ frequency spectrum available at HCS, toroid radiation of significant power was only achievable in the 1.0 to 20 MHZ range. Resonant frequencies selected for measuring field characteristics of each toroid were therefore within this range. Selected resonant frequencies for the large diameter toroid were 2.36 and 17.30 MHZ for current-adding and current-opposing modes of operation, while those for the medium diameter toroid were 2.36 and 18.30 MHZ. Selected resonant frequencies for the tear drop toroid was 5.66 and 3.94 MHZ for current-adding and current-opposing, while the current-opposing, resonant frequency selected for the loosely wound toroid was 19.70 MHZ.

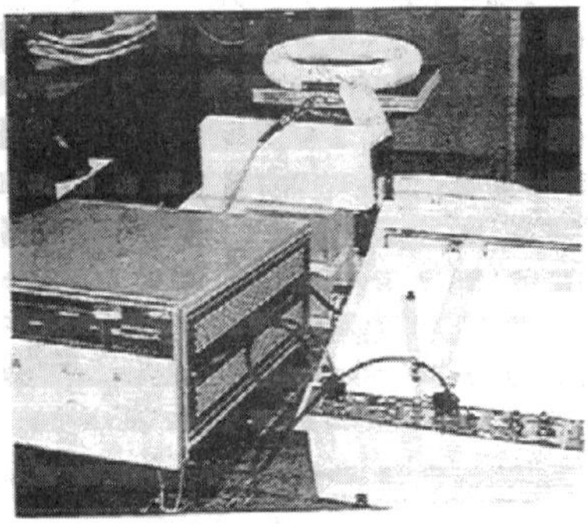

Fig. 5 – Resonance Sweep Set-Up

Toroid Field Intensity Measurements

Magnetic field intensity was measured out to 50 cm from each toroid center, and along the upper and lower surface of each toroid as well. For an applied power of 10W, the magnetic field component of each toroid's radiation was measured by a small magnetic pick-up coil shown in Figure 6, which converted the actual magnetic field intensity into an equivalent electric field strength (in micro volts per meter).

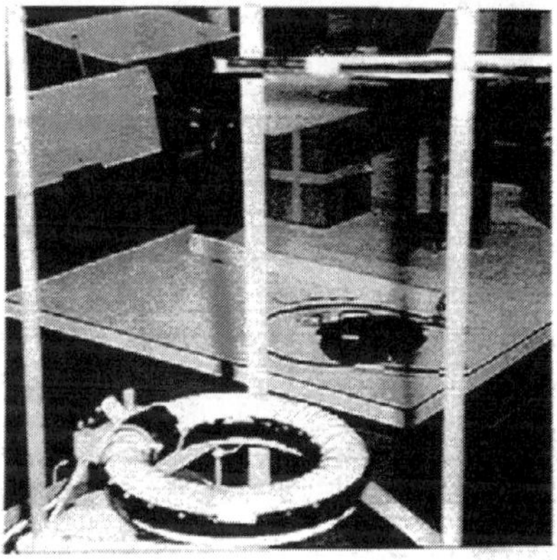

Fig. 6 – Magnetic Field Probe

Variation of the large diameter toroid's field strength with range (out to 10 meters) was measured with various types of antennas outside the HCS facility with the test set up as indicated in Figure 7. Data consistent with expected near-field signal strength variation with range was measured when the toroid was radiating in a current-adding resonant mode at 1.20 MHZ. But measurements in a current-opposing resonant mode were anomalous - in that no significant signal strength variation with range was detected.

Search For Gravity Modification

Final toroid testing activity was searching for gravitational field modifications within the specially conditioned em field regions surrounding toroids radiating at resonant frequencies. Gravitational disturbances were searched for with a "Prospector Model 420" gravitometer, manufactured by W. Sodin Ltd, which is capable of detecting changes as small as one-millionth of one percent of ambient gravity. This gravitometer's stainless steel shell and aluminum base does not provide complete magnetic field shielding. But its dewar-enclosed, all-quartz mechanical balance system is not influenced by ordinary em emissions. Unfortunately, preceding test activities took longer than expected, leaving time to search for gravity modification for only one of the toroids (the large diameter one) at only one location with respect to the gravitometer. The limited time remaining also required a very rapid toroid/gravitometer set up. This was achieved by the positioning shown in Figure 8.

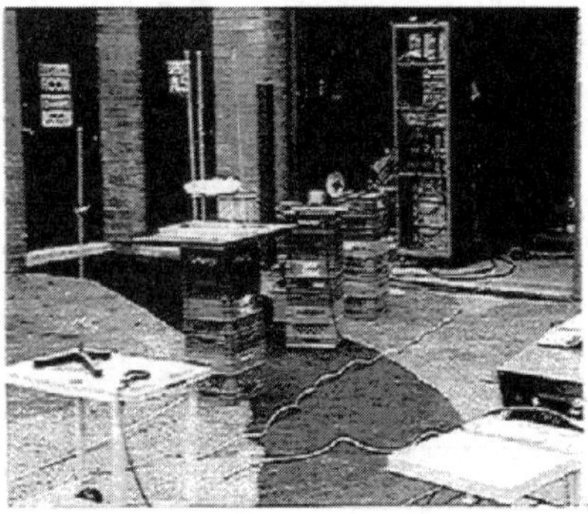

Fig. 7 –Toroid Range Test Set-Up

Fig. 8 – Toroid/Gravitometer Set-Up

TOROID TESTING RESULTS

Resonant frequencies between 400 KHZ and 110 MHZ were obtained for each toroid. And, for the purposes of mapping magnetic field intensity in each toroid's vicinity, one resonant frequency was selected for each operating mode for each toroid. Field intensity out to 10 m from the large diameter toroid was also measured together with the influences of magnetically shielded structures on its field intensity. Finally, the effect of large diameter toroid field intensity on gravity modification was explored. The results of these efforts are summarized in the following sections.

Toroid Resonance Determination

Resonant frequencies for current-adding and current-opposing operating modes were obtained for each toroid. Figures 9 and 10 show examples of the resonances obtained for the large diameter and tear-drop toroids throughout the 400 KHZ-110 MHZ radio frequency spectrum available at HCS.

Toroid Field Patterns

Magnetic field intensity variation with radial distance for the 3 circular toroids was similar with intensity maximizing near the inner surface of each toroid's ring. And, as would be expected, intensity diminished rapidly with increasing distance above and outside each toroid. Figure 11a and 11b show no definite trend with respect to the influence of toroid diameter. Higher magnetic field intensity is achieved by the smaller diameter toroid in a current-opposing mode of operation while higher magnetic field intensity is achieved by the larger diameter toroid in a current-aiding mode. Figure 11c shows a definite trend - with increased windings over a given toroid geometry resulting in increased magnetic field intensity.

Significant focusing of the electromagnetic energy radiated from the asymmetrical "tear drop" toroid was accomplished. Figure 11d shows that magnetic field intensity is enormously greater at given distances forward of the center of the toroid's hole than for the same distances aft of the hole center. Figure 9d also shows a top and front view of the tear drop toroid field pattern for a given magnetic field intensity. It is seen that more electromagnetic energy is focused into the forward direction than into the aft or side directions and that the toroids flattened shape (its reduced thickness) causes less radiation to be dissipated in directions transverse to the toroid plane.

One interesting discovery was formation, in the circumferential direction, of standing em waves along the upper and lower surfaces of transmitting toroids. No standing wave measurements were made on the loosely wound toroid. However, numbers of magnetic field peaks and nodes measured circumferentially on the top and bottom surfaces of the other circular toroids were 8 for the medium diameter toroid and 10 for the large diameter one. And at least 6 magnetic field peaks and nodes were measured on the top and bottom surface of the tear drop toroid.

Attenuation of Toroid Field Intensity

Field strength attenuation with range from the center of the large diameter toroid is shown in Figure 12. It was about as expected when radiating at a resonant frequency of 1.2 MHZ in a current adding mode of operation, with a steep signal drop (greater than $(1/r)^3$ out to about 1.0 meter and with an expected near-field $(1/r)^2$ variation between 3 and 10 meters from the toroid. But measurements made with the toroid radiating at a resonant frequency of 17.3 MHZ in a current opposing mode indicated no significant variation in signal strength at distances 3 to 10 meters from the toroid. After continual measurements and re-measurements with various types of antennas, we have no definitive explanation for lack of signal strength reduction with increasing range - other than the possibility of operating slightly off resonance and a significant drop in signal strength.

Additional anomalous behavior may also have been observed for the large-diameter toroid - in that almost identical signal strength was measured at a given location and distance from the toroid when radiating in free space and when radiating from within a magnetically shielded (mu metal) enclosure. These results are considered inconclusive because stray signals were detected from power supply leads which were outside the shielded enclosure. Since measured signals were almost identical for both shielded and un-shielded conditions, and since it is unlikely that almost all of the measured free space signal was from the power supply leads, it is conceivable that some of the toroid's em wave energy was propagated through the magnetically shielded mu metal walls.

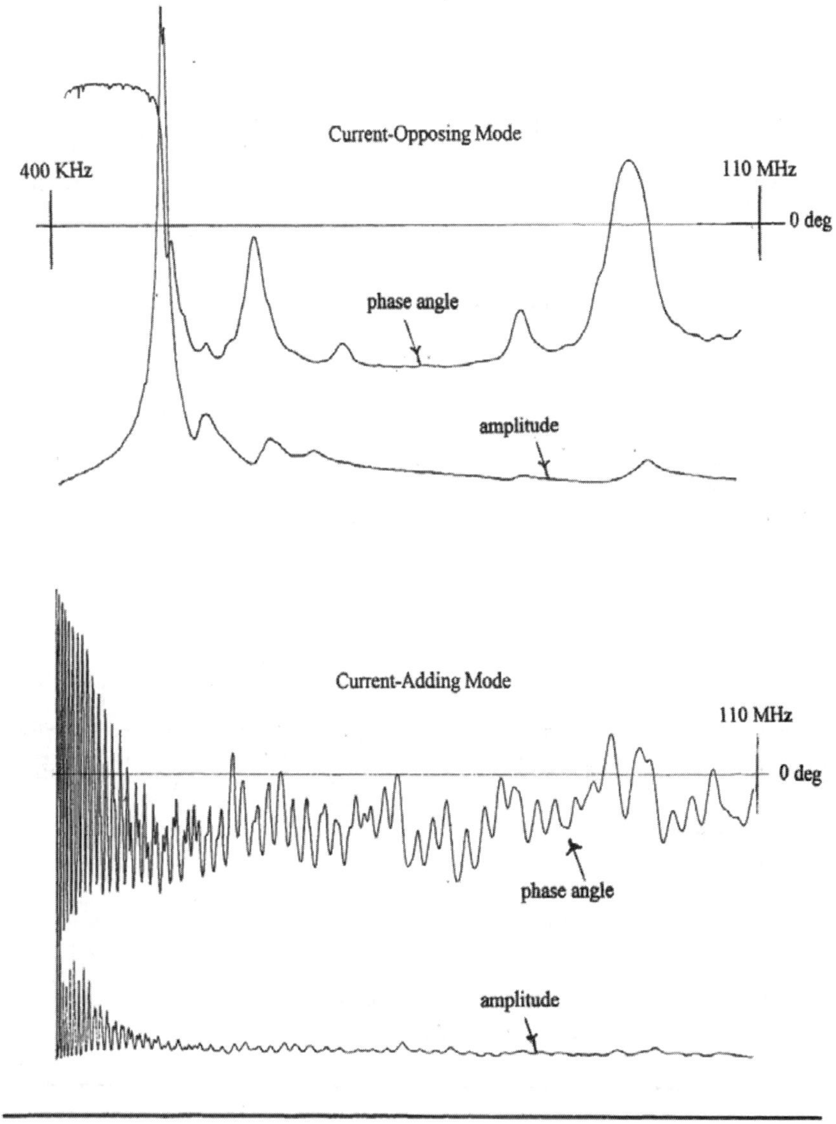

Fig. 9 – Resonances for Large Diameter Toroid

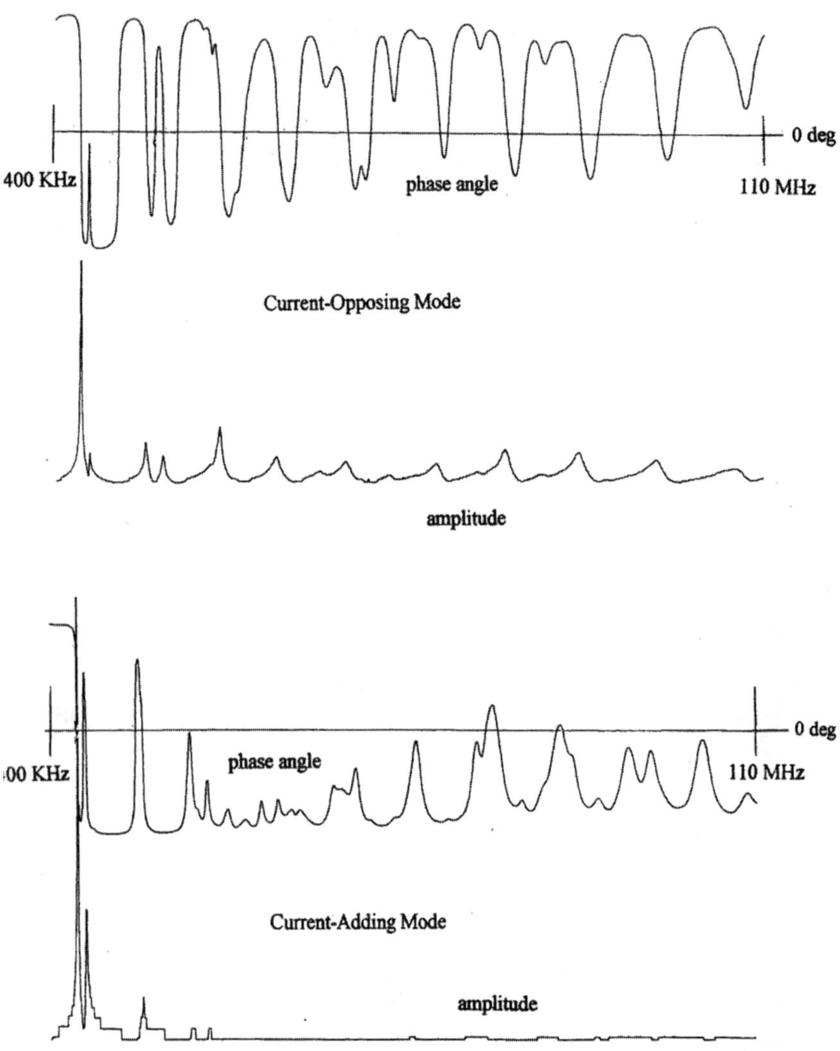

Fig. 10 – Resonances for "Tear-Drop" Toroid

Fig. 11 – Magnetic Field Patterns in Toroid Vicinity

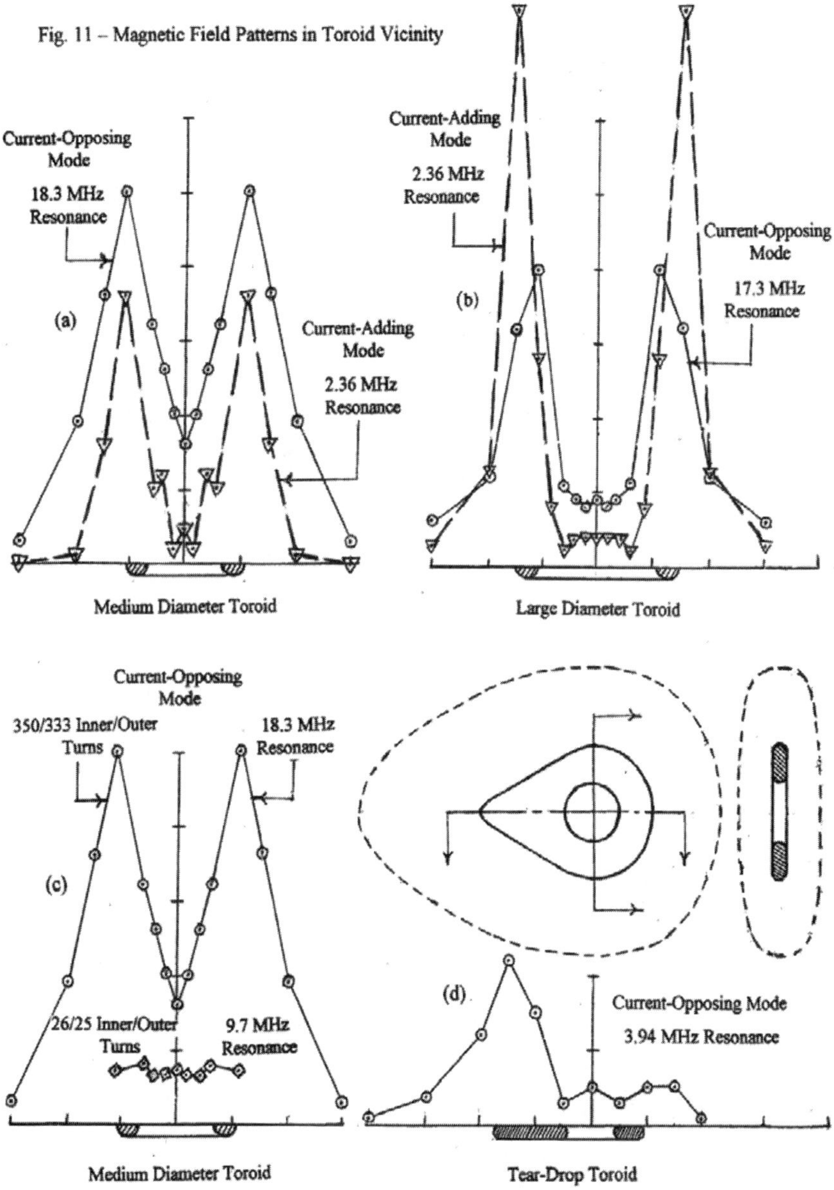

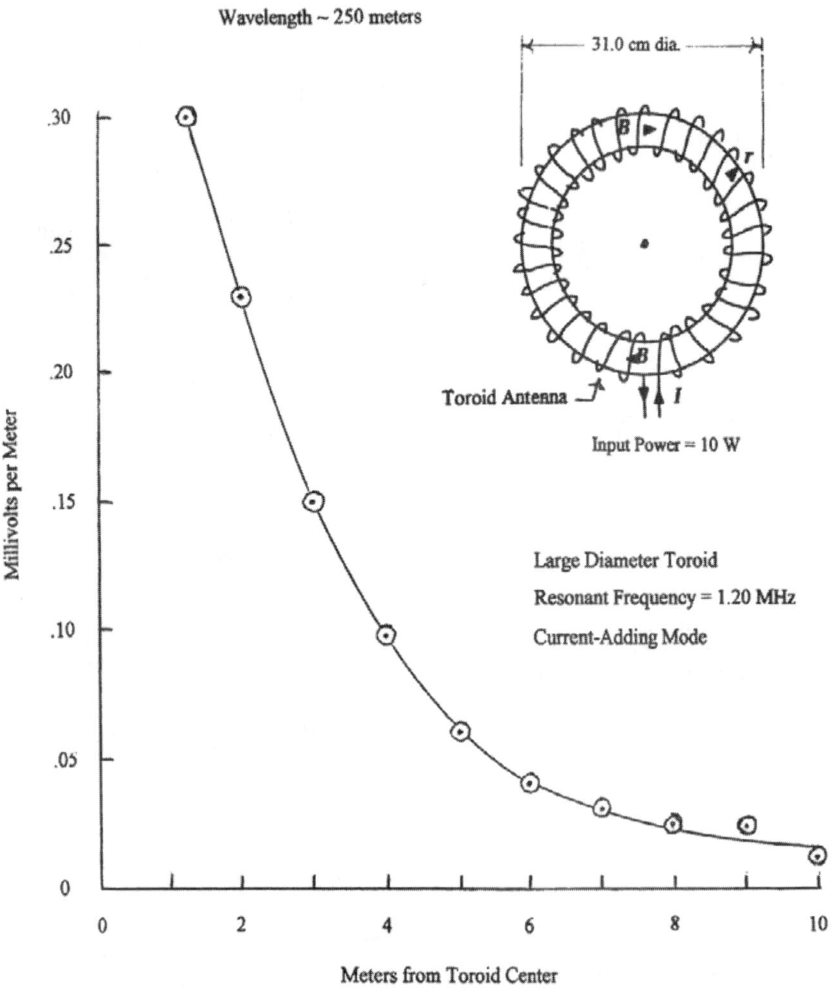

Fig. 12 - Attenuation of Toroid Signal with Range

Search for Gravity Modification

The possibility of gravity modification in the vicinity of transmitting toroid antennas was briefly investigated by use of the Prospector 420 gravitometer and the large diameter toroid radiating up to 0.5 KW of average power at the resonant frequencies associated with current-aiding and current-opposing operation modes. For these powers and operating modes, no discernable gravity modification was detected for the single toroid/gravitometer positioning that time allowed.

As previously mentioned, time limitations required a rapid toroid/gravitometer test set-up which resulted in the gravitational mass being located in a magnetic field region whose intensity was subsequently found to be much less than magnetic field intensity existing in other locations. For the current opposing mode of operation, measured magnetic field intensity at the gravitometer test mass location was only about 15 percent of the maximum intensity measured near the toroids inner diameter. And toroid magnetic field intensity at the gravitometer test mass location was only about 2 percent of the maximum measured magnetic field intensity for the current-adding mode.

One conceivable reason for non-discernable gravity modification is, of course, 5 to 50 times less em field intensity at the single location probed by the gravitometer, as compared to locations of maximum intensity. But another reason could be dissimilarity in field topologies associated with toroid em emanations and gravity. And still another reason could be enormous possible differences in the frequencies and wavelengths characterizing gravitational fields and those that characterize electromagnetic fields created by transmitting toroid antennas.

SUMMARY AND CONCLUSIONS

Although interesting phenomenon are associated with em fields created by alternating current flowing at resonant frequencies through toroid coils, no discernable gravity modification (caused by coupling of these fields with those of gravity) was detected.

Interesting electromagnetic phenomenon were: (a) standing em waves along toroid surfaces; (b) em wave energy focused into more intense beams by asymmetrical toroid shapes; and (c) possibly, em wave propagation through magnetically shielded enclosures.

There might have been increased probability of detecting a discernable gravity modification if there had been time for gravitometer measurements in regions where toroid field strength was much greater.

The possibility of anomalous wave propagation should be confirmed or refuted by re-testing the large diameter toroid within a magnetically shielded structure that encloses both the toroid and its power leads.

Zero gravity modification within the radio frequency em fields surrounding transmitting toroid antennas should be confirmed by a gravitometer search throughout the entire vicinity of the large diameter toroid.

ACKNOWLEDGMENTS

This experimental work was motivated by theoretical work by Dr. Terence Barrett with respect to specially conditioned em fields created by transmitting toroid antennas. Dr. Barrett also contributed useful suggestions as to recommended toroid configurations and test procedures. The most recent experimental work was conducted with the considerable assistance of Mr. Blair Cleveland, who was intimately involved in every aspect of the test preparations conduction and data gathering portions of this work.

REFERENCES

(1) Yang, C.N., "Gauge Theory", McGraw-Hill Encyclopedia of Physics, 2nd Edition, p483, (1993)

(2) Barrett, T.W., "Electromagnetic Phenomenon not Explained by Maxwell's Equations", Essays on Formal Aspects of Electromagnetic Theory, p6, World Scientific Publ. Co., (1993)

(3) Barrett, T.W., " Toroid Antenna as Conditioner of Electromagnetic Fields into (Low Energy) Gauge Fields", Proceedings of the: Progress in Electromegnetic Research Symposium 1998, (PIERS '98) 13-17 July, Nantes, France (1998)

(4) Froning, H.D., Barrett, T.W., "Theoretical and Experimental Investigations of Gravity Modification by Specially Conditioned EM Radiation", Space Technology and Applications International Forum 2000, Editor: Mohamed S. El-Genk, Published by the American Institute of Physics (2000)

14th AIAA/AHI Space Planes and Hypersonic Systems and Technologies Conference AIAA 2006-8100

Combining MHD Air breathing and Aneutronic Fusion for Aerospace Plane Power and Propulsion

H. David Froning
PO Box 180, Gumeracha, South Australia, 5233, Australia

An aerospace plane concept that combines the extraordinary propulsion and power generating capabilities of MHD air breathing engines and aneutronic (neutron-free) fusion rocket engines is summarized. This vehicle, using liquid hydrogen as coolant, fuel, and working fluid, for power and propulsion, generated: air breathing propulsion from takeoff to Mach 7; MHD air breathing propulsion and power from Mach 7 to Mach 14; combined MHD airbreathing and aneutronic fusion power and propulsion for Mach 14 cruise; or for fusion power and propulsion from Mach 14 to orbital speed. This resulted in a reusable air breathing SSTO launch vehicle with less than 1/3 the propellant of chemical airbreathing designs. This enabled rapid space access with aircraft-like operations and takeoff weights.

I. Introduction

Recent Air Force and NASA air breathing and rocket propulsion developments have focused on two-stage-to-orbit (TSTO) vehicles to improve military and civil access to space because propulsion and airframe technology developments for more revolutionary single-stage-to-orbit (SSTO) vehicles are not deemed achievable in the foreseeable future. In this respect, studies by Murthy and Froning[1] indicated that SSTO air breathing vehicles powered by "clean"aneutronic fusion rocket reactions would have about ½ the takeoff mass and 1/3 the propellant consumption of chemical air breathing SSTO vehicles. And Froning and Bussard, in a study for NASA[2], estimated about 17 years to develop aneutronic fusion to a NASA Technical Readiness Level (TRL) of 7. Since this 17 year span measured from the current epoch would enable start of full-scale vehicle engineering development in about 2025,the Air Force Research Laboratory (AFRL) initiated a study performed in 2003 and 2004 by Flight Unlimited to determine what a revolutionary military SSTO aerospace vehicle might be like if it could be powered and propelled by aneutronic fusion and commence development in about 2025. This study was performed with the assistance of the University of Illinois Fusion Physics Laboratory. The military SSTO vehicle work is described in some detail in Froning[3], and the fusion work inThomas[4, 5]. Both works are summarized in less detail in this paper.

II. MHD Air breathing and Fusion Rocket Vehicle Concept

The general requirement for the vehicle that will be described, is to :(a) provide the US Air Force with global-reach and rapid space-access with military aircraft-like vehicles that are no heavier than Air Force combat and transport aircraft such as the B1-B, B-2 and C-17; (b) deliver up to 18t of payload to space; and (c) be capable of commencing full-scale development in about 2025. Aircraft-like operations required aneutronic fusion power and propulsion to eliminate destructive neutron emissions and residual radioactivity. "Dense Plasma Focus"(DPF) fusion reactors were deemed by AFRL to show the most promise of achieving the plasma dynamics needed for aneutronic fusion and of having the lightness and compactness to achieve needed power and thrust-to-weight. Rocket and turbine based combined cycle air breathing engines were considered for accelerating the vehicle to Mach 14 - the highest possible fusion system ignition speed and altitude (which also minimizes fusion rocket thrust and mass). MHD power generation is used during Mach 7-14 air breathing flight because it enables the hundreds of megawatts of needed electrical power for DPF fusion system ignition. The DPF fusion system provides propulsion and power at speeds of Mach 12 or greater. And, a thrust vectoring chemical rocket system provides additional thrust and control any time during vehicle flight. The total consumption of liquid oxygen oxidizer depends upon whether Rocket or turbine-based air breathing propulsion is eventually selected. But liquid hydrogen, used as: regenerative coolant; working fluid; and propellant; constitute most of the consumed fluid during air breathing and fusion rocket flight.

* Visiting Lecturer, University of Adelaide, AIAA Associate Fellow

III. MHD Air breathing Power and Propulsion

Enormous amounts of electrical current are created within air breathing engines by very strong MHD (J x B) interactions within ionized and magnetized airflow at hypersonic (Mach 7 to Mach 14) flight speeds. Such current is a consequence of flow-slowing J x B interactions within airflow that is ionized to about 10^{13} electrons/cm^3 and subjected to magnetic fields of about 7 tesla. This current is: extracted from airflow by electrodes; conditioned to needed power and voltage within an MHD generator; and distributed to appropriate vehicle subsystems, - which, of course, include the MHD flow ionizing and magnetizing components. MHD components are located around and embedded within engine walls, and significantly increase air breathing engine mass. They include: superconducting coils-to create magnetic fields; electron beams for flow ionization; and electrodes to extract MHD-created current.

Figure 1, taken from an MHD air breathing aerospace plane study for NASA by Chase[6], shows a typical embodiment of MHD elements, such as "saddle magnets" (built up from superconducting coils) wrapped around the air-breathing engine walls. Also shown is flow path direction and the direction of the B-field generated by the magnets.

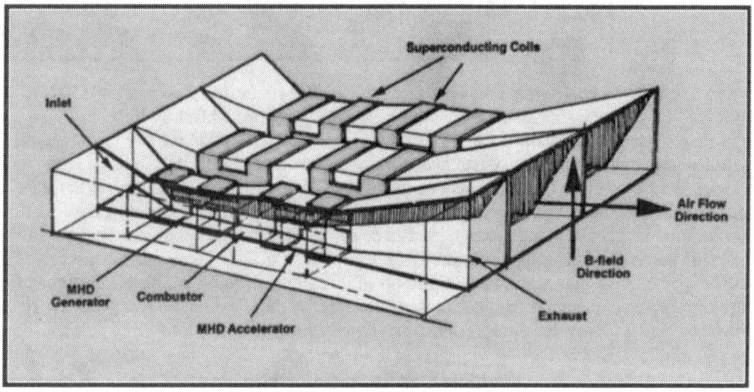

Figure 1. Typical MHD Airbreathing Engine Systems

The enormous amount of MHD-generated electrical power extracted from hypersonic airflow (of the order of 1.0 gigawatt) is vital for providing the large amount of electrical power needed for high altitude ignition of the DPF fusion system at Mach 12. After fusion system ignition, fusion energy is immediately injected into the air breathing engine flow by electron beams – where intense ohmic heating of nozzle airflow. increases air breathing thrust with negligible addition of mass. CFD computations indicate that this energizing of airflow also cause higher pressures to act along the vehicle aft-expansion surface to significantly decrease vehicle drag. And this increased thrust and decreased drag results in more than a 2-fold increase in effective air breathing ISP between Mach 12 and Mach 14.

IV. Aneutronic DPF Fusion Power and Propulsion

Figure 1 shows the DPF propulsion and power system selected for the 2025 vehicle. It fuses boron 11 and hydrogen nuclei to yield 3 helium 4 ions and 5.68 MeV of energy during each fusion. Very light shielding is needed to shield against soft x-rays, but elimination of neutron emissions eliminates neutron impacts upon vehicle structure and resulting radioactivity. And this enables aircraft-like maintenance and servicing of vehicle engines and structure.

The DPF consists of a coaxial electrode configuration in which a capacitor bank is discharged across the electrodes, ionizing injected nuclear fuels into a gas that forms into a plasma sheath. Current flows radially between electrodes, inducing an azmuthal magnetic field. "Rundown" then occurs wherein the plasma sheath is accelerated by J x B MHD interactions down the length of the anode as it proceeds rearward within the annular volume inside the concentric cylinder of the DPF. And finally, "Focus", wherein the accelerated plasma sheath collapses towards the central axis of the anode, compressing to fusion temperatures and pressures by enormously strong magnetic fields within a small pinch region immediately downstream of the DPF device. The released fusion energy then accelerates and expands injected hydrogen to high exhaust velocity and specific energy within a magnetic nozzle.

This aneutronic DPF fusion cycle is repeated about 10 times every second – with injected charges of nuclear fuel being vaporized by electrical pulses that are approximately: 400 kV in voltage; 20 MJ in energy; 200 MW in power.

Figure 2. Typical Plasma Focus Device

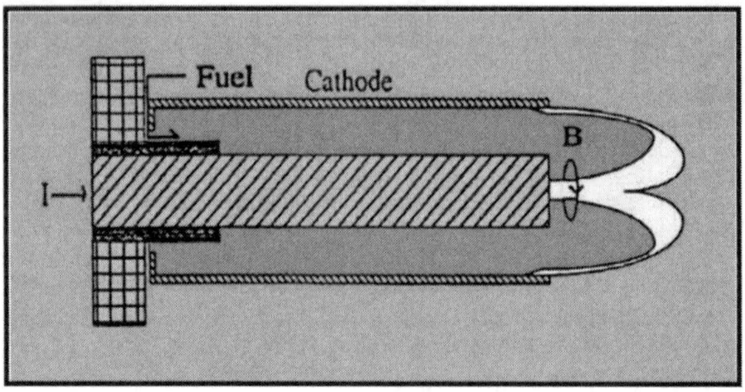

An attractive fusion vehicle design required fusion rocket ISP (thrust/fuel flow rate) of 1500s to 2000s for thrust levels in the 750 kN to 500 kN range, enabling vehicle acceleration from Mach 14 to Mach 25 speed in about 15 minutes of time. Thomas[4] indicate that the energies and power levels needed to achieve these thrust levels with satisfactory fusion power and propulsion system mass require capacitor energy storage/discharge capabilities would have to be about 15 kJ/kg. This is beyond current capacitor state of the art, but sources such as Sargeant[7] describe metalized capacitor developments that should yield capacitor efficiencies of 15-20 kJ/kg during the next 10 years.

Above the atmosphere, fusion must, supply needed power to vehicle subsystems, and Thomas[4] has identified an attractive MHD power generation approach. Here, nozzle expansion of the DPF plasma exhaust causes magnetic flux trapping between the plasma jet and a stator pick-up coil, and flux diffusion into both. Then rapid field compression transforms plasma kinetic energy into magnetic pressure, forcing field lines out of the stator ring, and inducing high voltage pulses propagated through appropriate transduction equipment for electric power distribution.

V. Vehicle Configuration Definition

MHD air breathing integrated reasonably well with DPF fusion, sharing some propulsive flow paths and having similar components- such as electron beams and superconducting magnets. The resulting vehicle, shown in Figure 3, had less than 40 % the takeoff mass and less than 20 % the propellant mass of an ordinary SSTO airbreathing plane.

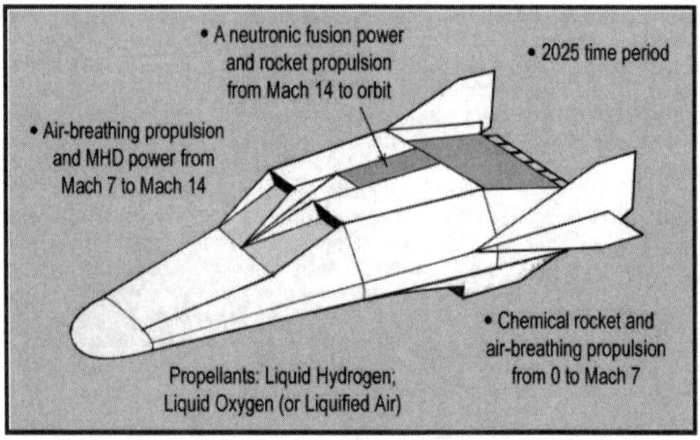

Figure 3. MHD Airbreathing and Fusion Rocket Aerospace Plane

Since the vehicle's takeoff mass is only 175 t - less than current Air Force bombers and long range jet airliners - it could use available runways and much of the other infrastructure at many military and/or civil airports in the USA.

Significant improvement in structural efficiency (vehicle surface area / structural+TPS mass) was assumed to be achievable by 2025 with use of high temperature (~950 deg C) metal matrix materials – such as rapidly solidified titanium and silicon carbide – over most of the vehicle. This resulted in about 25 percent improvement in structural efficiency over that of air breathing vehicles studied and documented in reports such as "NASA Access to Space Final Report Volume 1, July, 1993". A 10 percent improvement in superconducting magnet mass over that used in a previous NASA MHD study by Chase[5] was assumed. But recent YBCO and carbon nanotube materials work by Putnam[8] and Chapman[9] indicate that even lighter superconducting magnet weights than those which were assumed in this study may be possible. Shown in Table 1are estimated masses of the various subsystems of the 2025 vehicle

Table 1. Estimated Masses of the 2025 Vehicle

Takeoff Weight	174 t
Payload	18 t
Dry mass	82 t
Structure	27 t
Systems	12 t
Propulsion (chemical 22t, fusion 21t)	43 t
Propellant (airbreathing 47t, rocket 27t)	74 t

VI Critical Issues for DPF Fusion Power and Propulsion

Although many formidable challenges are associated with MHD air breathing power and propulsion, most technical uncertainties are associated with development of the aneutronic DPF fusion power and propulsion. system. One uncertainty is achievement of the high ignition temperatures needed for fusion of protons and Boron11 ions with acceptable input energy and power. Bremsstrahlung radiation from high temperature excitation of the electrons within greated plasma can result in excessive energy dissipation during fusion power build-up unless electron temperatures are lower than ion temperatures. This appears possible for the non-equilibrium dynamics of plasmas created within DPF devices, and for the narrow non-Maxwellian energy distributions within the DPF plasmas - which minimize wasteful heating of particles that have a fairly low probability of participating in the fusion reactions.

An associated challenge is retaining a significant amount (at least half) of the bremsstrahlung radiation inevitably emitted during aneutronic fusion - by reflecting and re absorbing it within the inner flow. Single-film Hohlraum-like reflecting cavities are under study by Murkami[10] and thin film multi-layer reflectors by Joenson[11]. . Neither concept has yet to be found entirely satisfactory, but they are believed capable of about 10 reflections of photons before being lost. If so, power and flux densities of about 10^5 J/cm^2 and 10^6 W/cm^2 can be reabsorbed in flow.

Critical DPF engineering issues include: conversion of very high fractions of fusion power into electrical power; and conversion of a high fraction of generated electrical power into useful jet power in the rocket exhaust.

References

[1]Murthy, S.N, Froning, H.D, "Combining Chemical-Electric-Nuclear Propulsion for High-Speed Flight", Proceedings of the International Society of Airbreathing Engines (ISABE), Nottingham, England, 1991

[2]Froning, H.D, Bussard, R.W, "Roadmap for QED Fusion Engine Research and Development", NASA Purchase Order H28027D for the George C. Marshall Space Flight Center, Huntsville, AL, USA, 1997

[3]Froning, H.D, Czysz, P.A, "Advanced Technology and Breakthrough Propulsion Physics for 2025 and 2050 Aerospace Vehicles", Proceedings of the Space Technology and Applications Forum (STAIF2005), Albuquerque, NM. USA, 2006

[4]Thomas, R, Yang, Y, Miley, G.H, Mead, F.B, "Advancements in Dense Plasma Focus (DPF) for Space Propulsion", Proceedings of the Space Technologies and Applications Forum (STAIF 2005), Albuquerque, NM, 2005

[5]Thomas, R., Miley, G.H., Mead, F.B., "An Investigation of Bremsstrahlung Radiation in a Dense Plasma Focus (DPF) Propulsion Device", Proceedings of the Space Technologies and Applications Forum (STAIF2007), Albuquerque, NM, 2006

[6]Chase, R.L, et al, "An Advanced Highly Reusable Space Transportation System", NASA Cooperative Agreement No. NCC8-104 Final Report, ANSER, Arlington, VA 1997

[7]Sargeant, W.J, et al, IEE Transactions on Plasma Science, **26**, p 1368, 19998

[8]Putman, P, et al, "Superconducting Permanent Magnets for Advanced Propulsion Applications", Proceedings of the Space Technologies and Applications Forum (STAIF2005), Albuquerque, NM, 2005

[9]Chapman, J.N, et al, "Flightweight Magnets for Space Applications Using Carbon Nanotubes", AIAA 2003-330, 41st Aerospace Sciences Meeting and Exhibition, Reno NV, 2003

[10]Murakami, M., et al, "Indirectly Driven Targets for Inertial Confinement Fusion", Nuc Fusion, **31**, p 1315,1991

[11]Joenson, K.D, "Broad-band Hard X-Ray Reflectors", Nuc. Inst. and Methods in Phys. Research B, **132**, p 221,1997

Available online at www.sciencedirect.com

SciVerse ScienceDirect

Physics Procedia 38 (2012) 77 – 86

Physics Procedia

Space, Propulsion & Energy Sciences International Forum - 2012

Specially conditioned EM fields to reduce nuclear fusion input energy needs

H. David Froning Jr[a] Terence W. Barrett[b] George H. Miley[c] *

[a]P.O. Box 1211, Malibu CA 90262 USA [b]BSEI, 1453 Beulah Rd. Vienna VA 22182 USA [c]912 West Armory, Champaign, IL 61821 USA

Abstract

Ordinary electromagnetic (EM) fields possess relatively simple U1gauge symmetry, and their angular momentum is analogous to that of spin1 particles whose likecharges attract and unlike charges repel. This manifests in coulomb repulsion between free electrons or ions and coulomb attraction between free electrons and ions. By contrast, angular momentum of SU(2) fields that describe the shortrange Weak Nuclear Force in atomic nuclei is analogous to that of spin2 particles whose likecharges attract. So, free ions that enter such small SU(2) field regions attract each other until their separation becomes so small that their fusion occurs. In this respect, Barrett has derived EM fields with the same SU(2) gauge symmetry and spin2 angular momentum as SU(2) matter fields in atomic nuclei. It is conceivable, therefore, that SU(2) EM fields might cause fuel ions inside nuclear fusion reactors to attract (rather than repel) each other. This paper, therefore, explores the possibility of SU(2) EM fields reducing the electrical compression energies these SU(2) EM fields must exert on fuel ions before fusion of the ions by the SU(2) matter fields of the weak nuclear force then occurs. A specific conditioning of U(1) EM field energy into SU(2) EM field energy was selected; a given type of fusion was assumed; and preliminary, parametric estimates of input electrical energy reductions were made.

PACS: 04.30, 41.20Jb, 41.60Bg Keywords:, A fields, aneutronic fusion, A vector potential, Inertial Electrostatic Confinement Fusion, , SU(1) fields, SU(2) fields

* Corresponding author. Tel.: +61883891949; fax: +61883891945. E-mail address: froning@infomagic.net

1. Specially conditioned electromagnetism

Ordinary electromagnetic (EM) fields possesses U(1) Lie group symmetry and angular momentum that is analogous to the angular momentum of spin1 particles like free ions, whose likecharges repel. By contrast, the matter fields associated with the weak nuclear force in atomic nuclei possess higher SU(2) symmetry; and their angular momentum is analogous to that of spin2 particles (like any ions confined in these fields – ions which attract each other until their fusions occur). In this respect, SU(2) EM fields with angular momentum analogous to spin1 particles have been derived by one of us (Barrett). It is, therefore, conceivable that SU(2) EM fields in reactors (like SU(2) matter fields in nuclei) could cause ions inside them to attract each other. If so, fuelcompressing, ionattracting SU(2) EM fields inside nuclear reactors would face less repulsive resistance and, thus, need to expend less EM field energy to accomplish fusion. In most cases, EM radiation fields are correctly and adequately described by the classical Maxwell equations which is a theory of U(1) symmetry form. However, in special situations, specially conditioned EM radiation fields can be produced that require an extension of Maxwell theory to higher symmetry. For such situations, Barrett [1] has used topology, group and gauge theory to derive specially conditioned SU(2) EM radiation fields. Even more complex EM fields are describable by more complex symmetry groups like SU(3) and even higher SU(5) groups. However, this paper only addresses SU(2) EM fields.

1.1 Maxwell equations for conventional and specially conditioned EM fields

Using group theoretic methods, EM radiation fields of SU(2) symmetry can be created by special conditioning of conventional U(1) EM fields. U(1) EM fields are described by Maxwell's equations in Table 1. They describe electric field strength (**E**), magnetic flux density (**B**) and current density (**J**). The **E** and **B** fields of force can be related to a vector potential" (**A**) and a "scalar electric potential" (φ) that, themselves, are not physical. However, they are of mathematical convenience in U(1) EM field theory.

Table 1. The four Maxwell vector field equations for conventional U(1) symmetry electromagnetic radiation fields

Gauss' Law	$\nabla \bullet \boldsymbol{E} = \boldsymbol{J}_0$
Ampere's Law	$\frac{\partial \boldsymbol{E}}{\partial t} - \nabla \times \boldsymbol{B} - \boldsymbol{J} = 0$
Coulomb's Law	$\nabla \bullet \boldsymbol{B} = 0$
Faraday's Law	$\nabla \times \boldsymbol{E} + \frac{\partial \boldsymbol{B}}{\partial t} = 0$

$$E = -\frac{\partial A}{\partial t} - \nabla\phi, \quad B = \nabla \times A$$

In the SU(2) field theory, Barrett [1] shows that the potentials **A** and φ have actual physicality. Table 2 shows extended Maxwell equations that describe propagation of specially conditioned SU(2) EM fields. These Maxwell equations are described by tensor, rather than vector quantities. SU(2) Maxwell equations include **E** and **B** fields just as U(1) Maxwell equations do. But they also include added tensor field terms that include imaginary number i (viewed as either square root of 1 or as an orthogonal rotation occurring in x, y, z, ct spacetime) and electron charge (q). These added tensor field terms describe added **AxE** and **AxB** and **A•E** and **A•B** interactions (Barrett [1] on p 145147). All tensors (matrices) function as operators that obey noncommutative, nonAbelian algebra. So, unlike vector multiplication, the product **(A•B)** doesn't equal **(B•A)** and **(AxB)** doesn't equal **(BxA)** in the matrix algebra of SU(2) fields.

H. David Froning Jr et al. / Physics Procedia 38 (2012) 77 – 86

Table 2. The four Maxwell tensor EM field equations for conditioned SU(2) symmetry electromagnetic radiation

$\nabla \bullet E = J_0 - iq(A \bullet E - E \bullet A)$
$\frac{\partial E}{\partial t} - \nabla \times B - J + iq[A_0, E] - iq(A \times B - B \times A) = 0$
$\nabla \bullet B + iq(A \bullet B - B \bullet A) = 0$
$\nabla \times E + \frac{\partial B}{\partial t} + iq[A_0, B] = iq(A \times E - E \times A) = 0$

The Lorentz force (**F**) plays important roles in the plasma dynamics of many nuclear fusion processes. It arises from an electromagnetic interaction that involves **E** and **B** fields and the velocity (**υ**) of moving particles with a charge (e). Table 3 shows Lorentz force equations for U(1) EM vector fields in terms of magnetic vector potentials and electric scalar potentials. It also shows Lorentz force equations for SU(2) tensor fields, in terms of these vector and scalar potentials. SU(2) Lorentz forces contain extra terms that include these potentials. So, they can be of different magnitude and direction than U(1) Lorentz forces.

Table 3. Comparison of Lorentz force equations for U(1) symmetry and SU(2) symmetry EM fields from Barrett [1].

U(1) Lorentz Force	$\mathscr{F} = e\mathbf{E} + ev \times \mathbf{B} = e\left(-\frac{\partial \mathbf{A}}{\partial t} - \nabla\phi\right) + ev \times \left((\nabla \times \mathbf{A})\right)$
SU(2) Lorentz Force	$\mathscr{F} = e\mathbf{E} + ev \times \mathbf{B} = e\left(-(\nabla \times \mathbf{A}) - \frac{\partial \mathbf{A}}{\partial t} - \nabla\phi\right) + ev \times \left((\nabla \times \mathbf{A}) - \frac{\partial \mathbf{A}}{\partial t} - \nabla\phi\right)$

1.2 Example of the special conditioning of an ordinary U(1) EM field into an SU(2)symmetry EM field

An example of ordinary U(1) EM field energy being transformed into specially conditioned SU(2) EM field energy is described on pages 46 and 61 of Barrett [2]. This example uses a wave guide system paradigm to portray oscillating U(1) EM wave energy being transformed into SU(2) EM wave energy by phase and polarization modulation. Figure 1 shows a completely adiabatic system where oscillating wave energy: enters from the left; divides into 3 parts; is modulated and recombined; and exits from the right. One part of the input wave energy is unchanged, another part of the input wave energy provides phase modulation ($\partial\varphi/\partial t$) and then combines with an orthogonally polarized part that has passed through a "polarization rotator". Two orthogonally polarized oscillating wave forms (with one being the unchanged fraction of oscillation wave energy) result. The two wave forms are then superimposed at an output where they are combined into a single EM beam of emitted SU(2) radiation. Owing to phase modulation of one waveform with respect to the other, and their initial orthogonal polarization, the output SU(2) radiation is of continuously varying polarization during one cycle of wave oscillation. This is symbolized by the diagram in the lower right hand portion of Figure 1. This diagram represents timevarying E and Bfield polarizations that traverse continuously through: horizontallinear; rightverticalelliptical; rightcircular; righthorizontalelliptical; horizontallinear; lefthorizontalelliptical; leftcircular; leftverticalelliptical polarization during a polarization modulation period Δt. And, here: $\partial\varphi/\partial t$ = constant and $0 < \varphi < 360^\circ$.

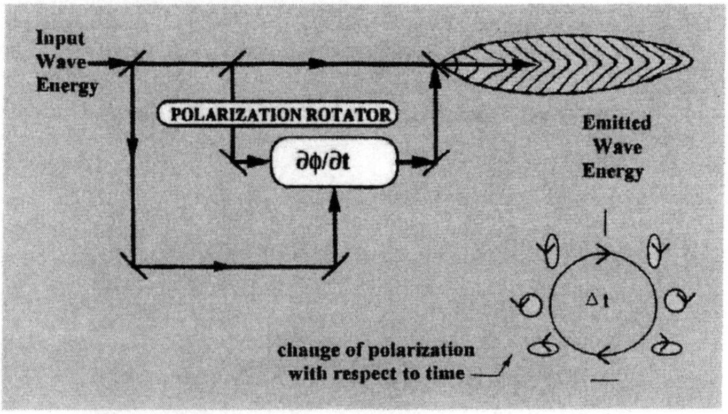

Figure 2, from Barrett [1] shows the magnitude and direction of the electric field E within a beam of SU(2) EM radiation during a cycle of its phase and polarization modulation. It is seen that magnitude and direction of the electric field can vary rapidly during one cycle of phase and polarization modulation and that many E field rotations can occur during a very short time and over a very short length of beam travel.

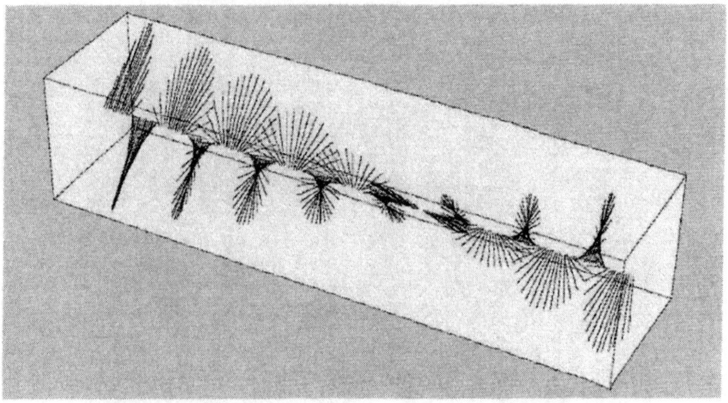

Figure 2. Rapid change in electric field vector direction and magnitude over a very short time and very short distance

Rapid electric and magnetic field rotations in SU(2) EM beams, combined with rapid change in field intensity result in different angular dynamics than that of U(1) EM beams with fixed linear or circular or elliptical polarization. And different angular dynamics of U(1) and SU(2) EM beams could possibly be reflected in repulsion of spin1ions in U(1) EM beams and attraction of spin2 ions in SU(2) EM beams.

H. David Froning Jr et al. / Physics Procedia 38 (2012) 77 – 86

2.0 Polarizationmodulated SU(2) EM fields for Inertial Electrostatic Confinement fusion systems

We explored the possibility of SU(2) EM fields lowering electrical input energy needs for fusion in certain "Inertial Electrostatic Confinement" (IEC) nuclear fusion systems. One IEC nuclear fusion system uses a central, negatively charged electrode (a grid) that draws many converging streams of positively charged ions from plasma injectors (RF ion guns) located around the reaction chamber periphery. Figure 3 shows such an IEC system at the University of Illinois accomplishing hundreds of fusions per second.

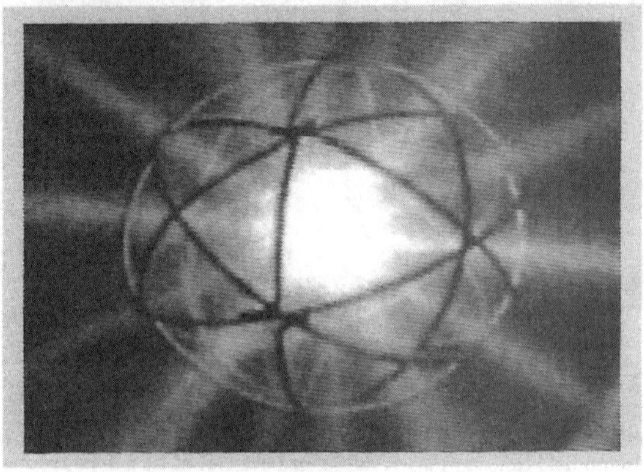

Figure 3. Operating IEC Fusion Reactor accomplishing hundreds of DD fusions per second at University of Illinois

Each radial ion stream in Figure 3 is emitted from a plasma generator that contains: nuclear fusion fuels; needed electrical power; and radiofrequency (RF) antenna discharges to transform fusion fuels into plasma and transform plasma into a focused ion beam. Figure 4 shows an IEC system installation at the University of Illinois with one of many plasma generators (one of many RF ion guns) that are needed.

Figure 4. IEC fusion reactor chamber with helicon plasma injector (RF ion gun) operated at the University of Illinois

Each circumferentiallydistributed injector contains a radio frequency (RF) antenna system that deposits pulsed energy into nuclear fuel material in the injector, transforming the material into flowing plasma. A magnetic field is also induced by current flow in windings. Such a plasma injector, developed by Miley at the University of Illinois [3], is shown in Figure 5. High power (13.5 MHz) discharges are caused by RF wave propagation through a coaxial copper helical resonator consisting of a singlelayer coil inside a copper shield. The RF system's helical coil antenna is magnet wire that is wound directly to a glass tube.

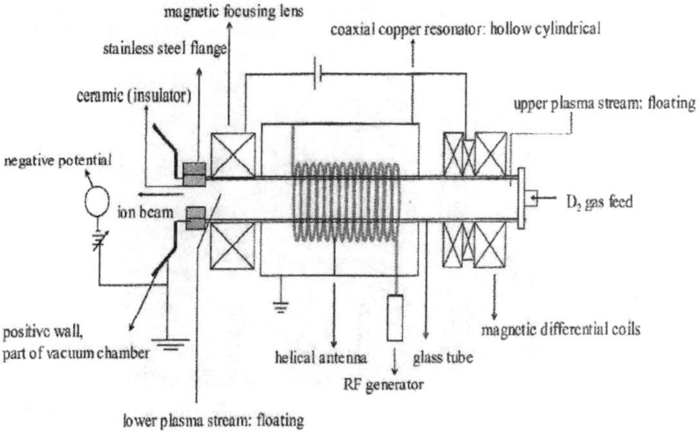

Figure 5. Typical schematic of one of the plasma injectors that are circumferentiallydistributed about an IEC reactor

Barrett's transforming of ordinary U(1) EM waves into SU(2) EM waves is extended to EM plasma waves emitted from plasma injectors. Windings and antenna elements of each injector would be modified to modulate both phase and polarization of the created SU(2) EM beam and favorably effect electrons and ions inside it. If SU(2) field content causes some ion attractions, coulomb resistance to applied EM input power will lessen, So, less input energy need be expended by ionattracting SU(2) EM fields in bringing ions close enough for SU(2) weak force matter fields in the ion nuclei, themselves, to cause their fusion.

We are now exploring possibilities for adding another set of helicon windings to the existing set. Such a circuit would "wiggle" RF discharges somewhat like an undulator (wiggler) in a free electron laser (FEL) accelerates and decelerates electrons. Electron accelerations and decelerations in a FEL can give off enormously complex and varied bursts of EM radiation as FEL undulator is 'played' like a very highly tuned musical instrument. Therefore, a critical issue is helicon windings (or selective driving of helicon windings) to provide control precision that may have to approach that of very large and complex FELs.

One concern is interaction of SU(2) ion beams from modified plasma generators with the U(1) electric field of the negativelycharged grid. Such interactions are not yet understood. Thus, it is possible that the grid's negativelycharged U(1) electric field could: (a) repel rather than attract ions in the SU(2) beam; or (b) quench the SU(2) field itself. One alternative might be to reverse the polarity of the grid. Another might be to remove the grid entirely and to let the mutual attraction of the merging nuclear fuel ions from the many converging SU(2) ion beams coalesce themselves naturally into a central region of increasing ion compactification. Hopefully, this would create a deep potential well of increasing ion attraction a deep well that would enable final fuel ion fusion by the SU(2) matter fields in the fuel ions, themselves.

3. Recommended early experiments involving SU(2) EM Field generation in a helicon plasma gun

Theoretical explorations of possibilities for ionattracting SU(2) EM plasmas will, of course, require much work. And early experiments, to augment and guide early analysis, are needed. Early experiments would involve modification of a helicon plasma generator for SU(2) ion beam emission. This need not involve expensive nuclear fuel plasmas. Instead, more ordinary, inexpensive plasmas from RF discharges in gases such as Argon or Xenon can be created in a wellinstrumented helicon plasma generator. Figure 6 shows an instrumented helicon plasma generator that is located at the Australian National University.

Figure 6. Helicon plasma generator in the Plasma Dynamics Laboratory of the Australian National University

Figure 7, illustrates some of the plasma properties that can be measured inside the Australian National University's helicon plasma generator shown in Figure 6. Typical measurements, described in [4] include plasma voltage and current mapping throughout the plasma generator chamber; ion speeds; voltage and energy variation (with high spatial and temporal resolution) within the beam. Such kinds of measurements and resolutions are needed to detect ionion attractions and ionelectron repulsions in SU(2) plasmas.

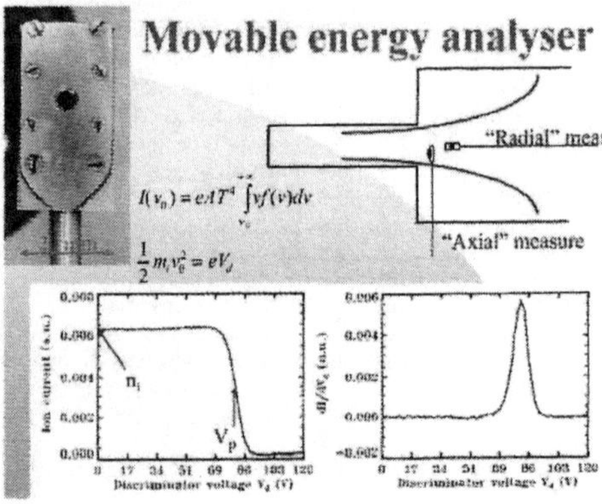

Figure 7. Energy analyzer that maps plasma properties and flows throughout the interior of the plasma generator

4. Preliminary analyses of potential reductions in fusion input energy needs with SU(2) EM fields

Figure 8, from Duncan [5], shows the effect of input electrical power (in thousands of electron volts) on the crosssectional areas of regions where ion fusions can occur. These areas are usually described in units called "barns", where a barn is $1.5 \times 10^{28} m^2$, the approximate area of a uranium nucleus. It follows that fusion crosssection sizes are associated with numbers of ion fusions and amounts of fusion power.

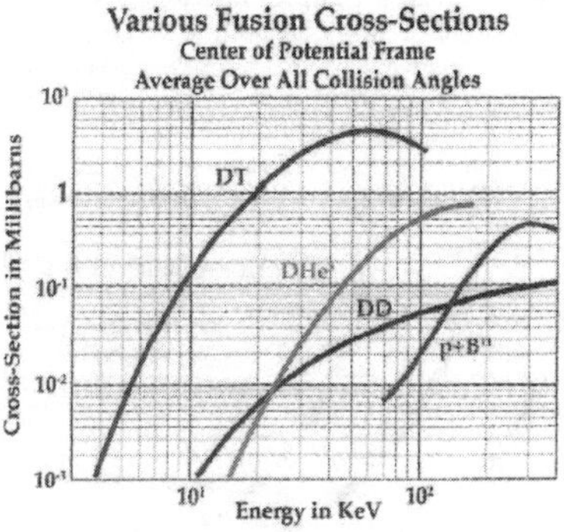

Figure 8. Crosssections of reaction regions where nuclear fusions occur as a function of input electrical field energy

Since we cannot yet claim that SU(2) EM fields can induce attractions between ions, it is certainly not possible to predict any input energy reductions these fields can provide. Possible input energy reductions, if any, would have to be ascertained by experiments with existing IEC reactors and ion guns. However, preliminary parametric explorations are possible with fusion cross section and input energy information from sources like Figure 8 and by assuming certain SU(2) ion attracting and electron repelling scenarios.

SU(2) EM field creation and its accomplishment of ion attracting states can be viewed as the product ($\alpha\beta$) of two efficiencies – with α: the efficiency in creating SU(2) EM field energy by phase and polarization modulation of input U(1) EM field energy in IEC ion guns; and β: the efficiency of created SU(2) EM field energy in mulling fuel ion repulsions in IEC fusion reactors. It follows that: (1α) is the fraction of input energy dissipated during the phase and polarization modulation processes that create SU(2) field energy; while (1β) is wasted field action from EM couplings that hamper achievement of ion attractions in SU(2) EM beams. Such couplings can begin as soon as RF discharges create electronion clouds and screening from intervening electrons hamper initial attractions of ions inside these clouds.

Increased input field energy force ions into closer proximity to increase probability and numbers of their fusions. And increased cross sectional areas reflect increased numbers of ion fusions. For example, Figure 8 shows a modest 4 KeV of EM input energy enabling only a modest amount of DT fusion inside a small (10^{3} millibarn) cross section. But, a 24fold increase in input energy to 200 KeV results in 10^{3} fold increase in fusion crosssection to 1 millibarn and a 10^{3} fold increase in fusion output energy. Thus, if ionattracting SU(2) EM fields could null all fuel ion repulsions within a 1.0 millibarn crosssectional area area by the time 4 KeV of EM field energy is spent, no more energy expenditure would be needed, For all ions in this region would now be attracting and moving ever closer until their fusion by the SU(2) fields of the weak nuclear force occurs at distances of about a 10^{3} millibarn area. Hence, 4 KeV of input

SU(2) EM field energy – if generated with negligible losses by polarization modulation of U(1) EM field energy would achieve the same output fusion power as 200 KeV of ordinary U(1) EM energy provides.

Figure 9 illustrates this. It indicates that if SU(2) EM fields could cause ion attractions over distances $10^{34}$to ^{15}10times longer than the 10cm range of the ionattracting, fusioncausing weak nuclear force, input energies could be reduced by factors of 10 to 20 for achieving a given DT fusion power. In this scenario, ionattracting SU(2) fields cause fuel ions to be drawn closed and closer by mutual attraction until their fusions are accomplished by the SU(2) matter fields associated with the weak nuclear forces in the ions themselves. Significant reductions in input energy are seen to be achievable, even for fairly poor field efficiencies (αβ of 1.0 % to 10 %.), if ionattractions in SU(2) EM fields can occur over distances that are $10^{24}$$10^{15}$times the 10cm distance where fuel ion fusion can occur. It must also be mentioned that Figure 9 assumes all fuel ions transition from repelling to attracting at the same time. But, of course, ions actually reach attracting state at different times, so ion resistance to compression is reducing long before all ions become attracting. Figure 9 calculated input energy is, thus, more than is actually needed.

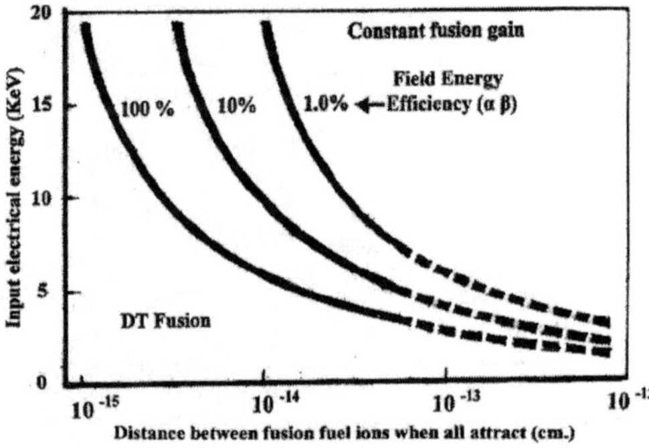

Figure 9. Influence of distance between fusion fuel ions when all become attracting on reduction in input energy to achieve a given DT fusion output power. Dashes denote extrapolation of Figure 8 reaction crosssection information.

A very clean nuclear fusion reaction is fusion of Hydrogen and Boron 11 nuclei. Figure 10, from [5], shows this reaction culminating in: the energy of 3 charged helium ions (which can be converted directly to electricity); no harmful, radioactivitycausing neutron emissions; and, hence, little radiation shielding.

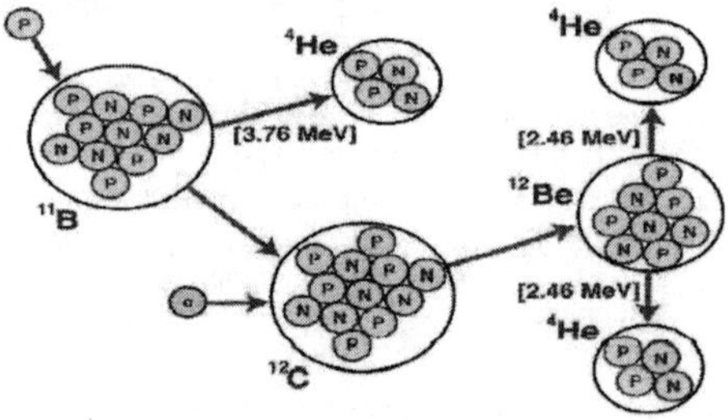

Figure 10. Aneutronic pB11fusion resulting in 8.68 MeV of energy and the electricity of 3 Helum 4 Ions

Effective, ionattracting SU(2) EM field interactions would generate fusion power with less electrical input energy for any of the fusion reaction in Figure 8. However, "clean" pB11 reactions, which result in very low radioactivity because very few neutrons are emitted, might benefit most from SU(2) EM fields.

Figure 8 shows pB11 fusion reactions requiring about 15times more input energy than lessclean DT fusion reactions to achieve a given fusion power. This results in higher ignition temperatures which can cause phenomenon such as bremsstrahlung radiation losses that certain fusion systems find very difficult to prevent or to cope with. Figure 11 compares reduced input energy needs for DT and pB11 fusion for a given fusion output power and for a fairly modest SU(2) EM field efficiency (αβ) of only 10 percent.

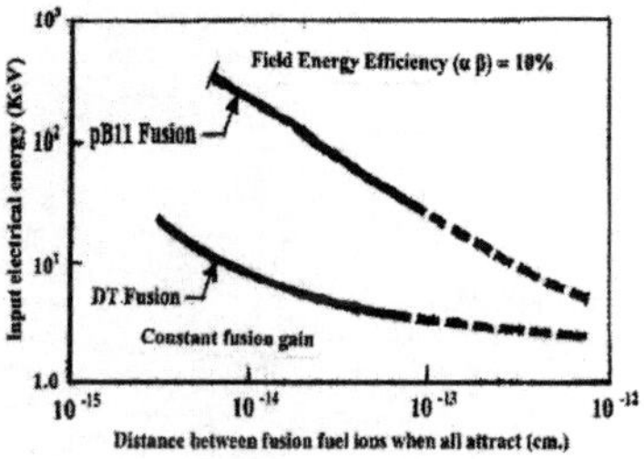

Figure 11. Obtaining longer ion-attracting range in SU(2) EM fields reduces DT and p-B11 fusion input energy needs

Aneutronic pB11 fusion is seen to requires more input energy than DT fusion for a given reaction distance. But, ionattracting SU(2) EM fields are seen to enable larger input energy reductions for pB11 fusion than for DT fusion. If so, these reductions should avoid things such as bremsstrahlung radiation.

Conclusions

Barrett has shown the possibility of EM radiation fields with the same SU2 gauge symmetry as the ion-attracting matter fields associated with ionattracting, weak nuclear forces in nuclei that cause fusion. If so, ionattracting SU(2) EM radiation fields in fusion reactors could conceivably attract (rather than repel) fusion fuel ions to reduce the coulomb resistance and electrical compression energy needed for fusion of these ions. So, the authors have begun exploring this seemingly bizarre possibility theoretically. However, an experiment to modify a helicon plasma generator to create SU(2) EM discharges and ion beams, and to search for ionion attractions and ionelectron repulsions inside the generator is also a needed first step.

References

[1] Barrett TW, Topological Foundations of Electromagnetism, World Scientific, 2008
[2] Barrett TW, On the distinction between fields and their metric, Annales de la Foundation Louis de Broglie, Volume 14, no. 1, 1989
[3] Miley, G.,H, Shaban, Y., Yang, Y., RF Gun Injector in Support of Fusion Ship II Research and Development, Proceedings of Space Technology and Applications Forum STAIF 2005, Edited by MS El Genk, American Institute of Physics, Melville, New York, 2005
[4] Charles, C., and Boswell, R., Laboratory evidence of a supersonic ion beam generated by a currentfree "helicon"doublelayer, Physics of Plasmas, Volume 11, Number 6, April 5, 2004
[5] Duncan M, "Should Google Go Nuclear", askmar.com, video of talk at http://video.google.com/videoplay?docid=1996321846673788606

IAA-98-IAA.3.3.03
POSSIBLE REVOLUTIONS IN ROCKET PROPULSION
49th International Astronautical Congress

H.D. Froning, Jr.
Flight Unlimited
5450 Country Club Dr.

Flagstaff, AZ USA

Roger E. Lo
Berlin Univ.of Technology
Marchstraße 12
D-10587
Berlin, GERMANY

Abstract

This paper documents the work of Group 3 of the IAA Advanced Propulsion Working Groups in identifying possible revolutions in rocket propulsion during the next 40 years. The work was accomplished at the IAA Workshop on Advanced Space Propulsion Concepts, held at the Aerospace Corporation in El Segundo, California, on January 20 and 21, 1998, and this paper describes the revolutions in rocket propulsion that Working Group 3 deemed possible during the next 40 years of spaceflight.

These possibilities included: (a) solid cryogenic propellants that could be stored at higher densities and release greater amounts of energy during combustion, or pulsed detonations to achieve higher effective combustion pressures and specific impulse with less overall engine pressures and heating; (b) nuclear fission rockets that embody solid core or gas core reactors to enable a 2 to 8-fold increase in specific impulse over that of current chemical rockets; and (c) nuclear fusion rockets for an 8 to 80-fold specific impulse increase over that of current chemical rockets for missions throughout the solar system - and, perhaps beyond. Most of the fusion rocket concepts embody nuclear fuels whose fusion reactions result in neutron emission and radioactivity. However, one concept embodies "aneutronic" fuels that do not result in neutron emissions or residual radioactivity. Thus, it could be used for earth-to-orbit missions in addition to outer space exploration.

This paper shows the preliminary estimates made by the Working Group for plausible time scales over which each of the revolutionary rocket propulsion concepts could be developed for earth-to-orbit and/or beyond-earth-orbit missions, together with estimated rocket performance advances that could be achieved in terms of: propellant density, specific impulse, and engine thrust-to-weight improvement as a function of time. Some of the critical technologies that the working group deemed necessary to enable development of the revolutionary rocket propulsion concepts are also described, together with certain environmental and public policy issues that would be encountered with some of the concepts.

Introduction

The theme of the IAA workshop on Advanced Propulsion Concepts. held in 1998 at the Aerospace Corporation in El Segundo, California on January 20 and 21, was to identify possible advances in space propulsion technology between year 2000 and 2040 - the next 40 years of spaceflight. Since another Working Group was tasked to identify near-term rocket propulsion advances in the 2000-2040 time period, Working Group 3 was directed to identify those more revolutionary rocket propulsion advances that were deemed possible between 2010 and 2040. Deliberations within our working group took place over a two day period and resulted in the creation of a series of charts which reflected the general consensus within our group. Thus, these exact charts (instead of specially prepared tables and figures derived from them) will be shown in this paper. Table 1 includes the composition of APW Working Group 3.

Revolutionary Technology Options

Chart 1 shows the revolutionary technical advances in rocket propulsion which our working group deemed possible between 2010 and 2040, during the last 3 decades of the next 40 years of spaceflight. These possibilities, categorized as 3 options, are: revolutionary advances in chemical rocket engines; revolutionary advances in solid core and gas core fission rocket engines; and revolutionary advances in fusion rocket engines that accomplish nuclear fusion by different reactions and confinement techniques.

Revolutionary advances in chemical rocket engines included chemical rockets that embody energetic solid cryogenic propellants in order to achieve higher propellant density and/or energy. Concepts studied in both the United States and Germany were discussed at the workshop. German solid cryogenic propellant concepts are described in reference (15). They embody solid fuel and solid oxidiser cast into separate modular solid grains. Propellant research at the U.S. Air Force Research Laboratory is described in (1). It includes atomic hydrogen stored in solid molecular hydrogen (so called matrix isolation) and exotic propellants using high-nitrogen molecules. Previous work sponsored by the Air Force Research Laboratory has included work on very dense materials such as metallic hydrogen and very energetic materials such as metastable helium (References 2,3). But such materials are proving to be very difficult to manufacture and to store. Another revolutionary possibility for improving chemical rockets is use of specially shaped pulsed detonations within combustion chambers to achieve higher effective combustion pressures and specific impulse with less overall engine pressures and heating.

Revolutionary advances in solid and gas core fission rocket engines would build upon past research and development of such systems in the United States and former Soviet Union. Promising recent work in solid core fission rockets is described in References (4,5) and recent work in gas-core fission rockets is described in references (6,7).

Revolutionary advances in fusion rocket engines would require breakthroughs in at least one of the fusion concepts that are being studied today. These concepts include: antiproton catalyzed fusion systems that simultaneously confine antiprotons and heated plasma within a nested ion trap (references 8,9), and dense plasma focus systems which accelerate and compress fusion fuel plasma between concentric electrodes until fusion temperature and pressure is reached within a "pinch region" at the end of the electrodes (10). Another system being studied by NASA, which was inadvertently left off Chart 1, is the "Gas Dynamic Mirror Fusion Rocket", in which a hot dense plasma is confined within a simple magnetic mirror geometry (11). The most ambitious system is the aneutronic Electrostatic Confinement Concept proposed by Bussard (12) which involves fusion of Boron 11 ions and protons. Such fusion emits no neutrons and does not result in any radioactivity.

Rocket Performance Revolutions

Chart 2 shows our best estimates of when rocket performance revolutions, in terms of increased specific impulse, might occur. Specific impulse of current LOX/LH2 systems are seen to increase slightly during the next 10 years until pulsed detonation systems or solid cryogenic systems with much higher propellant densities become available in about 2010. At about this same time, 10 years of advancement in solid core fission technologies between 2000 and 2010 could enable a specific impulse jump to about 1,000 seconds (for an engine thrust-to-mass in the 5 to 10g range).

As shown in Chart 2, specific impulse could then increase slightly during the next 10 years until gas-core fission systems (with engine thrust-to-mass in the 0.5 to 1.0g range) are perfected in about 2020. And, it is seen that High Energy Density Materials (HEDM), such as either atomic or metallic hydrogen could enable chemical rocket engine specific impulse in the 1,000 sec range with large increase in propellant density, as well.

It is seen that development of neutronic and aneutronic fusion, estimated to be achievable by 2040, would enable increase in fusion rocket specific impulse to values of the order of 30,000 sec--or even more for outer planet colonization and exploration by humans. Aneutronic fusion would enable specific impulse of the order of 3,000 sec for reusable earth-to-orbit transportation.

Revolutionary Rocket Roadmap

The roadmap produced at the workshop marks the earliest possible time of availability for the concepts discussed. Each block of time for each development represents total time for exploratory and advanced development, and for full-scale engineering development for a specific spaceflight application. All chemical rocket options are applicable for earth-to-orbit and beyond-earth-orbit flight. Nuclear fission and nuclear fusion options (except for the one aneutronic electrostatic confinement option) are, of course, limited to flight beyond earth orbit. This is indicated in Chart 3 which shows a roadmap for revolutionary rocket development, in terms of preliminary time estimates to accomplish revolutionary breakthroughs in rocket technology for earth-to-orbit and beyond-earth-orbit flight.

It is seen that earth-to-orbit missions with solid cryogenic rocket propellants (or pulsed detonation rocket engines) could begin in about 2010, while interplanetary missions with solid core fission could also begin as early as that. It is also seen that breakthroughs in high energy density matter for earth-to-orbit missions could occur sometime between 2010 and 2040, while interplanetary spacecraft powered by gas core fission propulsion could begin flying by 2020.

It is not obvious what fusion propulsion systems (if any) are most likely to be developed. But aneutronic fusion systems, which would not emit neutrons or cause radioactivity, would obviously be the most desirable, especially if their reduced shielding mass would result in needed specific impulse with attractive ratios of engine thrust-to-weight. Studies performed for NASA indicate that aneutronic fusion propulsion could be developed for earth-to-orbit and interplanetary flight in times of the order of 15 years after aneutronic fusion power breakeven is demonstrated.

Enabling Technologies

Chart 4 shows some of the critical technologies that are needed to enable revolutionary chemical, fission, and fusion systems. It is seen that refrigeration / insulation technology advancements are needed for revolutionary chemical propellant advances, while advances that significantly reduce the size and mass of superconducting magnets are needed for some of the chemical and nuclear rocket propulsion system possibilities. Supersonic combustion of LOX in the nozzle exhaust of solid core fission rockets is mentioned in (4,5) and is expected to significantly increase exhaust velocity and thrust, while compact and lightweight superconducting magnets and advances in electromagnetics and plasma dynamics would be required for most fusion propulsion systems.

Advances in artificial intelligence (knowledge based systems) are needed for all revolutionary propulsion systems to achieve a high degree of automated check-out and health management for high mission reliability and significant reduction in operations and support costs. And significant increase in the strength-to-weight of materials is needed for significant vehicle and propulsion system dry mass reduction. One example would be high temperature silicon nitride for rocket engine components. A more revolutionary possibility would be materials made by means of "molecular manufacturing" - whereby materials with very few structural defects and enormous strength-to-weight would be "grown" by nanoscale mechanical systems to guide the placement of reactive molecules, building complex structures with atom-by-atom control" (13). Such materials could be composed of carbon structures, such as either thin diamond films or carbon 60 crystals or nanotubes.

Technology and Policy Issues

Chart 5 indicates that some revolutionary rocket concepts have mainly technology issues, while others have issues with respect to both technology and policy. For example, achievement of higher rocket specific impulse by means of higher exhaust velocity at low altitude could result in excessive noise at takeoff. Thus, specific impulse at takeoff and initial ascent would have to be limited to lower than achievable values until higher altitude is reached. And solid core and gas core fission rockets not only have technical challenges, such as engine thrust-mass, but policy challenges due to emotional resistance to any nuclear power or propulsion system by anti-nuclear activists. Antiproton catalyzed fusion and dense plasma focus fusion was deemed to be somewhat more politically acceptable, because fusion is generally perceived as "cleaner" than fission. Also, such systems would only be used for space journeys to remote destinations. Aneutronic fusion, which would emit no neutrons nor cause any radioactivity, would be most politically acceptable of all.

Our group believed that the technology and policy issues indicated in Chart 5 can be addressed by international collaborations, involving scientists, engineers and policy makers from around the world. Such cooperation is believed to be most possible during early research and exploratory development phases, when proprietary considerations and competitiveness between companies, corporations and countries tend to be the least.

Breakthrough Propulsion Physics

Although not specifically considered by this or any other IAA Propulsion Working Group at the workshop, an activity entitled "Breakthrough Propulsion Physics" has been recently initiated by NASA and is currently underway at a modest level of effort. This effort is exploring various physics breakthrough possibilities that could conceivably enable: new sources of propulsive power; much higher vehicle speeds; or minimization of propellant consumption by thrust generation from action and reaction of fields (rather than by combustion and expulsion of mass). Reference (14) summarizes current Breakthrough Propulsion Activity. Although this activity probably deserves a category of its own in the future, this IAA Working Group and the breakthrough propulsion activity (lead Investigator Marc Millis, NASA-LERC) are informally affiliated and coordinating activities at the present time. It appears that propulsion physics breakthroughs may be achievable during the 2010 to 2040 time period that is being considered by the IAA Advanced Propulsion Working Group.

Conclusions

The conclusions listed in Chart 6 summarize the material described by our group in Charts 1 through 5, plus several additional observations that our group wished to make with respect to rocket propulsion revolutions.

Hydrocarbon and cryogenic propellants that will be used in currently envisioned chemical rockets are approaching a plateau in the performance that they can provide. Thus, chemical rockets will soon be limited in the vacuum specific impulse that they can generate. On the other hand, chemical rockets embodied within combined-cycle (rocket/air breathing) engines have significantly higher effective specific impulse because airflow is captured for oxidizer during atmospheric flight.

Significantly higher rocket specific impulse would result in significantly higher exhaust velocity than that of current rockets and result in excessive noise at low altitudes. Thus, the exhaust velocity of high energy rockets at lift-off must be about the same as that of current rockets. Then, specific impulse can be increased along with decreasing engine mass flow as higher altitude is reached. Another option is to use the high specific impulse of airbreathing propulsion for takeoff and ascent through the lower atmosphere, followed by high energy rocket propulsion in the upper atmosphere and space. The airbreathing portion of this option is being addressed by the IAA Airbreathing Propulsion Working Group.

Chemical rockets that embody solid cryogenic propellants and nuclear rockets of the solid core fission reactor type could be the first chemical and non-chemical revolutionary rocket systems for future spaceflight. Gas core fission reactors could be a mid-term option alternative to solid core fission, while some form of fusion propulsion is a possible long-term option. There are common technology links for many of the revolutionary rocket propulsion options, and any breakthroughs in propulsion physics that would significantly diminish rocket thrust and propellant needs would further revolutionize space propulsion.

References

(1) Carrick, P.G., "Theoretical Performance of High Energy Density Cryogenic Solid Rocket Propellants" AIAA 95-2892, 31st AIAA Joint Propulsion Conference, San Diego, CA July, 1995

(2) Forward, R.L., "Alternate Propulsion Energy Sources", Air Force Research Laboratory Report: AFRPL-83-039, Air Force Research Laboratory, Edwards Air Force Base, 1983

(3) Sivera, I.F., New phases of hydrogen at megabar pressures and metallic hydrogen ", pp. 41-46, in Proceedings of the High Energy Density Materials Contract or Conference, Long Beach, CA, February, 1990, L.P. Davis and F.J. Wodarezyk, Editors, Air Force Office of Scientific Research, Bolling AFB., DC, 20332-6448

(4) Borowski, S., etal., "Nuclear Thermal Rocket Vehicle Design Options for Future NASA Missions to the Moon and Mars", AIAA 93-4170, Space Programs and Technology Conference, Huntsville, AL, September, 1993

(5) Borowski, S., etal "A Reusable Mars Space Transportation Architecture Enabled by IRSU and Lox-Augmented NTR Propulsion" AIAA-98-3885, AIAA Joint Propulsion Conference, Cleveland, OH, July, 1998

(6) Howe, S.D., etal,. "Reducing the Risk to Mars: The Gas Core Nuclear Rocket", AIP Conf Proceedings 420 of the Space Technology and Applications International Forum, Albuquerque, NM, January, 1998

(7) Howe, S.D., etal,. "Gas-Core Nuclear Rocket Feasibility Project," AIAA-98-3887, AIAA Joint Propulsion Conference, Cleveland, OH, July, 1998

(8) Lewis, R.A., etal,. "Antiproton-Catalized Microfission/Fusion Propulsion Systems for Exploration of the Outer Solar Systems and Beyond", AIP Conf. Proceedings 387 of the Space Technology & Applications and International Forum, Albuquerque, NM, January, 1997

(9) Smith,G., "Antiproton-Catalyzed Microfission/fusion Propulsion System for Exploration of the Outer Solar System", AIAA-98-3589, AIAA Joint Propulsion Conference, Cleveland, OH, July 1, 1998

(10) Choi, C.K., "Engineering Considerations for the Self-Energized MPD-Type Fusion Plasma Thruster", Phillips Laboratory Report No. PL-TR-91-3087, February, 1992

(11) Kamish, T., etal, "Physics Basis for the Gas Dynamic Mirror (GDM) Fusion Rocket", AIP Conf. Proceedings 420 of the Space Technology and Applications International Forum, Albuquerque, NM, January, 1998

(12) Froning, H.D., etal., "Aneutronic Fusion Propulsion for Earth-to-Orbit and Beyond", AIP Conf. Proceedings 420 of the Space Technology and Applications International Forum, Albuquerque, NM, January, 1998

(13) Drexler, K.E., "Nanosystems" John Wiley & Sons Inc., New York, NY., ISBN 0-471-57547-X, 1992

(14) Millis, M.G., "Breakthrough Propulsion Physics Workshop Preliminary Results", AIP Conf. Proceedings 420 of the Space Technology and Applications International Forum, Albuquerque, NM, January, 1998

(15) R.E.Lo:
"Modular Fragmented Cryogenic Solid-Rocket Propellant Grains", 49th Int.Astronautical Congress, Melbourne, Austr., Sept.29.1998, Session S.3, Space Propulsion Technology, IAF-98-S.3.10

Name	Affiliation	Expertise
Brady, Brian B.	The Aerospace Corporation	Propulsion combustion
Culver, Don	Aerojet Propulsion- Gencomp	Liquid & Nuclear Rocket Eng. & Power
Froning, H. David	Flight Unlimited	nuclear and field pro-pulsion: Vehicle synthesis
Johnson, Ray F.	The Aerospace Corporation	Propulsion
Lo, Roger E.	Berlin University of Technology	Advanced Space Propulsion
Martin, L. Robbin (Bob)	The Aerospace Corporation	Chemical Physics
Olson, Glenn S.	US Air Force - AFRL/PRRM	Advanced Concepts
Pollard, James E.	The Aerospace Corporation	Spacecraft propulsion
Quinn, Lawrence P.	US Air Force - AFRL/PRR	Solid Liquid Electric Solar Propulsion
Thierschmann, Michael	DGLR	Super-High-Energy-Propellants

Table 1: Members of APW Subgroup - Revolutionary Rockets

Revolutionary Technology Options

- Chemical rockets
  - » Solid cryogenic
  - » Pulsed detonation
  - » Hydrogen atoms
  - » Metallic hydrogen
  - » Metastable helium
  - » Triatomic nitrogen
- Nuclear fission
  - » Solid core
  - » Gas core
- Nuclear fusion
  - » Antiproton catalyzed
  - » Dense plasma focus
  - » Electrostatic confinement
    - Neutronic
    - Aneutronic

Rocket Performance Revolutions

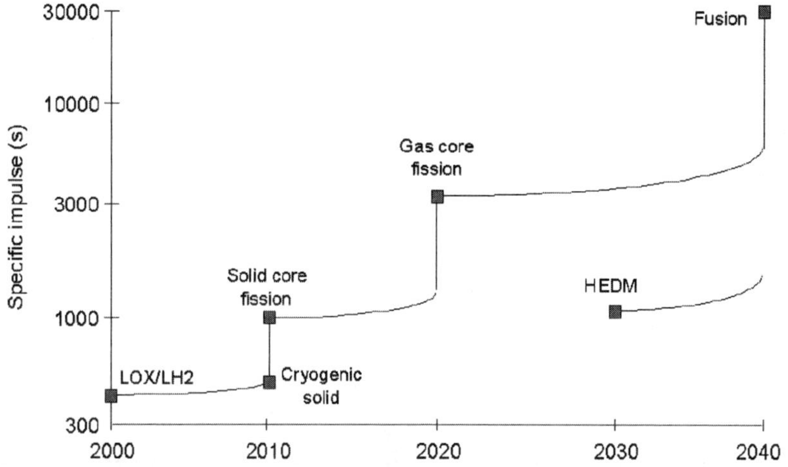

IAA **Advanced Space Propulsion Concepts Workshop** 2

Revolutionary Rocket Roadmap

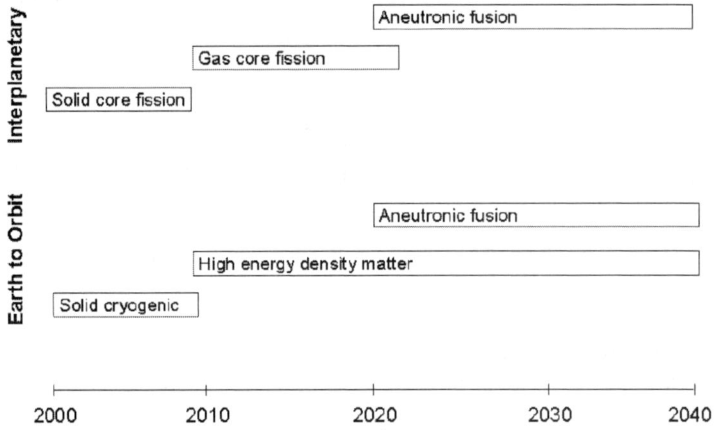

Advanced Space Propulsion Concepts Workshop

3

Enabling Technologies

Chemical	Fission	Fusion
Cryogenic solid Refrigeration Insulation	**Solid core** Supersonic combustion	**Antiproton catalyzed** Antimatter production Antimatter storage
High energy matter Refrigeration Insulation SC magnets Super high pressure	**Gas core** SC magnets	**Dense plasma focus** Lightweight capacitors Compact capacitors **Aneutronic** SC magnets

Artificial intelligence
Knowledge based systems

Ultralight weight structures

IAA Advanced Space Propulsion Concepts Workshop 4

Technology & Policy Issues

- Cryogenic solid rocket
  - Production and storage
- Solid core fission
  - Nuclear policy, thrust/weight ratio
- Gas core fission
  - Nuclear policy, thrust/weight ratio, magnetic confinement
- Antiproton catalyzed fusion
  - Antimatter storage, thrust/weight ratio
- Dense plasma focus fusion
  - Electrical energy storage (weight and volume), thrust/weight ratio
- Aneutronic fusion
  - Thrust/weight ratio, plasma dynamics, light weight magnets
- High energy density matter
  - Production and storage, materials

Conclusions

- Current rocket propellants have reached a plateau in performance
- Must combine existing rocket propellants with air breathing (rocket- or tubine-based combined cycle engines) for significant performance increase
- Higher I_{sp} rockets will be limited by noise and other environmental concerns at high launch rates
  - Use air breathing or rocket propulsion with lower I_{sp} at sea level
- Solid cryogenic propellants may be the best near-term option for revolutionary earth-to-orbit propulsion
- Nuclear fission is best medium-term option for planetary missions
- Nuclear fusion is possible long-term option
- Common technology links
  - Cryogenic refrigeration
  - Cryogenic insulation
  - Superconducting magnets

IAA Advanced Space Propulsion Concepts Workshop 6

Glossary of David Froning papers published through the American Institute of Aeronautics and Astronautics (AIAA)

https://arc.aiaa.org/action/showPreferences?menuTab=Articles&type=favorite%2Ctitle&type=subscribed%2Ctitle

2/5
Track Citations | Email | View Abstracts | Download Citations
Select All
Access Indicator: =Free =Full =Partial =No access
'Bank-to-turn steering' for highly maneuverable missiles
Guidance and Control Conference
Abstract | Remove
Aerodynamic design of slender missiles for back-to-turn flight at high angles of attack
19th Aerospace Sciences Meeting
Abstract | Remove
Aero-propulsion interactions of a modular missile
16th Joint Propulsion Conference
Abstract | Remove
Aerospace plane applications for heavy lift missions to the moon and Mars
4th Symposium on Multidisciplinary Analysis and Optimization
Abstract | Remove
Airframe-propulsion considerations for pulse-motor powered missiles
24th Aerospace Sciences Meeting
Abstract | Remove
Application of fluid dynamics to the problems of field propulsion and ultra high-speed flight
28th Aerospace Sciences Meeting
Abstract | Remove
Drag reduction, and possibly impulsion, by perturbing fluid and vacuum fields
31st Joint Propulsion Conference and Exhibit
Abstract | Remove
Drag reduction of transonic airfoils by freestream perturbations and heat addition
21st Atmospheric Flight Mechanics Conference
Abstract | Remove
Economic and technical challenges of expanding space commerce by RLV development
Space Programs and Technologies Conference
Abstract | Remove
Field propulsion for future flight
27th Joint Propulsion Conference
Field propulsion for future needs
28th Joint Propulsion Conference and Exhibit
Fusion-electric propulsion for aerospace plane flight
5th International Aerospace Planes and Hypersonics Technologies Conference

Fusion-electric propulsion for hypersonic flight
29th Joint Propulsion Conference and Exhibit
Abstract | Remove
Impact of dual fuel and dual expander rockets on RLVs
Space Programs and Technologies Conference
Abstract | Remove
Interstellar propulsion - A possible future derivative of air-breathing technology
4th Propulsion Joint Specialist Conference
Abstract | Remove
Investigation of a 'quantum ramjet' for interstellar flight
17th Joint Propulsion Conference
Abstract | Remove
Investigation of airbreathing antimatter propulsion for single-stage-to-orbit ships
24th Joint Propulsion Conference
Abstract | Remove
Rudi Beichel's unique dual fuel/dual expander reusable rocket engine
32nd Joint Propulsion Conference and Exhibit
Abstract | Remove
Use of an almost-developed propellant tank for flight demonstration of spaceplane art
International Aerospace Planes and Hypersonics Technologies
Abstract | Remove
VACUUM ENERGY FOR POWER AND PROPULSIVE FLIGHT?
30th Joint Propulsion Conference and Exhibit

Drag reduction, and possibly impulsion, by perturbing fluid and vacuum fields
31st Joint Propulsion Conference and Exhibit
Abstract | PDF | PDF Plus
Drag reduction, and possibly impulsion, by perturbing fluid and vacuum fields
31st Joint Propulsion Conference and Exhibit
Abstract | PDF | PDF Plus
Inertia reduction - and possibly impulsion - by conditioning electromagnetic fields
33rd Joint Propulsion Conference and Exhibit
Abstract | PDF | PDF Plus
Preliminary simulations of vehicle interactions with the zero point vacuum by fluid dynamic approximations

36th AIAA/ASME/SAE/ASEE Joint Propulsion Conference and Exhibit
Abstract | PDF | PDF Plus
Specially conditioned EM radiation research with transmitting toroid antennas
37th Joint Propulsion Conference and Exhibit
Abstract | PDF | PDF Plus
Use of an almost-developed propellant tank for flight demonstration of spaceplane art
International Aerospace Planes and Hypersonics Technologies
Abstract | PDF | PDF Plus

AIAA
12700 Sunrise Valley Drive, Suite 200
Reston, VA 20191-5807
703.264.7500

Ed. Note: The last article in this Appendix is a special related report courtesy of Dr. Harold Puthoff to whom we are indebted.

UNCLASSIFIED//FOR OFFICIAL USE ONLY

Defense Intelligence Reference Document

Acquisition Threat Support

29 March 2010

ICOD: 1 December 2009

DIA-08-1003-015

Advanced Space Propulsion Based on Vacuum (Spacetime Metric) Engineering

UNCLASSIFIED//FOR OFFICIAL USE ONLY

UNCLASSIFIED//FOR OFFICIAL USE ONLY

Defense Intelligence Reference Document

[redacted]

Acquisition Threat Support

29 March 2010

ICOD: 1 December 2009

DIA-08-1003-015

Advanced Space Propulsion Based on Vacuum (Spacetime Metric) Engineering

UNCLASSIFIED//FOR OFFICIAL USE ONLY

Advanced Space Propulsion based on Vacuum (Spacetime Metric) Engineering

Prepared by:

Acquisition Support Division (DWO-3)
Defense Warning Office
Directorate for Analysis
Defense Intelligence Agency

Author:

H.E. Puthoff, Ph.D.
EarthTech International, Inc.
11855 Research Blvd.
Austin, Texas 78759

Administrative Note

This product is one in a series of advanced technology reports produced in FY 2009 under the Defense Intelligence Agency, Defense Warning Office's Advanced Aerospace Weapon System Applications (AAWSA) Program. Comments or questions pertaining to this document should be addressed to James T. Lacatski, D.Eng., AAWSA Program Manager, Defense Intelligence Agency, ATTN: CLAR/DWO-3, Bldg 6000, Washington, DC 20340-5100.

UNCLASSIFIED//FOR OFFICIAL USE ONLY

Contents

Figures

Tables

UNCLASSIFIED//FOR OFFICIAL USE ONLY

Advanced Space Propulsion Based on Vacuum (Spacetime Metric) Engineering

Preface and Introduction

A theme that has come to the fore in advanced planning for long-range space exploration in the future is the concept that empty space itself (the quantum vacuum, or spacetime metric) might be engineered to provide energy/thrust for future space vehicles. Although far reaching, such a proposal is solidly grounded in modern physical theory, and therefore the possibility that matter/vacuum interactions might be engineered for spaceflight applications is not a priori ruled out (Reference 1). Given the current development of mainstream theoretical physics on such topics as warp drives and traversable wormholes that provides for such vacuum engineering possibilities (References 2-6), provided in this paper is a broad perspective of the physics and consequences of the engineering of the spacetime metric.

The concept of "engineering the vacuum" found its first expression in the mainstream physics literature when it was introduced by Nobelist T. D. Lee in his textbook *Particle Physics and Introduction to Field Theory* (Reference 7). There he stated, "The experimental method to alter the properties of the vacuum may be called vacuum engineering.... If indeed we are able to alter the vacuum, then we may encounter new phenomena, totally unexpected." This legitimization of the vacuum engineering concept was based on the recognition that the vacuum is characterized by parameters and structure that leave no doubt that it constitutes an energetic and structured medium in its own right. Foremost among these are that (1) within the context of quantum theory, the vacuum is the seat of energetic particle and field fluctuations and (2) within the context of general relativity, the vacuum is the seat of a spacetime structure (metric) that encodes the distribution of matter and energy. Indeed, on the flyleaf of a book of essays by Einstein and others on the properties of the vacuum, there is the statement, "The vacuum is fast emerging as the central structure of modern physics" (Reference 8). Perhaps the most definitive statement acknowledging the central role of the vacuum in modern physics is provided by 2004 Nobelist Frank Wilczek in his book *The Lightness of Being: Mass, Ether and the Unification of Forces* (Reference 9):

> "What is space? An empty stage where the physical world of matter acts out its drama? An equal participant that both provides background and has a life of its own? Or the primary reality of which matter is a secondary manifestation? Views on this question have evolved, and several times have changed radically, over the history of science. Today the third view is triumphant."

Given the known characteristics of the vacuum, one might reasonably inquire why it is not immediately obvious how to catalyze robust interactions of the type sought for spaceflight applications. For starters, in the case of quantum vacuum processes, uncertainties regarding global thermodynamic and energy constraints remain to be clarified. Furthermore, it is likely that energetic

components of potential utility involve very-small-wavelength, high-frequency field structures and thus resist facile engineering solutions. With regard to perturbation of the spacetime metric, the required energy densities predicted by present theory exceed by many orders of magnitude values achievable with existing engineering techniques. Nonetheless, one can examine the possibilities and implications under the expectation that as science and its attendant derivative technologies mature, felicitous means may yet be found that permit the exploitation of the enormous, as-yet-untapped potential of engineering so-called "empty space," the vacuum.

This paper introduces the underlying mathematical platform for investigating spacetime structure, the metric tensor approach. It then outlines the attendant physical effects that derive from alterations in the spacetime structure. Finally, the paper examines these effects as they would be exhibited in the presence of advanced aerospace craft technologies based on spacetime modification.

I. Spacetime Modification – Metric Tensor Approach

Despite the daunting energy requirements to restructure the spacetime metric to a significant degree, one can investigate the forms that such restructuring would take to be useful for spaceflight applications and determine their corollary attributes and consequences. Thus we embark on a "Blue Sky," general-relativity-for-engineers approach, as it were.

As a mathematical evaluation tool, the *metric tensor* that describes the measurement of spacetime intervals is used. Such an approach, well known from studies in general relativity (GR), has the advantage of being model independent—that is, it does not depend on knowledge of the specific mechanisms or dynamics that result in spacetime alterations but rather only assumes that a technology exists that can control and manipulate (that is, engineer) the spacetime metric to advantage. Before discussing the predicted characteristics of such engineered spacetimes, beginning in Section III, a brief mathematical digression for those interested in the mathematical structure behind the discussion to follow is introduced.

As a brief introduction, the expression for the four-dimensional line element ds^2 in terms of the metric tensor $g_{\mu\nu}$ is given by

$$ds^2 = g_{\mu\nu} dx^\mu dx^\nu \qquad (1)$$

where summation over repeated indices is assumed unless otherwise indicated. In ordinary Minkowski flat spacetime, a (four-dimensional) infinitesimal interval *ds* is given by the expression (in Cartesian coordinates)

$$ds^2 = c^2dt^2 - (dx^2 + dy^2 + dx^2) \qquad (2)$$

where the identification $dx^0 = cdt$, $dx^1 = dx$, $dx^2 = dy$, $dx^3 = dz$ is made, with metric tensor coefficients $g_{00} = 1$, $g_{11} = g_{22} = g_{33} = -1$, $g_{\mu\nu} = 0$ for $\mu \neq \nu$.

For spherical coordinates in ordinary Minkowski flat spacetime

$$ds^2 = c^2dt^2 - dr^2 - r^2d\theta^2 - r^2\sin^2\theta d\varphi^2 \qquad (3)$$

where $dx^0 = cdt$, $dx^1 = dr$, $dx^2 = d\theta$, $dx^3 = d\varphi$, with metric tensor coefficients $g_{00} = 1$, $g_{11} = -1$, $g_{22} = -r^2$, $g_{33} = -r^2\sin^2\theta$, $g_{\mu\nu} = 0$ for $\mu \neq \nu$.

As an example of spacetime alteration, in a spacetime altered by the presence of a spherical mass distribution *m* at the origin (Schwarzschild-type solution), the above can be transformed into (Reference 10)

$$ds^2 = \left(\frac{1 - Gm/rc^2}{1 + Gm/rc^2}\right)c^2dt^2 - \left(\frac{1 - Gm/rc^2}{1 + Gm/rc^2}\right)^{-1} dr^2 - \left(1 + Gm/rc^2\right)r^2\left(d\theta^2 + \sin^2\theta d\varphi^2\right) \qquad (4)$$

I. Spacetime Modification – Metric Tensor Approach

Despite the daunting energy requirements to restructure the spacetime metric to a significant degree, one can investigate the forms that such restructuring would take to be useful for spaceflight applications and determine their corollary attributes and consequences. Thus we embark on a "Blue Sky," general-relativity-for-engineers approach, as it were.

As a mathematical evaluation tool, the *metric tensor* that describes the measurement of spacetime intervals is used. Such an approach, well known from studies in general relativity (GR), has the advantage of being model independent—that is, it does not depend on knowledge of the specific mechanisms or dynamics that result in spacetime alterations but rather only assumes that a technology exists that can control and manipulate (that is, engineer) the spacetime metric to advantage. Before discussing the predicted characteristics of such engineered spacetimes, beginning in Section III, a brief mathematical digression for those interested in the mathematical structure behind the discussion to follow is introduced.

As a brief introduction, the expression for the four-dimensional line element ds^2 in terms of the metric tensor $g_{\mu\nu}$ is given by

$$ds^2 = g_{\mu\nu} dx^\mu dx^\nu \tag{1}$$

where summation over repeated indices is assumed unless otherwise indicated. In ordinary Minkowski flat spacetime, a (four-dimensional) infinitesimal interval *ds* is given by the expression (in Cartesian coordinates)

$$ds^2 = c^2 dt^2 - (dx^2 + dy^2 + dx^2) \tag{2}$$

where the identification $dx^0 = cdt$, $dx^1 = dx$, $dx^2 = dy$, $dx^3 = dz$ is made, with metric tensor coefficients $g_{00} = 1$, $g_{11} = g_{22} = g_{33} = -1$, $g_{\mu\nu} = 0$ for $\mu \neq \nu$.

For spherical coordinates in ordinary Minkowski flat spacetime

$$ds^2 = c^2 dt^2 - dr^2 - r^2 d\theta^2 - r^2 \sin^2 \theta d\varphi^2 \tag{3}$$

where $dx^0 = cdt$, $dx^1 = dr$, $dx^2 = d\theta$, $dx^3 = d\varphi$, with metric tensor coefficients $g_{00} = 1$, $g_{11} = -1$, $g_{22} = -r^2$, $g_{33} = -r^2 \sin^2 \theta$, $g_{\mu\nu} = 0$ for $\mu \neq \nu$.

As an example of spacetime alteration, in a spacetime altered by the presence of a spherical mass distribution *m* at the origin (Schwarzschild-type solution), the above can be transformed into (Reference 10)

$$ds^2 = \left(\frac{1 - Gm/rc^2}{1 + Gm/rc^2}\right) c^2 dt^2 - \left(\frac{1 - Gm/rc^2}{1 + Gm/rc^2}\right)^{-1} dr^2 - \left(1 + Gm/rc^2\right) r^2 \left(d\theta^2 + \sin^2 \theta d\varphi^2\right) \tag{4}$$

with the metric tensor coefficients $g_{\mu\nu}$ modifying the Minkowski flat-spacetime intervals dt, dr, and so forth, accordingly.

As another example of spacetime alteration, in a spacetime altered by the presence of a *charged* spherical mass distribution (Q, m) at the origin (Reissner-Nordstrom-type solution), the above can be transformed into (Reference 11)

$$ds^2 = \left(\frac{1 - Gm/rc^2}{1 + Gm/rc^2} + \frac{Q^2 G/4\pi\varepsilon_0 c^4}{r^2 \left(1 + Gm/rc^2\right)^2} \right) c^2 dt^2 - \left(\frac{1 - Gm/rc^2}{1 + Gm/rc^2} + \frac{Q^2 G/4\pi\varepsilon_0 c^4}{r^2 \left(1 + Gm/rc^2\right)^2} \right)^{-1} dr^2$$
$$- \left(1 + Gm/rc^2\right)^2 r^2 \left(d\theta^2 + \sin^2\theta d\varphi^2\right) \quad (5)$$

with the metric tensor coefficients $g_{\mu\nu}$ again changed accordingly. Note that the effect on the metric due to charge Q differs in sign from that due to mass m, leading to what in the literature has been referred to as *electrogravitic repulsion* (Reference 12).

Similar relatively simple solutions exist for a spinning mass (Kerr solution) and for a spinning electrically charged mass (Kerr-Newman solution). In the general case, appropriate solutions for the metric tensor can be generated for arbitrarily engineered spacetimes, characterized by an appropriate set of spacetime variables dx^μ and metric tensor coefficients $g_{\mu\nu}$. Of significance now is to identify the associated physical effects and to develop a table of such effects for quick reference.

We begin by simply cataloging metric effects—that is, physical effects associated with alteration of spacetime variables—saving for Section IV the significance of such effects within the context of advanced aerospace craft technologies.

II. Physical Effects as a Function of Metric Tensor Coefficients

In undistorted spacetime, measurements with physical rods and clocks yield spatial intervals dx^μ and time intervals dt, defined in a flat Minkowski spacetime, the spacetime of common experience. In spacetime-altered regions, dx^μ and dt are still chosen as *natural* coordinate intervals to represent a coordinate map, but now *local* measurements with physical rods and clocks yield spatial intervals $\sqrt{-g_{\mu\nu}}dx^\mu$ and time intervals $\sqrt{g_{00}}dt$, so-called *proper* coordinate intervals. From these relationships a table of associated physical effects to be expected in spacetime regions altered by either natural or advanced technological means can be generated. Given that, as seen from an unaltered region, alteration of spatial and temporal intervals in a spacetime-altered region result in an altered velocity of light, from an engineering viewpoint such alterations can in essence be understood in terms of a variable refractive index of the vacuum (see Section III below) that affects all measurement.

TIME INTERVAL, FREQUENCY, ENERGY

Begin by considering the case where $\sqrt{g_{00}} < 1$, typical for an altered spacetime metric in the vicinity of, say, a stellar mass, as expressed by the leading term in Equation (4). Local measurements with physical clocks within the altered spacetime yield a time interval $\sqrt{g_{00}}\,dt < dt$; thus an interval of time dt between two events in an undistorted spacetime remote[1] from the mass—say, 10 seconds—would be judged by local (proper) measurement from *within* the altered spacetime to occur in a lesser time interval, $\sqrt{g_{00}}\,dt < dt$—say, 5 seconds. From this one can rightly infer that, relatively speaking, clocks (atomic processes and so forth) within the altered spacetime run slower. Given this result, a physical process (for example, interval between clock ticks, atomic emissions) that takes a time Δt in unaltered spacetime slows to $\Delta t \to \Delta t/\sqrt{g_{00}}$ when occurring within the altered spacetime. Conversely, under conditions (for example, metric engineering) for which $\sqrt{g_{00}} > 1$, processes within the spacetime-altered region are sped up. Thus the first entry for a table of physical effects (see Table 1) is made.

Given that frequency measurements are the reciprocal of time duration measurements, the associated expression for frequency ω is given by $\omega \to \omega\sqrt{g_{00}}$, our second entry in Table 1. This accounts, for example, for the redshifting of atomic emissions from dense masses where $\sqrt{g_{00}} < 1$. Conversely, under conditions for which $\sqrt{g_{00}} > 1$, blueshifting of emissions would occur. In addition, given that quanta of energy are given by $E = \hbar\omega$, energy scales with $\sqrt{g_{00}}$, as does frequency, $E \to E\sqrt{g_{00}}$, our third entry in the table. Depending on the value of $\sqrt{g_{00}}$ in the spacetime-altered region, energy states may be raised or lowered relative to an unaltered spacetime region.

[1] An observer at "infinity."

Table 1. Metric Effects on Physical Processes in an Altered Spacetime as Interpreted by a Remote (Unaltered Spacetime) Observer

Variable	Typical Stellar Mass $(g_{00} < 1,\ \lvert g_{11}\rvert > 1)$	Spacetime-Engineered Metric $(g_{00} > 1,\ \lvert g_{11}\rvert < 1)$
Time Interval $\Delta t \to \Delta t/\sqrt{g_{00}}$	Processes (for example, clocks) run slower	Processes (for example, clocks) run faster
Frequency $\omega \to \omega\sqrt{g_{00}}$	Redshift toward lower frequencies	Blueshift toward higher frequencies
Energy $E \to E\sqrt{g_{00}}$	Energy states lowered	Energy states raised
Spatial $\Delta r \to \Delta r/\sqrt{-g_{11}}$	Objects (for example, rulers) shrink	Objects (for example, rulers) expand
Velocity $v_L = c \to c\sqrt{g_{00}/-g_{11}}$	Effective $v_L < c$	Effective $v_L > c$
Mass $m = E/c^2 \to \left(-g_{11}/\sqrt{g_{00}}\right)m$	Effective mass increases	Effective mass decreases
Gravitational "force" $f(g_{00}, g_{11})$	"Gravitational"	"Antigravitational"

Spatial Interval

Again, by considering the case typical for an altered spacetime metric in the vicinity of, say, a stellar mass, then $\sqrt{-g_{11}} > 1$ for the radial dimension $x^1 = r$, as expressed by the second term in Equation (4). Therefore, local measurements with physical rulers *within* the altered spacetime yield a spatial interval $\sqrt{-g_{11}}dr > dr$; thus a spatial interval dr between two locations in an undistorted spacetime—say, remote from the mass—would be judged by local (proper) measurement from within the altered spacetime to be greater. From this one can rightly infer that, relatively speaking, rulers (atomic spacings and so forth) within the altered spacetime are shrunken relative to their values in unaltered spacetime. Given this result, a physical object (for example, atomic orbit) that possesses a measure Δr in unaltered spacetime shrinks to $\Delta r \to \Delta r/\sqrt{-g_{11}}$ when placed within the altered spacetime. Conversely, under conditions for which $\sqrt{-g_{11}} < 1$, objects would expand—thus the fourth entry for the table of physical effects.

Velocity of Light in Spacetime-Altered Regions

Interior to a spacetime region altered by, say, a dense mass (for example, a black hole), the locally measured velocity of light c in, say, the $x^1 = r$ direction is given by the ratio of locally measured (proper) distance/time intervals for a propagating light signal (Reference 13).

$$v_L^i = \frac{\sqrt{-g_{11}}dr}{\sqrt{g_{00}}dt} = c \qquad (6)$$

From a viewpoint *exterior* to the region, however, from the above one finds that the remotely observed *coordinate* ratio measurement yields a different value

$$v_L^e = \frac{dr}{dt} = \sqrt{\frac{g_{00}}{-g_{11}}}\,c \tag{7}$$

Therefore, although a local measurement with physical rods and clocks yields *c*, an observer in an exterior reference frame remote from the mass speaks of light "slowing down" on a radial approach to the mass owing to the ratio $\sqrt{g_{00}/-g_{11}} < 1$. Conversely, under (metric engineering) conditions for which $\sqrt{g_{00}/-g_{11}} > 1$, the velocity of light—and exotic-technology craft velocities that obey similar formulas—would appear superluminal in the exterior frame. This gives our fifth entry for the table of physical effects.

Refractive Index Modeling

Given that velocity-of-light effects in a spacetime-altered region, as viewed from an external frame, are governed by Equation (7), it is seen that the effect of spacetime alteration on light propagation can be expressed in terms of an optical refractive index *n*, defined by

$$v_L^e = \frac{c}{n}, \qquad n = \sqrt{\frac{-g_{11}}{g_{00}}} \tag{8}$$

where *n* is an effective refractive index of the (spacetime-altered) vacuum. This widely known result has resulted in the development of refractive index models for GR (References 14-17) that have found application in problems such as gravitational lensing (Reference 18). The estimated electric or magnetic field strengths required to generate a given refractive index change given by standard GR theory (the Levi-Civita Effect) can be found in (Reference 19).

In engineering terms, the velocity of light c is given by the expression $c = 1/\sqrt{\mu_0 \varepsilon_0}$, where μ_0 and ε_0 are the magnetic permeability and dielectric permittivity of undistorted vacuum space ($\mu_0 = 4\pi \times 10^{-7}$ H/m and $\varepsilon_0 = 8.854 \times 10^{-12}$ F/m). The generation of an effective refractive index $n = \sqrt{-g_{11}/g_{00}} \neq 1$ by technological means can from an engineering viewpoint be interpreted as manipulation of the vacuum parameters μ_0 and ε_0. In GR theory, such variations in μ_0, ε_0, and hence the velocity of light, c, are often treated in terms of a "$TH\varepsilon\mu$" formalism used in comparative studies of gravitational theories (Reference 20).

As discussed below, a number of striking effects can be anticipated in certain engineered spacetime regions.

Effective Mass in Spacetime-Altered Regions

In a spacetime-altered region, $E = mc^2$ still holds in terms of local ("proper coordinate") measurements, but now energy *E* and the velocity of light *c* take on altered values as observed from an exterior (undistorted) spacetime region. Reference to the definitions for *E* and *c* in Table 1 permits one to define an effective mass as seen from the exterior undistorted region as therefore taking on the value $m \to m(-g_{11})/\sqrt{g_{00}}$, providing a sixth entry for our table. Depending on the values of g_{00} and g_{11}, the effective mass may be seen from the viewpoint of an observer in an undistorted spacetime region to have either increased or decreased.

Gravity/Antigravity "Forces"

Strictly speaking, from the GR point of view, there are no gravitational "forces" but rather (in the words of GR theorist John Wheeler) "matter tells space how to curve, and space tells matter how to move." (Reference 21) As a result, Newton's law of gravitational attraction to a central mass is therefore interpreted in terms of the spacetime structure as expressed in terms of the metric tensor coefficients, in this case as expressed in Equation (4) above. Therefore, in terms of the metric coefficients, gravitational attraction in this case derives from the condition that $g_{00} < 1, |g_{11}| > 1$. As for the possibility for generating "antigravitational forces," noted in equation (5), inclusion of the effects of charge led to metric tensor contributions counter to the effects of mass—that is, to *electrogravitic repulsion*. This reveals that conditions under which, say, the signs of the coefficients g_{00} and g_{11} could be reversed would be considered (loosely) as antigravitational in nature. A seventh entry in Table 1 represents these features of metric significance.

III. Significance of Physical Effects Applicable to Advanced Aerospace Craft Technologies as a Function of Metric Tensor Coefficients

As in Section III, metric tensor coefficients define the relationship between locally and remotely observed (that is, spacetime-altered and unaltered) variables of interest as listed in Table 1, and in the process define corollary physical effects. Table 1 thereby constitutes a useful reference for interpreting the physical significance of the effects of the alteration of spacetime variables. The expressions listed indicate specific spacetime alteration effects, whether owing to natural causes (for example, the presence of a planetary or stellar mass) or as a result of *metric engineering* by advanced technological means as might be anticipated in the development and deployment of advanced aerospace craft.

TIME ALTERATION

With regard to the first table entry (time interval), in a spacetime-altered region, time intervals are seen by a remote (unaltered spacetime) observer to vary as $1/\sqrt{g_{00}}$ relative to the remote observer. Near a dense mass, for example, $\sqrt{g_{00}} < 1$, and therefore time intervals are seen as lengthening and processes as running slower,[2] one consequence of which is redshift of emission lines. Should such a time-slowed condition be engineered in an advanced aerospace application, an individual who has spent time within such a temporally modified field would, when returned to the normal environment, find that more time had passed than could be experientially accounted for.

Conversely, for an engineered spacetime associated with an advanced aerospace craft in which $\sqrt{g_{00}} > 1$, time flow within the altered spacetime region would appear sped up to an external observer, while to an internal observer external time flow would appear to be in slow motion. A corollary would be that within the spacetime-altered region, normal environmental sounds from outside the region might cease to be registered, since external sounds could under these conditions redshift below the auditory range.

An additional implication of time speedup within the frame of an exotic craft technology is that its flightpath that might seem precipitous from an external viewpoint (for example, sudden acceleration or deceleration) would be experienced as much less so by the craft's occupants. From the occupants' viewpoint, observing the external environment to be in relative slow motion, it would not be surprising to consider that one's relatively modest changes in motion would appear abrupt to an external observer.

Based on the second entry in Table 1 (frequency), yet another implication of an accelerated timeframe due to craft-associated metric engineering that leads to $\sqrt{g_{00}} > 1$, frequencies associated with the craft would for a remote observer appear to be blueshifted. Corollary to observation of such a craft is the possibility that there would be a brightening of luminosity due to the heat spectrum blueshifting up into the visible portion of the spectrum (see Figure 1).

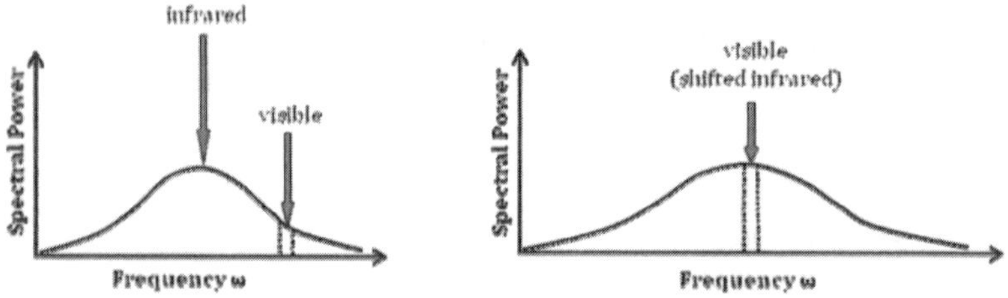

Figure 1. Blueshifting of Infrared Heat Power Spectrum

With regard to the third entry in Table 1 (energy), in a spacetime-altered region, energy scales as $\sqrt{g_{00}}$ relative to a remote observer in an undistorted spacetime. In the vicinity of a dense mass where $\sqrt{g_{00}} < 1$, the consequent reduction of energy bonds correlates with observed redshifts of emission. For engineered spacetimes associated with advanced craft technology in which $\sqrt{g_{00}} > 1$ (accelerated timeframe case), a

[2] *In the case of approach to a black hole, to stop altogether.*

craft's material properties would appear "hardened" relative to the environment owing to the increased binding energies of atoms in its material structure. Such a craft could, for example, impact water at high velocities without apparent deleterious effects.

SPATIAL ALTERATION

The fourth entry in Table 1 (spatial measure) indicates the size of an object within an altered spacetime region as seen by a remote observer. The size of, say, a spherical object is seen to have its radial dimension, r, scale as $1/\sqrt{-g_{11}}$. In the vicinity of a dense mass $\sqrt{-g_{11}} > 1$, in which case an object within the altered spacetime region appears to a remote observer to have shrunk. As a corollary, metric engineering associated with an advanced aerospace craft to produce this effect could in principle result in a large craft with a spacious interior appearing to an external observer to be relatively small. Additional dimensional aspects, such as potential dimensional changes, are discussed below in "Refractive Index Effects."

VELOCITY OF LIGHT/CRAFT IN SPACETIME-ALTERED REGIONS

Interior to a spacetime-altered region, the locally measured velocity of light, $v_L^i = c$, is given by the ratio of (locally measured) distance/time intervals for a propagating light signal, as expressed in Equation (6) above. From a viewpoint exterior to the region, however, the observed coordinate ratio measurement can yield a different value v_L^e greater or less than c as given by the fifth entry in Table 1 (velocity). As an example of a measurement less than c, one speaks of light "slowing down" as a light signal approaches a dense mass (for example, a black hole.) In an engineered spacetime in which $g_{00} > 1, |g_{11}| < 1$, however, the effective velocity of light v_L^e as measured by an external observer can be > c.

Given that velocities in general in different coordinate systems scale as does the velocity of light—that is, $v \to \sqrt{g_{00}/-g_{11}}\,v$ —for exotic propulsion an engineered spacetime metric can in principle establish a condition in which the trajectory of a craft approaching the velocity of light in its own frame would be observed from an exterior frame to exceed light speed—that is, exhibit motion at superluminal speed. This opens up the possibility of transport at superluminal velocities (as measured by an external observer) without violation of the velocity-of-light constraint within the spacetime-altered region, a feature attractive for interstellar travel. This is the basis for discussion of warp drives and wormholes in the GR literature (References 2-6). Therefore, although present technological facility is far from mature enough to support the development of warp drive and wormhole technologies (Reference 22), the possibility of developing such technologies in the future cannot be ruled out. In other words, effective transport at speeds exceeding the conventional speed of light could occur in principle, and therefore the possibility of reduced-time interstellar travel is not fundamentally ruled out by physical principles.

REFRACTIVE INDEX EFFECTS

When considering metric-engineered spacetime associated with exotic propulsion, a number of corollary side effects associated with refractive index changes of the vacuum structure emerge as possibilities. Expected effects would mimic known refractive index effects in general and can therefore be determined from known phenomena. Indistinct boundary definition associated with "waviness" as observed with heat waves off a desert floor is one example. As another, a light beam may bend (as in the GR example of the bending of starlight as it grazes the sun; see Figure 2) or even terminate in mid-space. Such an observation would exhibit features that under ordinary circumstances would be associated with a high-refractive index optical fiber in normal space (well-defined boundaries, light trapped within, bending or termination in mid-space). Additional observations might include apparent changes in size or shape (changes in lensing magnification parameters). Yet another possibility is the sudden "cloaking" or "blinking out," which would at least be consistent with strong gravitational lensing effects that bend a background view around a craft, though other technical options involving, for example, the use of metamaterials, exist as well.

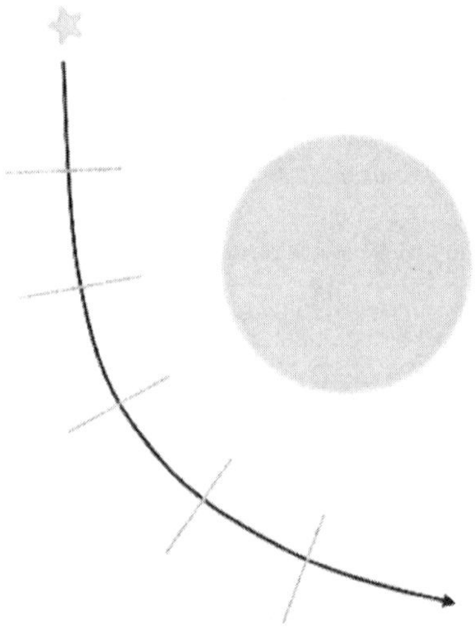

Figure 2. Light-Bending in a Spacetime-Altered Region

EFFECTIVE MASS IN SPACETIME-ALTERED REGIONS

As noted in the preceding sections, spacetime alteration of energy and light-speed measures leads to an associated alteration in the effective mass of an object in a spacetime-altered region as viewed from an external (unaltered) region. Of special interest is the case in which the effective mass is decreased by application of spacetime metric engineering principles as might be expected in the case of metric engineering for spaceflight applications (reference last column in Table 1). Effective reduction of inertial mass as viewed in our frame of reference would appear to mitigate against untoward effects on craft occupants associated with abrupt changes in movement. (The physical principles involved can also be understood in terms of associated coordinate transformation properties as discussed above.) In any case, changes in effective mass associated with engineering of the spacetime metric in a craft's environs can lead to properties advantageous for spaceflight applications.

GRAVITY/ANTIGRAVITY/PROPULSION EFFECTS

In the GR ansatz gravitational-type forces derive from the spacetime metric, whether determined by natural sources (for example, planetary or stellar masses) or by advanced metric engineering. Fortunately for our consideration of this topic, discussion can be carried out solely based on the form of the metric, independent of the specific mechanisms or dynamics that determine the metric. As one exemplar, consider Alcubierre's formulation of a "warp drive," a spacetime metric solution of Einstein's GR field equation (References 2, 22). Alcubierre derived a spacetime metric motivated by cosmological inflation that would allow arbitrarily short travel times between two distant points in space. The behavior of the warp drive metric provides for the simultaneous expansion of space behind the spacecraft and a corresponding contraction of space in front of the spacecraft (see Figure 3). The warp drive spacecraft would thus appear to be "surfing on a wave" of spacetime geometry. By appropriate structuring of the metric, the spacecraft can be made to exhibit an arbitrarily large apparent faster-than-light speed as viewed by external observers without violating the local speed-of-light constraint within the spacetime-altered region. Furthermore, the Alcubierre solution showed that the proper (experienced) acceleration along the spaceship's path would be zero, and that the spaceship would suffer no time dilation—highly desirable features for interstellar travel. In order to implement a warp drive, one would have to construct a "warp bubble" that surrounded the spacecraft by generating a thin shell or surface layer of exotic matter—that is, a quantum field having negative energy and/or negative pressure. Although the technical requirements for such are unlikely to be met in the foreseeable future (Reference 22), the exercise nonetheless serves as a good example for showcasing attributes associated with manipulation of the spacetime metric at will.

The entire discussion of the possibility of generating a spacetime structure like that of the Alcubierre warp drive is based simply on assuming the form of a metric (that is, $g_{\mu\nu}$) that exhibits desired characteristics. In like manner, arbitrary spacetime metrics to provide gravity/antigravity/propulsion characteristics can in principle be postulated. What is required for implementation is to determine appropriate sources for their generation, a requirement that must be met before advanced spaceship technology based on vacuum engineering can be realized in practice. The difficulties, challenges, and options for meeting such requirements can be found in the relevant literature (Reference 22).

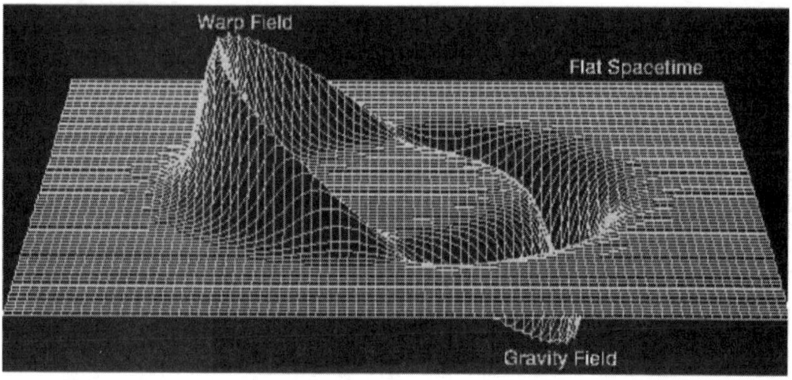

Figure 3. Alcubierre Warp Drive Metric Structure

IV. Discussion

This paper has considered the possibility—even likelihood—that future developments with regard to advanced aerospace technologies will trend in the direction of manipulating the underlying spacetime structure of the vacuum of space itself by processes that can be called vacuum engineering or metric engineering. Far from being simply a fanciful concept, a significant literature exists in peer-reviewed, Tier 1 physics publications in which the topic is explored in detail.[3]

The analysis presented herein, a form of general relativity for engineers, takes advantage of the fact that in GR a minimal-assumption, metric tensor approach can be used that is model-independent—that is, it does not depend on knowledge of the specific mechanisms or dynamics that result in spacetime alterations but rather only assumes that a technology exists that can control and manipulate (that is, engineer) the spacetime variables to advantage. Such an approach requires only that the hypothesized spacetime alterations result in effects consonant with the currently known GR physics principles.

In the metric engineering approach, the application of the principles gives precise predictions as to what can be expected as spatial and temporal variables are altered from their usual (that is, flat space) structure. Signatures of the predicted contractions and expansions of space, slowdown and speedup of time, alteration of effective mass, speed of light and associated consequences, both as occur in natural phenomena in nature and with regard to spacetimes specifically engineered for advanced aerospace applications, are succinctly summarized in Table 1.

Of particular interest with regard to innovative forms of advanced aerospace craft are the features tabulated in the right-hand column of Table 1, features that presumably describe an ideal craft for interstellar travel: an ability to travel at superluminal speeds

[3] See Reference 1 for a comprehensive introduction to the subject with contributions from lead scientists from around the globe.

relative to the reference frame of background space, energy bonds of materials strengthened (that is, hardened) relative to the background environment, a decrease in effective mass vis-à-vis the environment, an accelerated timeframe that would permit rapid trajectory changes relative to the background rest frame without undue internal stress, and the generation of gravity-like forces of arbitrary geometry—all on the basis of restructuring the vacuum spacetime variables. As avant garde as such features appear to be, they are totally in conformance with the principles of general relativity as currently understood. A remaining challenge is to develop insight into the technological designs by which such vacuum restructuring can be generated on the scale required to implement the necessary spacetime modifications.

Despite the challenges, sample calculations as presented herein indicate the direction of potentially useful trends derivable on the basis of the application of GR principles as embodied in a metric engineering approach, with the results constrained only by what is achievable practically in an engineering sense. The latter is, however, a daunting constraint. At this point in the consideration of such nascent concepts, given our present level of technological evolution, it is premature to even guess about an optimum strategy, let alone attempt to form a critical path for the engineering development of such technologies. Nonetheless, only through rigorous inquiry into such concepts can one hope to arrive at a proper assessment of the possibilities inherent in the evolution of advanced spaceflight technologies.

---

[1] See, for example, a series of essays in the compendium Frontiers of Propulsion Science, Eds. M. G. Millis and E. W. Davis, AIAA Press, Reston, Virginia (2009).
[2] M. Alcubierre, "The warp drive: Hyper-fast travel within general relativity," Class. Quantum Grav. 11, p. L73 (1994).
[3] H. E. Puthoff, "SETI, the velocity-of-light limitation, and the Alcubierre warp drive: An integrating overview," Physics Essays 9, p. 156 (1996).
[4] M. S. Morris and K. S. Thorne, "Wormholes in spacetime and their use for interstellar travel: A tool for teaching general relativity," Am. J. Phys. 56, pp. 395-412 (1988).
[5] M. Visser, Lorentzian Wormholes: From Einstein to Hawking, AIP Press, New York, 1995.
[6] M. S. Morris, K. S. Thorne and U. Yurtsever, "Wormholes, time machines, and the weak energy condition," Phys. Rev. Lett. 61, p. 1446 (1988).
[7] T. D. Lee, Particle Physics and Introduction to Field Theory, Harwood Academic Press, London (1988).
[8] The Philosophy of Vacuum, Eds. S. Saunders and H. R. Brown, Clarendon Press, Oxford (1991).
[9] F. Wilczek, The Lightness of Being: Mass, Ether and the Unification of Forces, Basic Books, New York (2008).
[10] *A. Logunov and M. Mestvirishvili, The Relativistic Theory of Gravitation, Mir Publ., Moscow (1989), p. 76.*
[11] Op. cit., p. 83.
[12] S. M. Mahajan, A. Qadir and P. M. Valanju, "Reintroducing the concept of 'force' into relativity theory," Il Nuovo Cimento 65B, 404 (1981).
[13] R. Klauber, "Physical components, coordinate components, and the speed of light," www.arXiv:gr-qc/0105071 v1 (18 May 2001).
[14] F. de Felice, "On the gravitational field acting as an optical medium," Gen. Rel. and Grav. 2, 347 (1971).
[15] K. K. Nandi and A. Islam, "On the optical-mechanical analogy in general relativity," Am. J. Phys. 63, 251 (1995).
[16] H. E. Puthoff, "Polarizable-vacuum (PV) approach to general relativity," Found. Phys. 32, 927 (2002).
[17] P. Boonserm et al., "Effective refractive index tensor for weak-field gravity," Class. Quant. Grav. 22, 1905 (2005).
[18] X.-H. Ye and Q. Lin, "A simple optical analysis of gravitational lensing," J. Modern Optics 55, no. 7, 1119 (2008).
[19] H. E. Puthoff, E. W. Davis and C. Maccone, "Levi-Civita effect in the polarizable vacuum (PV) representation of general relativity," Gen. Relativ. Grav. 37, 483 (2005).
[20] A. P. Lightman and D. P. Lee, "Restricted proof that the weak equivalence principle implies the Einstein equivalence principle," Phys. Rev. D 8, 364 (1973).
[21] C. W. Misner, K. S. Thorne and J. A. Wheeler, Gravitation, Freeman, San Francisco (1973), p. 5.
[22] E. W. Davis, "Chapter 15: Faster-than-Light Approaches in General Relativity," Frontiers of Propulsion Science, Progress in Astronautics and Aeronautics Series, Vol. 227, eds. M. G. Millis and E. W. Davis, AIAA Press, Reston, VA, pp. 473 (2009).

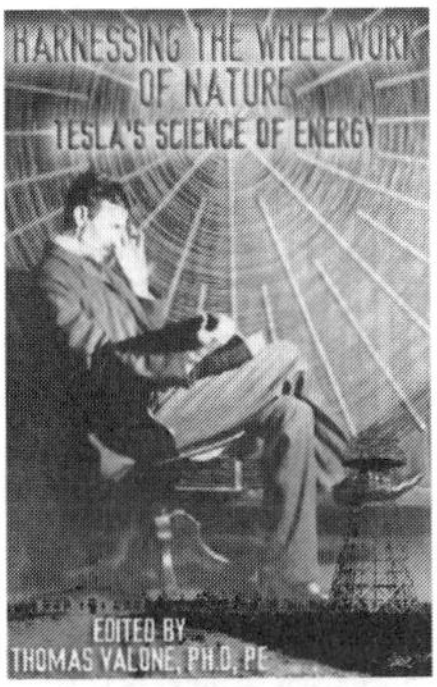

HARNESSING THE WHEELWORK OF NATURE: Tesla's Science of Energy
By Thomas Valone, Ph.D., P.E.

Chapters include:Non-Hertzian Waves: True Meaning of the Wireless Transmission of Power by Toby Grotz; On the Transmission of Electricity Without Wires by Nikola Tesla; Tesla's Magnifying Transmitter by Andrija Puharich; Tesla's Self-Sustaining Electrical Generator and the Ether by Oliver Nichelson; Modification of Maxwell's Equations in Free Space... Scalar Electromagnetic Waves; Disclosures Concerning Tesla's Operation of an ELF Oscillator; Electric Weather Forces: Tesla's Vision by Charles Yost; The New Art of Projecting Concentrated Non-Dispersive Energy Through Natural Media; Tesla's Death Ray" plus Selected Tesla Patents; more.

288 pages. 6x9 Paperback. Illustrated. $16.95. Code: HWWN

TECHNOLOGY OF THE GODS
The Incredible Sciences of the Ancients
by David Hatcher Childress

Childress looks at the technology that was allegedly used in Atlantis and the theory that the Great Pyramid of Egypt was originally a gigantic power station. He examines tales of ancient flight and the technology that it involved; how the ancients used electricity; megalithic building techniques; the use of crystal lenses and the fire from the gods; evidence of various high tech weapons in the past, including atomic weapons; ancient metallurgy and heavy machinery; the role of modern inventors such as Nikola Tesla in bringing ancient technology back into modern use; impossible artifacts; and more.

356 PAGES. 6x9 PAPERBACK. $16.95. CODE: TGOD

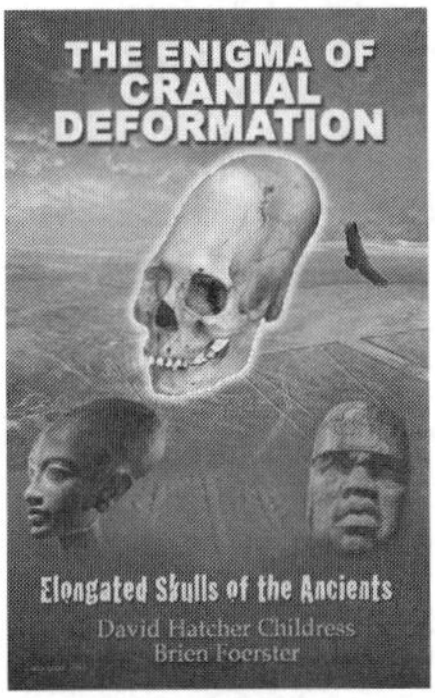

THE ENIGMA OF CRANIAL DEFORMATION
Elongated Skulls of the Ancients
By David Hatcher Childress and Brien Foerster

In a book filled with over a hundred astonishing photos and a color photo section, Childress and Foerster take us to Peru, Bolivia, Egypt, Malta, China, Mexico and other places in search of strange elongated skulls and other cranial deformation. The puzzle of why diverse ancient people—even on remote Pacific Islands—would use head-binding to create elongated heads is mystifying. Where did they even get this idea? Did some people naturally look this way—with long narrow heads? Were they some alien race? Were they an elite race that roamed the entire planet? Why do anthropologists rarely talk about cranial deformation and know so little about it? Color Section.

250 Pages. 6x9 Paperback. Illustrated. $19.95. Code: ECD

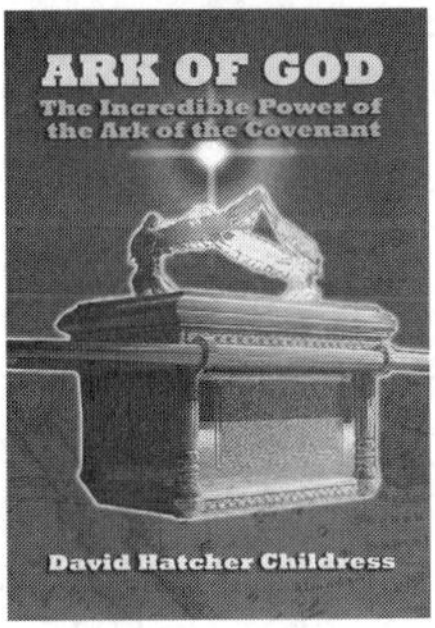

ARK OF GOD
The Incredible Power of the Ark of the Covenant
By David Hatcher Childress

Childress takes us on an incredible journey in search of the truth about (and science behind) the fantastic biblical artifact known as the Ark of the Covenant. This object made by Moses at Mount Sinai—part wooden-metal box and part golden statue—had the power to create "lightning" to kill people, and also to fly and lead people through the wilderness. The Ark of the Covenant suddenly disappears from the Bible record and what happened to it is not mentioned. Was it hidden in the underground passages of King Solomon's temple and later discovered by the Knights Templar? Was it taken through Egypt to Ethiopia as many Coptic Christians believe? Childress looks into hidden history, astonishing ancient technology, and a 3,000-year-old mystery that continues to fascinate millions of people today. Color section.

420 Pages. 6x9 Paperback. Illustrated. $22.00 Code: AOG

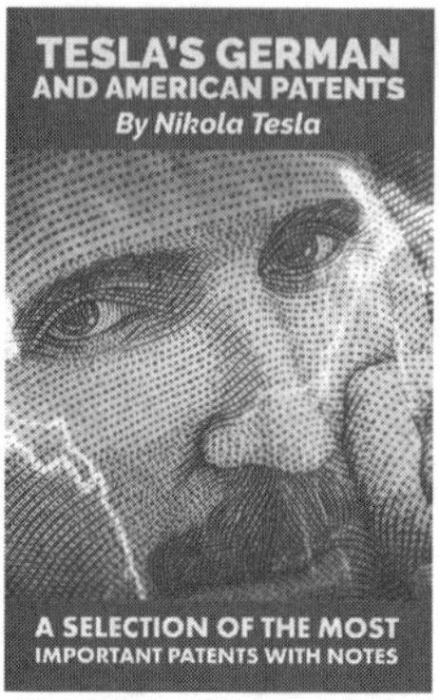

TESLA'S GERMAN AND AMERICAN PATENTS
A Selection of the Most Important Patents with Notes
By Nikola Tesla

In Tesla's own words, are such topics as wireless transmission of power, his towers for transmitting electrical power, death rays, and radio-controlled airships. The many patents include the electric-arc lamp, the dynamo-electro machine, armature for electric machines, electrical transformer induction device, apparatus for electrical conversion and distribution, system of electric lighting, electric incandescent lamp, electrical condenser, coil for electro magnets, electric generator, electric meter, electric-circuit controller, means for increasing the intensity of electrical oscillators, apparatus for the utilization of radiant energy, speed indicator, Tesla's water fountain, method of aerial transportation, tons more! A great visual compilation of all of Tesla's best inventions with text by Nikola Tesla himself in both English and German (in connection with the German patents). Tons of detailed drawings and patent notes!

802 Pages. 6x8 Paperback. Illustrated.. $24.95. Code: TGAP

TAPPING THE ZERO POINT ENERGY
Free Energy & Anti-Gravity in Today's Physics
by Moray B. King

King explains how free energy and anti-gravity are possible. The theories of the zero point energy maintain there are tremendous fluctuations of electrical field energy imbedded within the fabric of space. This book tells how, in the 1930s, inventor T. Henry Moray could produce a fifty kilowatt "free energy" machine; how an electrified plasma vortex creates anti-gravity; how the Pons/Fleischmann "cold fusion" experiment could produce tremendous heat without fusion; and how certain experiments might produce a gravitational anomaly.

180 PAGES. 5x8 PAPERBACK. ILLUSTRATED. $12.95. CODE: TAP

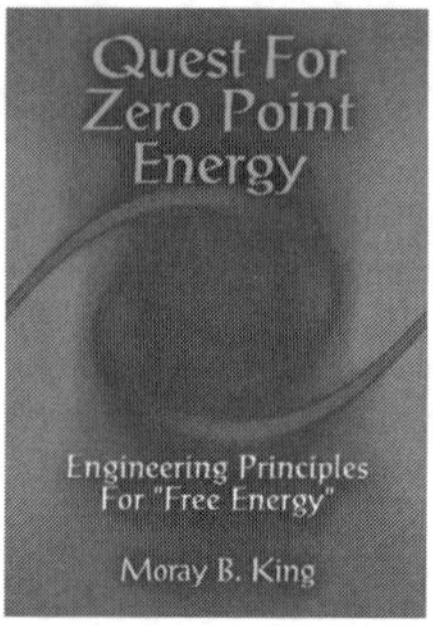

QUEST FOR ZERO-POINT ENERGY
Engineering Principles for "Free Energy"
by Moray B. King

King expands, with diagrams, on how free energy and anti-gravity are possible. The theories of zero point energy maintain there are tremendous fluctuations of electrical field energy embedded within the fabric of space. King explains the following topics: TFundamentals of a Zero-Point Energy Technology; Vacuum Energy Vortices; The Super Tube; Charge Clusters: The Basis of Zero-Point Energy Inventions; Vortex Filaments, Torsion Fields and the Zero-Point Energy; Transforming the Planet with a Zero-Point Energy Experiment; Dual Vortex Forms: The Key to a Large Zero-Point Energy Coherence. Packed with diagrams, patents and photos.

224 PAGES. 6x9 PAPERBACK. ILLUSTRATED. $14.95. CODE: QZPE

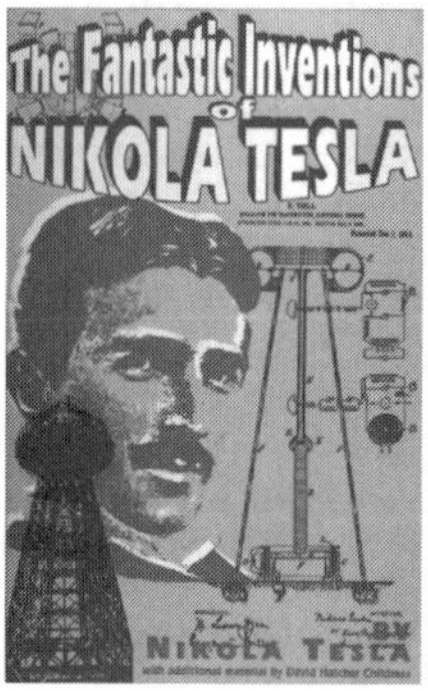

THE FANTASTIC INVENTIONS OF NIKOLA TESLA
by Nikola Tesla with David Hatcher Childress

This book is a readable compendium of patents, diagrams, photos and explanations of the many incredible inventions of the originator of the modern era of electrification. In Tesla's own words are such topics as wireless transmission of power, death rays, and radio-controlled airships. In addition, rare material on a secret city built at a remote jungle site in South America by one of Tesla's students, Guglielmo Marconi. Marconi's secret group claims to have built flying saucers in the 1940s and to have gone to Mars in the early 1950s! Incredible photos of these Tesla craft are included. •His plan to transmit free electricity into the atmosphere. •How electrical devices would work using only small antennas. •Why unlimited power could be utilized anywhere on earth. •How radio and radar technology can be used as death-ray weapons in Star Wars.

342 PAGES. 6x9 PAPERBACK. ILLUSTRATED. $16.95. CODE: FINT

SECRETS OF THE UNIFIED FIELD
The Philadelphia Experiment, the Nazi Bell, and the Discarded Theory
by Joseph P. Farrell

American and German wartime scientists determined that, while the Unified Field Theory was incomplete, it could nevertheless be engineered. Chapters include: The Meanings of "Torsion"; The Mistake in Unified Field Theories and Their Discarding by Contemporary Physics; Three Routes to the Doomsday Weapon: Quantum Potential, Torsion, and Vortices; Tesla's Meeting with FDR; Arnold Sommerfeld and Electromagnetic Radar Stealth; Electromagnetic Phase Conjugations, Phase Conjugate Mirrors, and Templates; The Unified Field Theory, the Torsion Tensor, and Igor Witkowski's Idea of the Plasma Focus; tons more.

340 pages. 6x9 Paperback. Illustrated. $18.95. Code: SOUF

THE ANTI-GRAVITY HANDBOOK
edited by David Hatcher Childress

The new expanded compilation of material on Anti-Gravity, Free Energy, Flying Saucer Propulsion, UFOs, Suppressed Technology, NASA Cover-ups and more. Highly illustrated with patents, technical illustrations and photos. This revised and expanded edition has more material, including photos of Area 51, Nevada, the government's secret testing facility. This classic on weird science is back in a new format!

230 pages. 7x10 paperback. Illustrated. $16.95. code: AGH

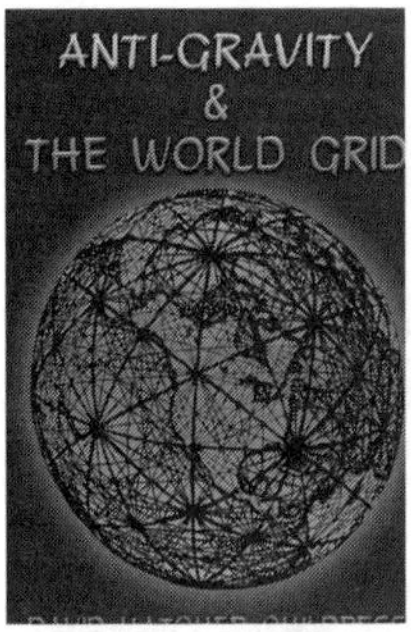

ANTI–GRAVITY & THE WORLD GRID

Is the earth surrounded by an intricate electromagnetic grid network offering free energy? This compilation of material on ley lines and world power points contains chapters on the geography, mathematics, and light harmonics of the earth grid. Learn the purpose of ley lines and ancient megalithic structures located on the grid. Discover how the grid made the Philadelphia Experiment possible. Explore the Coral Castle and many other mysteries, including acoustic levitation, Tesla Shields and scalar wave weaponry. Browse through the section on anti-gravity patents, and research resources.

274 pages. 7x10 paperback. Illustrated. $14.95. code: AGW

ANTI–GRAVITY & THE UNIFIED FIELD
edited by David Hatcher Childress

Is Einstein's Unified Field Theory the answer to all of our energy problems? Explored in this compilation of material is how gravity, electricity and magnetism manifest from a unified field around us. Why artificial gravity is possible; secrets of UFO propulsion; free energy; Nikola Tesla and anti-gravity airships of the 20s and 30s; flying saucers as superconducting whirls of plasma; anti-mass generators; vortex propulsion; suppressed technology; government cover-ups; gravitational pulse drive; spacecraft & more.

240 pages. 7x10 Paperback. Illustrated. $14.95. Code: AGU

THE TIME TRAVEL HANDBOOK
A Manual of Practical Teleportation & Time Travel
edited by David Hatcher Childress

The Time Travel Handbook takes the reader beyond the government experiments and deep into the uncharted territory of early time travellers such as Nikola Tesla and Guglielmo Marconi and their alleged time travel experiments, as well as the Wilson Brothers of EMI and their connection to the Philadelphia Experiment—the U.S. Navy's forays into invisibility, time travel, and teleportation. Childress looks into the claims of time travelling individuals, and investigates the unusual claim that the pyramids on Mars were built in the future and sent back in time. A highly visual, large format book, with patents, photos and schematics. Be the first on your block to build your own time travel device!

316 pages. 7x10 Paperback. Illustrated. $16.95. code: TTH

OBELISKS: TOWERS OF POWER
The Mysterious Purpose of Obelisks
By David Hatcher Childress

Some obelisks weigh over 500 tons and are massive blocks of polished granite that would be extremely difficult to quarry and erect even with modern equipment. Why did ancient civilizations in Egypt, Ethiopia and elsewhere undertake the massive enterprise it would have been to erect a single obelisk, much less dozens of them? Were they energy towers that could receive or transmit energy? With discussions on Tesla's wireless power, and the use of obelisks as gigantic acupuncture needles for earth, Chapters include: Megaliths Around the World and their Purpose; The Crystal Towers of Egypt; The Obelisks of Ethiopia; Obelisks in Europe and Asia; Mysterious Obelisks in the Americas; The Terrible Crystal Towers of Atlantis; Tesla's Wireless Power Distribution System; Obelisks on the Moon; more. 8-page color section.

336 Pages. 6x9 Paperback. Illustrated. $22.00 Code: OBK

NIKOLA TESLA'S ELECTRICITY UNPLUGGED
Wireless Transmission of Power as the Master of Lightning Intended
Edited by Tom Valone, Ph.D.

The immense genius of Tesla resulted from his ability to see an invention in 3-D, from every angle, within his mind before it was easily built. Tesla's inventions were complete down to dimensions and part sizes in his visionary process. Tesla would envision his electromagnetic devices as he stared into the sky, or into a corner of his laboratory. His inventions on rotating magnetic fields, creating AC current as we know it today, have changed the world—yet most people have never heard of this great inventor. Includes: Tesla's fantastic vision of the future, his wireless transmission of power, Tesla's Magnifying Transmitter, the testing and building of his towers for wireless power, tons more. The genius of Nikola Tesla is being realized by millions all over the world!

464 pages. 6x9 Paperback. Illustrated. Index. $21.95 Code: NTEU

THE TESLA PAPERS
Nikola Tesla on Free Energy & Wireless Transmission of Power
by Nikola Tesla, edited by David Hatcher Childress

David Hatcher Childress takes us into the incredible world of Nikola Tesla and his amazing inventions. Tesla's fantastic vision of the future, including wireless power, anti-gravity, free energy and highly advanced solar power. Also included are some of the papers, patents and material collected on Tesla at the Colorado Springs Tesla Symposiums, including papers on: •The Secret History of Wireless Transmission •Tesla and the Magnifying Transmitter •Design and Construction of a Half-Wave Tesla Coil •Electrostatics: A Key to Free Energy •Progress in Zero-Point Energy Research •Electromagnetic Energy from Antennas to Atoms

325 PAGES. 8x10 PAPERBACK. ILLUSTRATED. $16.95. CODE: TTP

COVERT WARS & THE CLASH OF CIVILIZATIONS
UFOs, Oligarchs and Space Secrecy
By Joseph P. Farrell

Farrell's customary meticulous research and sharp analysis blow the lid off of a worldwide web of nefarious financial and technological control that very few people even suspect exists. He elaborates on the advanced technology that they took with them at the "end" of World War II and shows how the breakaway civilizations have created a huge system of hidden finance with the involvement of various banks and financial institutions around the world. He investigates the current space secrecy that involves UFOs, suppressed technologies and the hidden oligarchs who control planet earth for their own gain and profit.

358 Pages. 6x9 Paperback. Illustrated. $19.95. Code: CWCC

HITLER'S SUPPRESSED AND STILL-SECRET WEAPONS, SCIENCE AND TECHNOLOGY
by Henry Stevens

In the closing months of WWII the Allies assembled mind-blowing intelligence reports of supermetals, electric guns, and ray weapons able to stop the engines of Allied aircraft—in addition to feared x-ray and laser weaponry. Chapters include: The Kammler Group; German Flying Disc Update; The Electromagnetic Vampire; Liquid Air; Synthetic Blood; German Free Energy Research; German Atomic Tests; The Fuel-Air Bomb; Supermetals; Red Mercury; Means to Stop Engines; more.

335 Pages. 6x9 Paperback. Illustrated. $19.95. Code: HSSW

PRODIGAL GENIUS
The Life of Nikola Tesla
by John J. O'Neill

This special edition of O'Neill's book has many rare photographs of Tesla and his most advanced inventions. Tesla's eccentric personality gives his life story a strange romantic quality. He made his first million before he was forty, yet gave up his royalties in a gesture of friendship, and died almost in poverty. Tesla could see an invention in 3-D, from every angle, within his mind, before it was built; how he refused to accept the Nobel Prize; his friendships with Mark Twain, George Westinghouse and competition with Thomas Edison. Tesla is revealed as a figure of genius whose influence on the world reaches into the far future. Deluxe, illustrated edition.

408 pages. 6x9 Paperback. Illustrated. Bibliography. $18.95. Code: PRG

HAARP
The Ultimate Weapon of the Conspiracy
by Jerry Smith

The HAARP project in Alaska is one of the most controversial projects ever undertaken by the U.S. Government. At at worst, HAARP could be the most dangerous device ever created, a futuristic technology that is everything from super-beam weapon to world-wide mind control device. Topics include Over-the-Horizon Radar and HAARP, Mind Control, ELF and HAARP, The Telsa Connection, The Russian Woodpecker, GWEN & HAARP, Earth Penetrating Tomography, Weather Modification, Secret Science of the Conspiracy, more. Includes the complete 1987 Eastlund patent for his pulsed super-weapon that he claims was stolen by the HAARP Project.

256 pages. 6x9 Paperback. Illustrated. Bib. $14.95. Code: HARP

THE TRIANGLE
The Truth Behind the World's Most Enduring Mystery
By Mike Bara

Hundreds of ships, planes and yachts have disappeared in the dark, mysterious waters between Bermuda and Florida. Ships have vanished without a trace only to magically reappear years later in good order but minus their crews, almost as if the intervening years had not even passed—for them. Pilots have reported bizarre problems with their instruments as compasses and guidance systems have spun inexplicably out of control over the shadowy waters of the Triangle. Entire squadrons of military aircraft have disappeared off of radarscopes in clear weather and with no forewarning. Explanations range from alien encounters to rogue waves to twisting unnatural funnel spouts caused by submerged civilizations left over from the days of Atlantis.

218 Pages. 6x9 Paperback. Illustrated. $19.95. Code: TRI

ORDER FORM

10% Discount When You Order 3 or More Items!

One Adventure Place
P.O. Box 74
Kempton, Illinois 60946
United States of America
Tel.: 815-253-6390 • Fax: 815-253-6300
Email: auphq@frontiernet.net
http://www.adventuresunlimitedpress.com

ORDERING INSTRUCTIONS

- ✓ Remit by USD$ Check, Money Order or Credit Card
- ✓ Visa, Master Card, Discover & AmEx Accepted
- ✓ Paypal Payments Can Be Made To: info@wexclub.com
- ✓ Prices May Change Without Notice
- ✓ 10% Discount for 3 or More Items

SHIPPING CHARGES

United States

- ✓ Postal Book Rate { $4.50 First Item; 50¢ Each Additional Item
- ✓ POSTAL BOOK RATE Cannot Be Tracked! Not responsible for non-delivery.
- ✓ Priority Mail { $6.00 First Item; $2.00 Each Additional Item
- ✓ UPS { $7.00 First Item; $1.50 Each Additional Item

NOTE: UPS Delivery Available to Mainland USA Only

Canada

- ✓ Postal Air Mail { $15.00 First Item; $2.50 Each Additional Item
- ✓ Personal Checks or Bank Drafts MUST BE US$ and Drawn on a US Bank
- ✓ Canadian Postal Money Orders OK
- ✓ Payment MUST BE US$

All Other Countries

- ✓ Sorry, No Surface Delivery!
- ✓ Postal Air Mail { $19.00 First Item; $6.00 Each Additional Item
- ✓ Checks and Money Orders MUST BE US$ and Drawn on a US Bank or branch.
- ✓ Paypal Payments Can Be Made in US$ To: info@wexclub.com

SPECIAL NOTES

- ✓ RETAILERS: Standard Discounts Available
- ✓ BACKORDERS: We Backorder all Out-of-Stock Items Unless Otherwise Requested
- ✓ PRO FORMA INVOICES: Available on Request
- ✓ DVD Return Policy: Replace defective DVDs only

ORDER ONLINE AT: www.adventuresunlimitedpress.com

10% Discount When You Order 3 or More Items!

Please check: ✓

☐ This is my first order ☐ I have ordered before

Name
Address
City
State/Province Postal Code
Country
Phone: Day Evening
Fax Email

Item Code	Item Description	Qty	Total

Please check: ✓

Subtotal ▶
Less Discount-10% for 3 or more items ▶

☐ Postal-Surface
Balance ▶

☐ Postal-Air Mail (Priority in USA)
Illinois Residents 6.25% Sales Tax ▶
Previous Credit ▶

☐ UPS (Mainland USA only)
Shipping ▶
Total (check/MO in USD$ only) ▶

☐ Visa/MasterCard/Discover/American Express

Card Number:
Expiration Date: Security Code:

✓ SEND A CATALOG TO A FRIEND: